Cyber-Physical, IoT, and Autonomous Systems in Industry 4.0

Cyber-Physical, IoT, and Autonomous Systems in Industry 4.0

Edited by
Vikram Bali
Vishal Bhatnagar
Deepti Aggarwal
Shivani Bali
Mario José Diván

CRC Press is an imprint of the
Taylor & Francis Group, an **informa** business

First edition published 2022
by CRC Press
6000 Broken Sound Parkway NW, Suite 300, Boca Raton, FL 33487-2742

and by CRC Press
2 Park Square, Milton Park, Abingdon, Oxon, OX14 4RN

CRC Press is an imprint of Taylor & Francis Group, LLC

Library of Congress Cataloging-in-Publication Data
Names: Bali, Vikram, editor. | Bhatnagar, Vishal, 1977- editor. | Aggarwal, Deepti, editor. | Bali, Shivani, editor.
Title: Cyber-physical, IoT, and autonomous systems in industry 4.0 / edited by Vikram Bali, Vishal Bhatnagar, Deepti Aggarwal, Shivani Bali, and Mario José Diván.
Description: First edition. | Boca Raton : CRC Press, 2022. | Includes bibliographical references and index.
Identifiers: LCCN 2021021790 (print) | LCCN 2021021791 (ebook) | ISBN 9780367705152 (hardback) | ISBN 9780367705169 (paperback) | ISBN 9781003146711 (ebook)
Subjects: LCSH: Industry 4.0.
Classification: LCC T59.6 .C93 2022 (print) | LCC T59.6 (ebook) | DDC 658.4/0380285574--dc23
LC record available at https://lccn.loc.gov/2021021790
LC ebook record available at https://lccn.loc.gov/2021021791

ISBN: 978-0-367-70515-2 (hbk)
ISBN: 978-0-367-70516-9 (pbk)
ISBN: 978-1-003-14671-1 (ebk)

DOI: 10.1201/9781003146711

Typeset in Times
by SPi Technologies India Pvt Ltd (Straive)

Contents

Preface ix
Editors xvii
Contributors xix

Chapter 1 Cyber Systems and Security in Industry 4.0 1

Bhogaraju Padma and Thalatam Mohana Naga Vamsi

Chapter 2 A Demand Response Program for Social Welfare Maximization in the Context of the Indian Smart Grid: A Review 25

Sachin Ramnath Gaikwad and R. Harikrishnan

Chapter 3 Cloud Computing Security Framework Based on Shared Responsibility Models: Cloud Computing 39

Umesh Kumar Singh and Abhishek Sharma

Chapter 4 Performance Analysis of a Hypervisor and the Container-based Migration Technique for Cloud Virtualization 57

Aditya Bhardwaj

Chapter 5 Segmentation of Fine-Grained Iron Ore Using Deep Learning and the Internet of Things 73

Ada Cristina França da Silva and Omar Andres Carmona Cortes

Chapter 6 Amalgamation of Blockchain Technology and Cloud Computing for a Secure and More Adaptable Cloud 87

Himanshu V. Taiwade and Premchand B. Ambhore

Chapter 7 Cloud, Edge, and Fog Computing: Trends 103

Siddhant Thapliyal, Lisa Gopal, Piyush Bagla, and Kuldeep Kumar

Chapter 8 Progression in Cyber Security Concerns for Learning Management Systems: Analyzing the Role of Participants 117

Tejaswini Apte and Sarika Sharma

Chapter 9 A Security Model for Cloud-computing-based E-governance Applications 133

Aditya Makwe, Anand More, Priyesh Kanungo, and Niranjan Shrivastava

Chapter 10 Automatic Time and Motion Study Using Deep Learning 147

Jefferson Hernandez, Sofia Lopez, Gabriela Valarezo, and Andres G. Abad

Chapter 11 Applications of IoT Based Frameworks in Industry 4.0: Applications of IoT Based Frameworks 163

Kaushik Ghosh, Sugandha Sharma, Piyush Bagla, and Kuldeep Kumar

Chapter 12 Impact of Deep Learning and Machine Learning in Industry 4.0 179

Umesh Kumar Lilhore, Sarita Simaiya, Amandeep Kaur, Devendra Prasad, Meenu Khurana, Deepak Kumar Verma, and Afsan Hassan

Chapter 13 IoT Applications and Recent Advances 199

T. Mohana Naga Vamsi and Pratibha Lanka

Chapter 14 A Spatio-temporal Model for the Analysis and Classification of Soil Using the IoT 221

M. Umme Salma, Subbaiah Rachana, and Narasegouda Srinivas

Chapter 15 A Critical Survey of Autonomous Vehicles 235

Nipun R. Navadia, Gurleen Kaur, Harshit Bhardwaj, Aditi Sakalle, Yashpal Singh, Taranjeet Singh, Arpit Bhardwaj, and Divya Acharya

Chapter 16 A Meta-learning Approach for Algorithm Selection for Capacitated Vehicle Routing Problems 255

Neha Sehta and Urjita Thakar

Chapter 17 Early Detection of Autism Disorder Using Predictive Analysis 269

T. Harshvardhan, G. Preethi, S. Aishwarya, T. Vinod, and K. S. Meghana

Chapter 18 Computing Technologies for Prognosticating the Emanation of Carbon Using an ARIMA Model 283

Shyla, Kapil Kumar, and Vishal Bhatnagar

Chapter 19 DWT and SVD-based Robust Watermarking Using Differential Evolution with Adaptive Optimization 297

Meenal Kamlakar, Chhaya Gosavi, Vaidehee Salunkhe, Priyanka Bagul, Aishwarya Keskar, and Shweta Barge

Chapter 20 A Novel Framework Based on a Machine Learning Algorithm for the Estimation of COVID-19 Cases 311

Rashmi Welekar, Sharvari Tapase, Shubhi Bajaj, Isha Pande, Abhishek Verma, Ashutosh Katpatal, and Vaibhav Mishra

Chapter 21 Assessing the Impact of Coronavirus on Pollutant Concentration: A Case Study in Malaysia 331

Imam Wahyu Amanullah, Sharifah Sakinah Syed Ahmad, and Emaliana Kasmuri

Chapter 22 A Comprehensive Review of SLAM Techniques 341

Karthi Mohan, Surya Bharath Achalla, and Arpit Jain

Chapter 23 A Novel Evolutionary Computation Method for Securing the Data in Wireless Networks 371

Inderpreet Kaur, Vibha Tripathi, and Nidhi

Index 389

Preface

Industry 4.0 is revolutionizing planning and production processes. It originated in a simple idea: merging an existing physical entity with advanced IT elements such as the Internet of Things (IoT), cloud-dased data, sensors, automators, and analytics – creating something that offers a surplus benefit over its original predecessor, Industry 3.0. Machine learning, cloud computing, cyber security, and the IoT are the topics of contemporary research interest. Cyber-physical systems (CPSs) refer to an Industry 4.0-enabled manufacturing environment that offers real-time data collection, analysis, and transparency across every aspect of a manufacturing operation. The evolution of Industry 4.0 and advancements in the field of the IoT have significantly increased the complexity and the level of risk to which all enterprises are subjected to. Industry 4.0 is a massive shift towards automation and digitization, utilizing the IoT and CPSs such as sensors to aid in data collection for manufacturing verticals. Industry 4.0 involves a hyperconnected system that includes the smarter use of robotics to effectively and efficiently move manufacturing to new heights. With the use of all these technological systems, it is imperative to ensure that cyber security plays a role during the rise of this digital industrial revolution. Advances in machine learning are changing the traditional manufacturing era into the smart manufacturing era of Industry 4.0.

This book addresses the topics related to the IoT, machine learning, CPSs, and cloud computing for Industry 4.0. It brings together researchers, developers, practitioners, and users who are interested in these areas to explore new ideas, techniques, and tools and to exchange their experiences. This book intends to encourage researchers to develop novel theories to enrich scholars' and firms' knowledge to achieve sustainable development and/or foster sustainability. The book investigates challenges across multiple sectors and industries and also considers Industry 4.0 for operations and supply chain management.

In Chapter 1, "Cyber Systems and Security In Industry 4.0," it is shown that integrating the operations of manufacturing companies in Industry 4.0 means exposing the company's IT infrastructure and data, which makes them more susceptible and vulnerable to cyber threats. Shielding the industrial IoT against cyber threats is crucial in order to control and stimulate the economic engine of the future. Protecting cyber systems from theft and adversaries is essential, since it includes everything that is appropriate for securing sensitive data, personal information, cerebral property, and industrial and governmental information systems. Cyber security refers to the umbrella of processes, technologies, and tools intended to protect networks, programs, devices, and data from various cyber attacks, damage, and unauthorized access. Cyber security ought to become an essential part of design, strategy, and operations, and should be focused from the start on any new, connected, Industry 4.0-driven approach. This chapter focuses on acquiring a clear understanding of cyber space, cyber systems, and the security challenges in Industry 4.0.

In Chapter 2, "A Demand Response Program for Social Welfare Maximization in the Context of the Indian Smart Grid: A Review," it is shown that the smart grid allows the bi-directional flow of energy and information between energy generation and the load side. As a demand response program is a key element in the present smart grid system, so the incentive and price-based demand response programs help prosumers and consumers to be actively involved in the smart grid system. This chapter proposes a thematic idea with a gray literature review on building a new business model on demand response programs for social welfare maximization in the context of the Indian smart grid. Moreover, this concise study also helps us to understand the challenges and solutions through pilot projects and case studies at the forefront of developing countries like India in smart grid deployments. The chapter deals with collecting and analyzing the relevant information and presenting it in a logical flow which provides more information to researchers who are working on smart grid development.

In Chapter 3, "A Cloud Computing Security Framework Based on Shared Responsibility Models," it is shown that in the present scenario, especially in the current pandemic situation, cloud computing enables new innovation in the IT industry and that more than 90% of IT enterprises are migrating their services towards virtualization using cloud computing platforms. In order to gain the maximum outcome from the advantages of such platforms, cloud users rely on cloud service providers (CSPs) for managing their data and IT services. So, for the smooth management of data and services for cloud users, CSPs need a strong cloud security framework. More than 50% of the market of cloud computing services are occupied by the top three CSPs: Amazon Web Services (AWS), Microsoft Azure, and Google Cloud. This chapter presents a comprehensive analysis and the findings of a security framework related to these top CSPs and proposes a cloud security shared responsibility model and its application in various internet computing services like healthcare and e-governance.

In Chapter 4, "Performance Analysis of a Hypervisor and the Container-based Migration Technique for Cloud Virtualization," it is shown that containers emerged as a lightweight alternative to a virtual machine (VM) and offer better microservice architecture support. The major limitation of a VM is that it comprises a large image size due to the fully installed guest OS, binaries, and library files. This disadvantage hinders its applicability to real-time applications. On the other hand, containers comprise only the required files, thus making them lighter in size compared to a VM. With the development of container technology, the issue of how to facilitate migration support needs to be addressed. Therefore, in this study an experimental setup is proposed to facilitate the migration of running containers and evaluate its performance compared to the existing hypervisor-based VM migration scheme. To provide a solution for this, the authors use LXD as a container technology and checkpoint/restore mechanism (CRIU) to facilitate the migration of running containers. Experimental results demonstrate that compared to the existing VM migration scheme, the proposed container-based migration technique shows improvement with a reduction in downtime by 72.08%, in migration time by 54.94%, and in the number of pages transferred by 97.54%.

In Chapter 5, "Segmentation of Fine-Grained Iron Ore Using Deep Learning and the Internet of Things," it is shown that sieving ore is a critical process in iron ore production. It reduces the iron ore size, improving the next stage's performance.

Failing to detect fine-grained ore in the initial steps can lead to a loss in production. Detecting fine-grained iron ore automatically is an essential task to improve the product's economic value in general. Hence, the chapter shows an automatic process based on deep learning and an IoT architecture that take decisions in real time. Using deep learninghas been inspired by the fact that traditional segmentation algorithms have proven to beslow; therefore, their use is not possible in an ore vibrating screen in real-time applications. In this context, this research proposes using a deep neural network calledU-Net for image segmentation. U-Net is a convolutional neural network developedfor biomedical image segmentation; consequently, it yields better performanceon fast image segmentation than other networks devised specially for regular objectdetection such as ResNet. In this context, U-Net was trained andtested in a Jetson TX1 architecture. Preliminary results show that U-Net canproperly segment around 3.46 fps, suitable for the authors application.

In Chapter 6, "Amalgamation of Blockchain Technology and Cloud Computing for a Secure and More Adaptable Cloud," the focus is on an existing cloud computing model along with an in-depth review of existing cloud computing security measures and related constraints. Through this chapter the authors also try to incorporate a systematic understanding of the basics of blockchain technology and how it can be integrated with cloud computing in order to enhance the latter's existing security mechanisms that can lead us towards a more secure and more adaptable cloud.

In Chapter 7, "Cloud, Edge and Fog Computing: Trends," it is shown that in recent years the emergence of artificial intelligence and the Internet of Things (IoT) has enhanced the working capability and digital powers of sectors like healthcare, banking, and education. The benefits of these technologies are faster transmission, security, and integrity of data. These features can be achieved through cloud computing. Cloud computing is the virtual storage concept for processing and storing data. But every second the amount of data is increasing, which will affect the analyzing and processing speed of data and will become more complicated for processing and analyzing in the coming years. As a solution to these problems, there are two enhancements to cloud computing – edge computing and fog computing. These computing technologies try to reduce the size of data to be transmitted to the cloud by performing processing closer to the entry point of the data. The strategy behind these technologies is to provide an enhancement such that only the necessary and important data which can be used frequently or which will affect real-time conditions in the future are stored on the cloud, and all temporary and intermediate data are analyzed and processed at the data feeding end.

In Chapter 8, "Progression in Cyber Security Concerns for Learning Management Systems: Analyzing the Role of Participants," it is shown that the educational sector, in adopting technologies and online deliveries of sessions, has increased the scope of discussion and knowledge sharing multifold. It has also opened up new ground for various opportunities in cyberspace. Educational bodies are facing several challenges as dependency on cyber-physical systems (CPSs) is increasing. A learning management system (LMS) is one of the CPSs used as a platform for online delivery of sessions. Many professional bodies are disseminating online education and are responsible for the respective assessments. However, a down side is that it tags and is vulnerable to many cyber threats, with which users or participants are not familiar.

The perception and knowledge of cyber security is predominantly required to address assessment and thereby the outcome of delivery by professional bodies supporting higher education. The infrastructure used to build and deploy online session delivery (i.e., LMS) plays a considerable role in implementing the security standards and guidelines, and thereby the policies, for an organization. The distributed nature of the online LMS with the greater support in advancement of scalability and functionality has increased the complexity of the system. This complexity has spawned various threads for non-legitimate users to exploit the vulnerabilities in the existing architecture of LMS. In addition, the invention of various mobile devices to access an online LMS serves as a catalyst for such exploitation. There has been a need to bring forward the essentials of cyber security education and its adoption by participants (i.e., instructors or learners). The authors analyze various vulnerabilities present in LMS and study the role and adoption intentions of the instructor towards cyber security measures.

In Chapter 9, "A Security Model For Cloud-computing-based E-governance Applications," it is shown that in a cloud environment, the real time output of the execution of E-governance applications depends on various factors like the time of applying security measures such as firewalls, IDS/IPS, encryption, deciding on the priority of jobs (filtering high priority jobs), and the execution time on virtual machines (VMs). This chapter addresses the problem of the secure execution of E-governance jobs in a cloud computing environment along with various factors concerning the security of the cloud and also proposes a security architecture. This architecture involves three layers, where the first layer describes the security aspects of user jobs, the second layer decides the priority of user jobs (filtering out the E-governance jobs from other jobs in the cloud), and a third layer is responsible for the execution of sorted jobs on VMs. In the study, the overall execution time of the model is determined. The execution time for layer one depends upon the processing time of the firewall, the second layer depends upon the number of parameters taken for comparing the tasks, and for layer three the shortest job first (SJF) and the first come first serve (FCFS) policy is applied and the time of execution is obtained.

In Chapter 10, "Automatic Time and Motion Study Using Deep Learning," it is shown that measuring the performance of manual-labor activities is a key element of work scheduling and resource management. This measurement is usually performed using a data-collection methodology called a time and motion study (TMS). Many industries still rely on humaneffort to execute TMS, which can be time-consuming, error-prone,and expensive. This chapter introduces an automatic alternativeto human-performed TMS that works at two levels of abstraction:micro-actions (defined as the movement of a specific body part) andmacro-actions (defined as the combination of successive micro-actions). Specifically, an encoder–decoder based classifier is employed to perform micro-action recognition, and a continuous-time hidden Markov model to perform macro-action recognition. The authors show how the proposed system can compute productivity indicators such as: workeravailability, worker performance, and overall labor effectiveness; andhand-usage indicators such as hand speed and handedness.

In Chapter 11, "Applications of IoT Based Frameworks in Industry 4.0," it is shown that the advent of sensor networks and technologies like microelectronics and microsensors has made data communication without an infrastructure not only a reality, but also, pervasive. The evolution of pervasive and ubiquitous computing has paved the way for the technology of the Internet of things (IoT). In fact, IoT is a buzzword for both industry and academia today. It is among one of the handful of innovative technologies which have risen sufficiently to step out of the walls of research laboratories and enter the realms of our daily lives. In fact, there is hardly any industry where the IoT has not found any considerable application. Agriculture, healthcare, and the automobile and oil industries are some of the areas where an extensive application of the IoT is to be found. In this chapter, the authors discuss how the IoT can be integrated for industries of different domains. They describe how a single framework of the IoT can be used in three different industries: agriculture, healthcare, and the oil industry.

In Chapter 12, "Impact of Deep Learning and Machine Learning in Industry 4.0," it is shown that Industry 4.0 provides emergence to what is called the "Smart Factory." Industry 4.0 (IR 4.0) is a growing phenomenon for automation and information sharing in industrial technology. Artificial-intelligence-based methods (i.e., deep learning and machine learning) play a crucial role within IR 4.0. The main component of IR 4.0 includes a sensor, robotics, smart cameras, and software integration to manage the central system. These sensors generate huge amounts of data. This chapter examines the effect of numerous machine learning and deep-learning technologies on IR 4.0. The work also proposes a hybrid machine learning and deep learning-based model (HMDL) for IR4.0. The proposed HMDL model utilizes the features of the random forest machine learning method and the multilayer perceptron networks of the deep learning method. The importance of machine and deep learning strategies for Industry 4.0 is addressed further, as well as their application to problems in industrial production. Finally, the recent state-of-the-art object-tracking process is described, including the possible outcomes and future growth prospects.

In Chapter 13, "IOT Applications and Recent Advances," it is shown that since research in IoT applications is the amalgamation of embedded electronics, sensor technology, networking, cloud computing, artificial intelligence, data analytics, and other technologies, it will become a good area for interdisciplinary research for both academics and industrial people. At the outset there are a lot of new methodologies and findings that have evolved with this IoT research, and these outcomes give better solutions in the areas of smart environments, IoT services, intelligent systems, smart health, smart homes, smart agriculture, networking protocols, communication protocols, and autonomous industrial applications. Hence the research direction in IoT applications is definitely a challenge for academic researchers and industrial developers. The major challenges are in identification technology, choosing an IoT architecture technology, and establishing networks and communication technologies for the specific research problem. In this chapter the authors attempt to depict all such challenges by focusing on these IoT-enabled topics and throw light on these issues for new researchers who are entering this domain.

In Chapter 14, "A Spatio-temporal Model for the Analysis and Classification of Soil using the IoT," it is shown that the IoT in agriculture transforms entities such as crops, soils, and livestock in a smart way by utilizing underlying technologies such as embedded systems, pervasive computing, sensor networks, ubiquitous computing, ad hoc networks, various wireless communication technologies, Internet protocols and other advanced technologies. The research here focuses on the most important agriculture entity – soil. It is the soil that determines the yield of a crop. The more fertile the soil, more qualitative is the yield. The main idea behind the research is to identify the soil most suitable for agriculture. Using a spatio-temporal model, the soil samples collected from various parts of the country are classified into agricultural soil and non-agricultural soil. This classification is done by the aid of features such as the pH of the soil, and its humidity, moisture, and temperature collected from IoT sensors.

In Chapter 15, "A Critical Survey of Autonomous Vehicles," it is shown that in this developing world of technologies where every domain is converting into the autonomous, the one domain going to transportation is transportation or automobiles. Many things are becoming autonomous, like manufacturing and transportation, including trains and airplanes. Vehicles are the most important aspect from the perspective of transportation, which in the upcoming period are going to become completely autonomous or run without any human aid or interaction. The automation of vehicles will change transportation: humans will not play any role in running the vehicles in the revolution in this domain. We have moved someway towards auto-driven vehicles but need to alter the entire transportation and automobile industry. This chapter discusses the autonomous technologies and the examples of autonomous vehicles that we have currently, and the challenges in terms of the practical aspects and current scenarios.

In Chapter 16, "A Meta-learning Approach for Algorithm Selection for Capacitated Vehicle Routing Problems," it is shown that optimization is a must for dealing with scarce resources. Vehicle routing is a major concern in an enormous number of fields involving transportation of goods and services. The vehicle routing problem is a widely studied optimization problem for handling this. A variety of heuristic and meta-heuristic algorithms exist for its solution. For a given instance of an optimization problem, it is challenging to select an algorithm that will perform better out of many existing algorithms. This task of selecting an algorithm to solve a given instance of a problem is known as the algorithm selection problem, which is recognized as a learning task. Learning is done using the performance data of algorithms collected from past experiments, hence it is termed meta-learning.

In Chapter 17, "Early Detection Of Autism Disorder Using Predictive Analysis," it is shown that autism in children is a kind of inability at social and communicative behavior. If not detected between 20 to 60 months, treatment is exceedingly difficult. The earlier the detection, the earlier the treatment and therapy can be employed, though they would still be challenging. With the upsurge in the use of machine learning (ML) techniques for research dimensions in medicine and diagnostics, we can attempt to make use of many different ML techniques, networks, and convolutional neural networks for predicting and analyzing ASD problems in a child. The authors use ML and neural network algorithms like SVM, random forest, and CNN to detect and predict autism in children.

In Chapter 18, "Computing Technologies for Prognosticating the Emanation of Carbon using an ARIMA Model," it is shown that the core responsibility towards environment protection is the adaptation of green computing technologies. The common practices of technology users releases a large amount of carbon dioxide (CO_2), contributing to environmental pollution. The regulated and controlled usage of peripherals for a limited period of time, the efficient usage of energy intensive peripherals, power management, hardware inactivity, the employment of alternate computing devices, and the implementation of efficient servers all impact the environment positively. In this chapter, a carbon emission dataset is used which includes the amount of CO_2 released from different peripherals, the different levels of electricity generation, and the amount of electricity consumption. The autoregressive integrated moving average (ARIMA) model is used on time series data to forecast CO_2 emanation.

In Chapter 19, "DWT and SVT-based Robust Watermarking using Differential Evolution with Adaptive Optimization," it is shown that digital technologies have been rapidly developing and low copy-cost and convenient temper make a huge impact. Digital watermarking plays a crucial role in copyright protection, even in IoT applications. It provides a good solution for copyright problems and content authentication to achieve data integrity. The objective behind this chapter is to obtain optimal, robust, and invisible image watermarking using differential evolution. A good watermark is robust against attacks, is invisible, and carries a good amount of information related to the owner. This chapter presents a digital watermarking algorithm that makes use of differential evolution along with adaptive optimization for choosing the most suitable scaling factors. It can be inferred from the results after experimentation that the proposed algorithm performs well with respect to invisibility and robustness.

In Chapter 20, "A Novel Framework Based on a Machine Learning Algorithm for the Estimation of COVID-19 Cases," it is shown that the outbreak of the coronavirus disease (COVID-19) in 2019, which was first reported in the Hubei province of China, spread over most of the world very quickly, with staggering economic and medical outcomes. India has also faced a very tough task to control the virus outbreak but has managed to do so through some strict measures; however, it has been a difficult task to predict when the number of cases will start to reduce. The authors attempt to predict, using the susceptible-infectious-removed machine learning model on the dataset for India, when the cases in India will start to reduce. The implementation started at a country level, which the authors continue to the state level and then to the district/city level. They develop a Web application, which maintains a dashboard of active, confirmed, recovered, and death cases in the form of graphs, tables, and hover-over-effect maps.

In Chapter 21, "Assessing the Impact of Coronavirus on Pollutant Concentrations: A Case Study in Malaysia," it is shown that the new variant of coronavirus presents many open questions as to how it affects human health. The exposure to polluted air might worsen the condition of a Covid-19 patient, especially in their respiratory organs. Most countries have implemented lockdowns as a step to minimize the transmission of Covid-19, which are predicted to reduce anthropogenic activity. The lockdown effort has not only reduced transmission of Covid-19, but also reduces the rate of air pollution. This study (1) describes the variation of air pollution concentration during Malaysian lockdown measures and (2) highlights the relationship between air pollutant variables and Covid-19 daily new cases and death cases.

In Chapter 22, "A Comprehensive Review of SLAM techniques," it is shown that robots are developed and programmed by replicating human or animal activities and functions in one way or another. In this chapter, the authors investigate mobile robots and how they navigate from an initial state to a goal state. For a mobile robot, navigation is essential for it to travel autonomously through any environment following certain functions like: obstacle avoidance; mapping and constructing the environment; and localization so as to move from its initial position to its goal state, or navigate through different waypoints. The chapter involves simultaneous localization, mapping, and 3D reconstruction using 2D laser scanners, which are installed on the top of the robot to obtain a 2D map of any area.

In Chapter 23, "A Novel Evolutionary Computation Method for Securing the Data in Wireless Networks," it is shown that today wireless networks are bringing a revolution in the field of communication. But, with development and utilization, the threat to data security shared on those networks has been growing. This chapter compares evolutionary computational algorithms to see which one is better when it comes to wireless network data security. Among the various evolutionary techniques, this one majorly focuses on four such techniques, namely particle swarm, ant colony, artificial bee colony, and cuckoo approach optimization. The chapter gives a comprehensive idea of the techniques and how they can be used to secure the data in wireless networks.

Vikram Bali
Vishal Bhatnagar
Deepti Aggarwal
Shivani Bali
Mario José Diván

Editors

Vikram Bali is Professor and Head of the Computer Science and Engineering Department at the JSS Academy of Technical Education, Noida, India. He graduated from REC, Kurukshetra with a B.Tech (CSE), an M.E. (CSE) from NITTTR, Chandigarh, and a PhD from Banasthali Vidyapith, Rajasthan. He has more than 20 years of rich academic and administrative experience. He has published more than 50 research papers in international journals and at conferences and has edited books. He has authored five text books. He has registered six patents. He is on the editorial board and on the review panel of many international journals. He is the series editor for three book series with CRC Press (Taylor & Francis Group). He is a lifetime member of the IEEE, ISTE, CSI, and IE. He won the Green Thinker Z-Distinguished Educator Award 2018 for a remarkable contribution to the field of Computer Science and Engineering, presented at the Third International Convention on Interdisciplinary Research for Sustainable Development (IRSD) at the Confederation of Indian Industry (CII), Chandigarh. He has attended faculty enablement programs organized by Infosys and NASSCOM. He is a member of the board of studies of different Indian universities and is a member of the organizing committee for various national and international seminars/conferences. He is working on four sponsored research projects funded by TEQIP-3 and Unnat Bharat Abhiyaan. His research interests include software engineering, cyber security, and automata theory.

Vishal Bhatnagar holds a B.E, Mtech, and PhD in the Engineering field. He has more than 21 years of teaching experience in various technical institutions. He is currently working as Professor in the Computer Science & Engineering Department at Netaji Subhash University of Technology East Campus (formerly the Ambedkar Institute of Advanced Communication Technologies & Research), Delhi, India. His research interests include data mining, social network analysis, data science, blockchain, and big data analytics. He has to his credit more than 130 research papers in various international/national journals, conferences, and book chapters. He is currently working as Associate Editor of a number of journals of IGI Global and Inderscience. He has experience of handling the special issues of the journals of Many Scopus, ESCI, and SCIE. He is the series editor for three book series with CRC Press (Taylor & Francis Group). He has also edited many books for Springer, Elsevier, IGI Global, and CRC Press. He is a life-time member of the Indian Society for Technical Education (ISTE).

Deepti Aggarwal is Assistant Professor in the Department of Computer Science and Engineering at the JSS Academy of Technical Education, Noida. She has more than 20 years of academic experience in teaching undergraduate and postgraduate computer science courses. She received her B. Tech. (CSE) from Maharishi Dayanand University, Rohtak, her M.Tech. (CSE) from Rajasthan Vidyapeeth, Udaipur, and her PhD from Jaipur National University, Jaipur. She has registered a patent for a smart dustbin: sanitation and solid-liquid waste management. She wrote a book *Computer Organisation* in 2011 and has published several papers in international journals and presented at conferences. Her areas of interest include machine learning, cyber security, the IoT, and data mining. She has been a reviewer and guest editor of several reputed journals and books.

Shivani Bali is an academic, trainer, researcher, and consultant with more than 17 years of experience in the field of Decision Sciences and Operations Management. She is working as Professor at the Jaipuria Institute of Management, Noida. She has published around 30 research papers in SCI/WoS/Scopus journals and registered four patents. She has been working on consulting assignments with many reputed companies and has delivered training programs in the area of data science and business analytics. Her areas of interest are business analytics, data science, operations, and supply chain analytics.

Mario José Diván is Full Professor and Head of the Data Science Research Group at the National University of La Pampa, Argentina. His scientific interests lie in the areas of data mining, data stream, stream mining, high-performance computing, big data, the IoT, data quality, measurement, and evaluation.

Contributors

Afsan Hassan
Chitkara University Institute of Engineering and Technology
Chitkara University
Punjab, India

Amandeep Kaur
Chitkara University Institute of Engineering and Technology
Chitkara University
Punjab, India

Andres G. Abad
Industrial Artificial Intelligence (INARI) Research Lab at ESPOL
Guayaquil, Ecuador

Surya Bharath Achalla
University of Petroleum and Energy Studies
Dehradun, India

Divya Acharya
Bennett University
Greater Noida, Uttar Pradesh, India

Sharifah Sakinah Syed Ahmad
Universiti Teknikal Malaysia Melaka

S. Aishwarya
Department of Information Science and Engineering
Jyothy Institute of Technology
Bangalore, India

Imam Wahyu Amanullah
Universiti Teknikal Malaysia Melaka

Premchand B. Ambhore
Government College of Engineering
Amravati

Tejaswini Apte
Symbiosis Institute of Computer Studies and Research
Symbiosis International (Deemed University)
Pune, India

Piyush Bagla
Dr. B.R. Ambedkar National Institute of Technology
Jalandhar

Priyanka Bagul
Cummins College of Engineering for Women
Pune, India

Shubhi Bajaj
Shri Ramdeobaba College of Engineering and Management
Nagpur, India

Shweta Barge
Cummins College of Engineering for Women
Pune, India

Aditya Bhardwaj
Bennett University (Times of India Group)
Greater Noida, Uttar Pradesh, India

Arpit Bhardwaj
Bennett University
Greater Noida, Uttar Pradesh, India

Harshit Bhardwaj
USICT
Gautam Buddha University
Greater Noida, Uttar Pradesh, India

Vishal Bhatnagar
Department of Computer Science and Engineering
NSUT East Campus
Formerly Ambedkar Institute of Advanced Communication Technologies and Research
India

Omar Andres Carmona Cortes
Instituto Federal do Maranhão (IFMA)
Brazil

Sachin Ramnath Gaikwad
Department of Electronics and Telecommunication
Symbiosis Institute of Technology (SIT)
Symbiosis International Deemed University (SIDU)
Pune, India

Kaushik Ghosh
UPES
Dehradun, India

Lisa Gopal
Graphic Era Hill University
Dehradun, India

Chhaya Gosavi
Cummins College of Engineering for Women
Pune, India

R. Harikrishnan
Department of Electronics and Telecommunication
Symbiosis Institute of Technology (SIT)
Symbiosis International Deemed University (SIDU)
Pune, India

T. Harshvardhan
Centre for Incubation, Innovation, Research and Consultancy
Jyothy Institute of Technology
Bangalore, India

Jefferson Hernandez
Industrial Artificial Intelligence (INARI) Research Lab at ESPOL
Guayaquil, Ecuador

Arpit Jain
University of Petroleum and Energy Studies
Dehradun, India

Meenal Kamlakar
Cummins College of Engineering for Women
Pune, India

Priyesh Kanungo
SCSIT DAVV
Indore MP

Emaliana Kasmuri
Universiti Teknikal Malaysia Melaka

Ashutosh Katpatal
Shri Ramdeobaba College of Engineering and Management
Nagpur, India

Gurleen Kaur
Department of Computer Science and Engineering
Dronacharya Group of Institutions
Greater Noida, Uttar Pradesh, India

Inderpreet Kaur
Galgotia College of Engineering and Technology

Aishwarya Keskar
Cummins College of Engineering for Women
Pune, India

Meenu Khurana
Chitkara University Institute of Engineering and Technology
Chitkara University
Punjab, India

Kapil Kumar
Department of Computer Science and Engineering
NSUT East Campus
Formerly Ambedkar Institute of Advanced Communication Technologies and Research
GGSIPU
India

Kuldeep Kumar
Dr. B.R. Ambedkar National Institute of Technology
Jalandhar, India

Pratibha Lanka
Gayatri Vidya Parishad College for Degree and PG Courses(A)
Visakhapatnam, Andhra Pradesh, India

Sofia Lopez
Escuela Superior Politecnica del Litoral (ESPOL)
Guayaquil, Ecuador

Umesh Kumar Lilhore
Chitkara University Institute of Engineering and Technology
Chitkara University
Punjab, India

Aditya Makwe
Institute of Engineering and Technology DAV
Indore, MP

K. S. Meghana
Department of Information Science and Engineering
Jyothy Institute of Technology
Bangalore

Vaibhav Mishra
Shri Ramdeobaba College of Engineering and Management
Nagpur, India

Karthi Mohan
University of Petroleum and Energy Studies
Dehradun, India

Anand More
Computer Centre DAVV
Indore MP

Nipun R. Navadia
Department of Computer Science and Engineering
Dronacharya Group of Institutions
Greater Noida, Uttar Pradesh, India

Nidhi
Galgotia College of Engineering and Technology

Bhogaraju Padma
Gayatri Vidya Parishad College for Degree and PG Courses
Visakhapatnam, Andhra Pradesh, India

Isha Pande
Shri Ramdeobaba College of Engineering and Management
Nagpur, India

Devendra Prasad
Chitkara University Institute of Engineering and Technology
Chitkara University
Punjab, India

G. Preethi
Department of Information Science and Engineering
Jyothy Institute of Technology
Bangalore

Sarita Simaiya
Chitkara University Institute of Engineering and Technology
Chitkara University
Punjab, India

Subbaiah Rachana
Christ (Deemed to be University)

Aditi Sakalle
USICT
Gautam Buddha University
Greater Noida, Uttar Pradesh, India

M. Umme Salma
Christ (Deemed to be University)

Vaidehee Salunkhe
Cummins College of Engineering for Women
Pune, India

Neha Sehta
Department of Computer Engineering
Shri G.S. Institute of Technology and Science
Indore

Sarika Sharma
Symbiosis Institute of Computer Studies and Research
Symbiosis International (Deemed University)
Pune, India

Sugandha Sharma
UPES
Dehradun, India

Niranjan Shrivastava
IMS DAVV
Indore MP

Shyla
Department of Computer Science and Engineering
NSUT East Campus
Formerly Ambedkar Institute of Advanced Communication Technologies and Research
GGSIPU
India

Ada Cristina França da Silva
Universidade Estadual do Maranhão (UEMA)
Brazil

Taranjeet Singh
Department of Computer Science and Engineering, GL Bajaj, Greater Noida, Uttar Pradesh, India

Umesh Kumar Singh
Institute of Computer Science
Vikram University
Ujjain

Yashpal Singh
Department of Computer Science and Engineering
MIET
Greater Noida, Uttar Pradesh, India

Narasegouda Srinivas
Jyoti Nivas College

Himanshu V. Taiwade
Priyadarshini Institute of Engineering & Technology
Nagpur, India

Sharvari Tapase
Shri Ramdeobaba College of Engineering and Management
Nagpur, India

Urjita Thakar
Department of Computer Engineering
Shri G.S. Institute of Technology and Science
Indore

Siddhant Thapliyal
Graphic Era University
Dehradun, India

Vibha Tripathi
Galgotia College of Engineering and Technology

Gabriela Valarezo
Escuela Superior Politecnica del Litoral (ESPOL)
Guayaquil, Ecuador

Thalatam Mohana Naga Vamsi
Gayatri Vidya Parishad College for Degree and PG Courses
Visakhapatnam, Andhra Pradesh, India

Abhishek Sharma
Shri Ramdeobaba College of Engineering and Management
Nagpur, India

Abhishek Verma
Shri Ramdeobaba College of Engineering and Management
Nagpur, India

Deepak Kumar Verma
Chhatrapati Shahu Ji Maharaj University
Kanpur, Uttar Pradesh, India

T. Vinod
Department of Information Science and Engineering
Jyothy Institute of Technology
Bangalore, India

Rashmi Welekar
Shri Ramdeobaba College of Engineering and Management
Nagpur, India

1 Cyber Systems and Security in Industry 4.0

Bhogaraju Padma and
Thalatam Mohana Naga Vamsi
Gayatri Vidya Parishad College for Degree and PG Courses,
Visakhapatnam, Andhra Pradesh, India

CONTENTS

1.1 Introduction to Cyberspace, Cyber Systems, and Cyber Security 2
1.1.1 Cyber Threats, Vulnerabilities, and Risks 3
1.2 The Internet of Things and Security Issues 4
1.2.1 IoT Architecture and Protocol Stacks 4
1.2.2 Applications of the IoT 6
1.2.3 Security of IoT Systems 7
1.2.4 Securing IoT Devices and Networks 8
1.3 Industrial Internet of Things (IIoT) 10
1.3.1 Industry 4.0: Overview 10
1.3.2 IIoT Applications 11
1.3.3 IIoT Architecture 12
1.3.4 Impact of IIoT on the Economy 13
1.4 Security Challenges in Industry 4.0 14
1.4.1 How Secure Is IIoT? 15
1.4.2 IIoT Security Challenges 16
1.4.2.1 Device Hijacking 16
1.4.2.2 Data Siphoning 17
1.4.2.3 Denial of Service Attacks 17
1.4.2.4 Data Breaches 17
1.4.2.5 Device Theft 17
1.4.2.6 Man-in-the-Middle or Device "Spoofing" 18
1.4.3 IIoT Security Solutions 18
1.4.3.1 Endpoint Security-by-Design 18
1.5 Conclusions and Future Work 21
1.5.1 Conclusions 21
1.5.2 Limitations 21
1.5.3 Future Work 22
References 22

DOI: 10.1201/9781003146711-1

1.1 INTRODUCTION TO CYBERSPACE, CYBER SYSTEMS, AND CYBER SECURITY

The term "cyber" stands for "computerized." The term "cyberspace" was first used by William Gibson in 1984, to describe any feature or facility that is associated with the Internet. Cyberspace is an electronic medium that facilitates online communication and data exchange activities. It encompasses the whole virtual computer world that explicitly employs transmission control protocol/Internet protocol (TCP/IP) for communication. Cyberspace is often referred to as the Internet but it is a symbolic space that exists within the scope of the Internet, whereas the Internet is a network of networks. The ever-increasing practice of using the Internet creates a cyberspace. The abundant use of desktop computers, laptops, and smart phones to access the Internet for gaming, streaming videos, social networks, language translation, e-banking, and so on means that, in a theoretical and practical sense, cyberspace is growing. Geopolitically, cyberspace is growing across borders, making the globe smaller.

A cyber system makes use of a cyberspace. It is any combination of equipment that facilitates infrastructure, procedures, personnel, and integrated communications to provide the services of the Internet that include control systems, business systems, and access control systems. A cyber-physical system (CPS) is a part of a cyber system that interacts with its physical surroundings (Boyes et al., 2018). CPS integrates cyberspace and physical elements. A CPS is made up of highly integrated communication, computation, control, and the physical world. It is the major current research area in academia, industry, and government. In recent decades, cyberspace has increased enormously. It has brought exceptional economic growth and opportunities.

The global, interconnected, and digital nature of the cyber infrastructure has unfortunately presented bad actors with new crime opportunities by means of exploiting computer and network vulnerabilities. A cyber attack is any deed that may modify, steal, or devastate a specified object by hacking into a susceptible system. Cyber attacks can vary from installing malware on a personal computer to destroying the infrastructure of a whole nation. Cyber attacks in business environments are deliberate and inspired by financial gain. Cyber crime is a criminal action that involves a cyber assault on a computer or a network. Cyber law can be described as that branch of law which deals with communications in information technology (Berrehili & Belmekki, 2016). Simply put, cyber law governs computers and cyberspace/the Internet. Cyber security is meant to protect Internet-connected systems, including software, hardware, and data from a range of cyber attacks. In other words, cyber security is the protection of cyberspace. It and information security are frequently considered as the same thing. Undoubtedly, this creates perplexity in the security world. The difference between these two lies in their respective scope. Cyber security is the capability of defending or protecting cyberspace against cyber attacks, but information security deals with protecting information and information systems to provide integrity, confidentiality, and availability (Thakur et al., 2015). Confidentiality ensures the information transmitted is not easily accessed by unauthorized users, including systems, applications, processes, or humans. Integrity is the capability of ensuring that data do not suffer from unauthorized modification. Availability ensures that systems, applications, and data are accessible to users, particularly when they need them. Non-repudiation is a term used in law that means one's intent to accomplish one's obligations in an agreement and that one party cannot refuse to send or receive a transaction.

1.1.1 Cyber Threats, Vulnerabilities, and Risks

The terms cyber threats, vulnerabilities, and risks are often used interchangeably, but this causes confusion. Organizations must understand the relationships between three components: threat, vulnerability, and risk. Although these three terms are employed interchangeably, they are different terms with special meanings and inferences. Through security vulnerabilities, a hacker finds his or her way into systems and networks, or pulls out sensitive information. The term cyber security vulnerability refers to any sort of exploitable weak spot that threatens the cyber security of an organization. The most frequent security vulnerabilities are phishing and ransom ware. Phishing refers to an attempt to obtain secret information while acting as a reliable contact, such as an online service or bank. Phishing is mostly targeted at gaining the confidential data of individuals. Phishing emails often look fully convincing, with ideal text and authentic logos. People use smart phones and tablets extensively. Ransom ware is a form of malware (malicious software) and most are brought in via malicious emails. Security vulnerabilities of these portable storage devices cause data leakages. Weak credentials are vulnerabilities which are always prone to dictionary and brute-force attacks. Vulnerabilities such as system misconfigurations, running needless services, or vulnerable settings can be exploited by attackers to violate networks. With attacks due to poor encryption, an enemy can interrupt the communication among connected systems and obtain information.

A cyber threat is a malevolent act that seeks to steal data, damage data, or disrupt digital life in general. The most common cyber threats are phishing, ransom ware, data leakage, and hacking. Hacking is achieving entrance to IT systems and networks from outside of an organization and offers an attractive alternative for criminals. There have been many successful attempts at admission to customer personal bank account data and also credit card numbers. By means of a social engineering threat, opponents trap users into disclosing their names and passwords. A Denial of Service (DoS) attack is an assault in which a computer is employed to send numerous TCP or user datagram protocol (UDP) data packets to servers. A Distributed Denial of Service (DDoS) attack is an assault where several systems aim at only one system with a DoS attack. The communication network that is targeted is showered with a huge number of packets directed from multiple locations. Cyber espionage is committed by state-sponsored or directed groups, government actors, or others on behalf of a government seeking to gain unauthorized admission to data and systems in order to develop their own country's economic competitiveness, national security, or military strength. Cyber terrorism refers to exploiting the Internet so as to perform violent and cruel actions that result in serious bodily harm or even loss of life. A cyber risk is the probability of a threatening agent successfully exploiting vulnerability. As the severity and frequency of cyber crimes are getting higher, there is a considerable need for cyber security management as a part of every business enterprise's risk profile. Irrespective of an organization's risk appetite, stakeholders need to include cyber security as part of a business's risk management process and regular business operations.

1.2 THE INTERNET OF THINGS AND SECURITY ISSUES

The Internet of Things (IoT) is an arrangement of connection and communication among computing devices, objects, mechanical and digital instruments, and people which are provided with the facility to transmit data over a communication network without human-to-human or human-to-computer interaction. An IoT platform can be implemented in many domain areas including healthcare, transportation, agriculture, energy production and distribution, and many other areas that necessitate things to interact with over the Internet to execute business projects intelligently without human participation. When IoT devices were invented, security concerns were not an issue. So generally, there is no way of dealing with security concerns on devices (Dean & Agyeman, 2018). Additionally, devices are sometimes shipped with malware on them; consequently, this may lead to infection of the other networks to which they get connected. Some of the network security systems don't have the capacity to detect which IoT devices are connected to them nor the visibility to know about other devices that are communicating through the present network. Since the IoT is applied to essential areas of the economy, healthcare, intelligent transportation, and security needs are higher in their dependability on the availability of IoT implementations. Machine-to-Machine communication (M2M) is said to be the predecessor technology, and generally refers to data exchange between various devices through the Internet without human involvement. All most all applications that implement the IoT need to include one or more sensors to accumulate data from the surrounding environment. Sensors and actuators are crucial and the integrated components of smart objects and context awareness is the most important feature of the IoT, which may not be achievable without sensor technology.

1.2.1 IoT Architecture and Protocol Stacks

There exists no standard consensus on the architecture of the IoT which is universally defined. Several IoT architectures diverge according to their functional areas and solutions (Al-Fuqaha et al., 2015). There are several functional layers of the IoT which are built upon the potentiality and performance of IoT elements and which offer optimal business solutions for enterprises as well as end users. IoT architecture is an elementary way for designing different elements of the IoT, so that it can bring about future needs in addition to services over the networks. The IoT requires a layered architecture for facilitating communication among trillions of diverged heterogeneous objects through the Web. Most of the earlier models used a layered architecture that included three layers: perception, application, and network. Physical sensors in the perception layer collect data in order to digitize it and send it through secure communications. The network layer administers the routing of data to IoT devices and hubs and also performs aggregation. The application layer transports

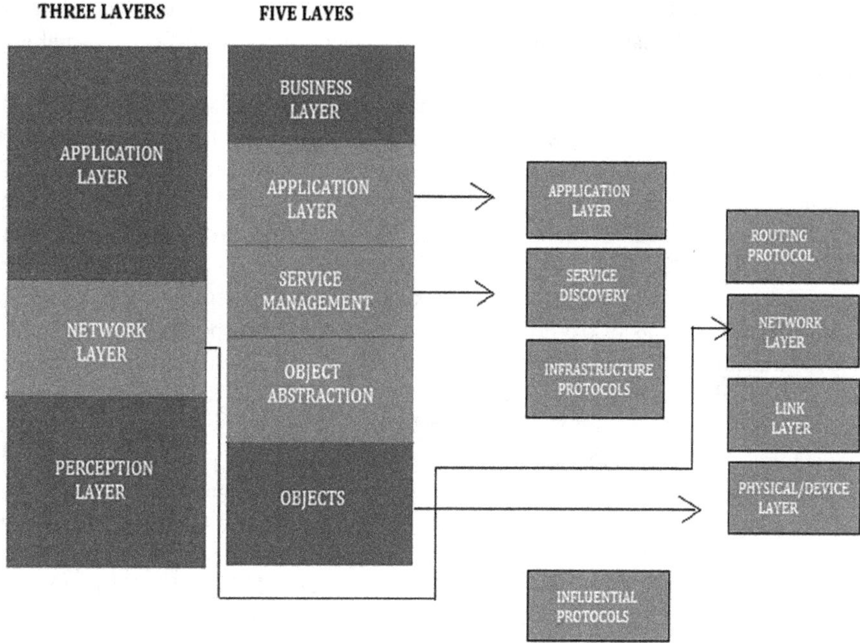

FIGURE 1.1 Five-layer architecture of the IoT.

data to end-user devices. The present architecture models include more views to the structure of the IoT. As a result, the five-layer framework was instigated which depends on the performance demand of IoT devices (Sicari et al., 2016), as shown in Figure 1.1.

Big data generated by "things" in the IoT are analyzed, and this process begins in the physical layer that contains devices such as sensors that gather and also analyze and process the data collected. The physical devices or objects include actuators as well as sensors for submitting queries on weight, location, temperature, acceleration, motion, humidity, and so on. The object abstraction layer encompasses communication technologies such as the global system for mobile communication (GSM), 3G, radio frequency identification (RFID), Wi-Fi, infrared, Bluetooth (Low Energy), and Zigbee for transferring information which is generated in the object layer to the service management layer (Jia et al., 2010). The abstraction layer controls various tasks such as data management and cloud-computation-related processes. The service management layer connects together services and requesters depending on names as well as addresses. The application designers of the IoT deal with a wide variety of objects that involve hardware platforms. The service management layer processes the collected data, takes decisions, and brings in obliged services according to network protocols. The application layer makes available smart quality devices and services for smart buildings, smart homes, industrial automation, transportation, and smart healthcare facilities. The business layer's responsibility is to manage the IoT system services, activities, graphs, the building of business models, and flow charts of the application layer. It plans, implements, evaluates, develops, and monitors IoT elements. Based on big data analysis, the business layer also supports decision-making processes.

The IoT protocol stack is designed as an enhancement of the TCP/IP protocol model. It consists of the following layers: physical/link layer, network layer, transport layer, application protocol layer, and application services layer. The most common physical layer protocols used by embedded systems are Ethernet, Bluetooth Low Energy, Wi-Fi, Wireless HART, Zigbee, Z-wave, RFID, Cellular (GPRS/2G/3G/4G/5G), and LoRAWAN. The network layer is accountable for the addressing and routing of information packets in the IoT. There are other network layer protocols such as 6LoWPAN, IPv6, and IPv4. The transport layer is accountable for transmitting and routing of data packets over a communication channel. The most commonly used transport layer protocols are TCP, UDP, and RPL (Routing Protocol for Low-Power and Lossy Networks). The application layer is accountable for data presentation and formatting, and is characteristically based on the Hypertext Transfer (HTTP) protocol. However, HTTP is not appropriate in a resource constrained environment since it is a heavy weight protocol and therefore involves a huge parsing overhead. Hence, there exist many application service protocols which have been invented for IoT surroundings; these include: MQTT, SMQTT, AMQP, DDS, CoAP, XMPP, REST, and SOAP.

1.2.2 Applications of the IoT

The IoT allows devices to interrelate, interact, collaborate, and be trained by each other's knowledge, as individuals do. In spite of the fear of cyber attacks and piracy, IoT applications will grow very rapidly in the coming years. Wearable technology is a trademark of IoT applications and is supposed to be the earliest IoT function. We commonly see Fitbits, smart watches, and other heart beat monitors everywhere these days. IoT applications have an ability to change reactive automated medical-based systems into proactive wellness-based systems in healthcare sectors. The sources that were used recently for medical research use lack real-world information, which is critical.

The ambition for optimized traffic control was a major motivation for the concept of a smart city (Zeadally & Khoukhi, 2018). Yet global issues, such as deteriorating air quality, unclean drinking water, and an ever-increasing urban density, take place in high intensity environments in many cities. Therefore, cities have been redeveloped in a different manner. There exist tremendous prospects for the study and field of IoT, such as smart greenhouses. Greenhouse farming is a technology which augments the yield of green crops by managing several environmental factors or parameters. Greenhouse manual handling results in energy loss, production loss, as well as labor charge, causing development to be less efficient. A greenhouse with IoT devices is built not only for flexible supervision, but also allows controlling the atmosphere, such as the internal temperature. Industrial automation is said to be one of the major fields in which faster development and product quality are crucial factors in returning a huge profit on investments. Using IoT platforms, we can reengineer products and package them to bring significant results in cost factors and customer satisfaction and experience. The IoT is proven to be game changing according to the requirements for quality control, product flow monitoring, inventory management, factory digitalization, packaging optimization, security, logistics, and supply chain optimization (Wunck & Bauman, 2017).

1.2.3 Security of IoT Systems

The explosion of connected devices in business industries, healthcare, and consumer companies has brought in-house security breaches and vulnerabilities, created by cyber criminals, such as a threat called zero-day attack that attacks devices such as printers, smart televisions, webcams, smart homes, and routers (Suo et al., 2012). There are several severe threats created by connected devices:

1. Smart security cameras: Recently, a home security systems supplier, owned by Amazon, was attacked by multiple hacking incidents that disabled security cameras that made victims upset. Security experts from cyber security firm Bitdefender found out and stated that there was a defect in Amazon's Ring Video Doorbell Pro that allows attackers unauthorized admission to users' Wi-Fi networks and other devices connected to them. Amazon has now sent a security patch to the Ring Doorbell cameras to alleviate the issue.
2. Hackers can "faxploit" connected fax machines: If fax machines have security vulnerabilities, they could permit an attacker to thieve information through an organization's network by making use of a fax number. The security loopholes that exist in fax devices allow the malicious code to decipher the information or data files and store them to its memory core that violates the sensitivity of the information and causes an interruption in the networks.
3. Smart TVs: As the FBI stated, smart televisions have a huge number of ignored and neglected security issues. Hackers can control not only our unprotected TV easily by changing the channels and volume controls, but also perusing our daily conversations and schedules using the microphone and integrated cameras.
4. Smart bulbs can be hacked: Many people have revealed smart bulb security weaknesses. Researchers at the University of Texas at San Antonio stated that attackers may target smart bulbs which are infrared-enabled by forwarding instructions through an invisible infrared light produced by these smart bulbs for exploiting the other IoT physical devices that exist within home networks.
5. Smart homes are vulnerable: A Milwaukee-based family experienced a terrible threat in their smart home unit that was compromised by anonymous hackers. The couple declared that attackers had captured their smart home by attacking the devices connected to it.
6. Stealing details of bank accounts from a coffee machine: Smart coffee machines are devices which are connected to the Internet by employing dedicated applications that are targeted to gain the owner's credit card details. The principal executive of security giant Avast, Vince Steckler, stated that IoT coffee machines permit their owners to monitor/control them distantly by using apps in their personal phones. Customers convey their oral commands while

they are connected to Amazon's effective virtual assistant software like Alexa. "Smart coffee machines are lagging in their security and it's very difficult to protect them," Steckler stated in the media.

7. Connected printers: A research firm Quocirca announced that smart printers that are connected to a company's set-up are the potential victims of cyber attacks. He provided a report which emphasized that around 60% of business industries in France, the USA, the UK, and Germany experienced data infringement that was printer related, resulting in monetary terms, on average, of more than US$400,000 in 2019.
8. Smart speakers can be hacked: Wu HuiYu and Qian Wenxiang are security specialists at Tencent Blade, who disclosed threats to smart speakers at a conference titled "DEFCON" on IoT security. These experts used Amazon Echo smart and stylish speakers to demonstrate the attacking process. The researchers easily compromised these speakers by using a malevolent tool embedded with a malicious code. They reported their results to Amazon prior to their presentation; later Amazon used a security patch for preventing the issue.
9. Gas stations on the IoT are vulnerable: Privacy and security are the major decision factors in the process of developing and designing IoT infrastructure and devices, and dealing with these concerns should be of the highest priority (Jindal et al., 2018). A 2019 survey done in Canada, France, Australia, Japan, the USA, and the UK disclosed that 63% of consumers found their connected devices were vulnerable. This demonstrates many aspects of IoT security challenges. Trust is vital to realizing the complete potentiality of the IoT. Therefore, cyber security must be implemented in IoT devices at all points in the ecosystem to thwart vulnerabilities in one part from jeopardizing the security of the whole (Vignesh & Samydurai, 2017).

1.2.4 Securing IoT Devices and Networks

From the viewpoint of modern technology, we are witnessing a remarkable increase in connecting heterogeneous devices and managing and controlling them remotely. The IoT is one of the emerging technologies that enable this perception. Considering the historical evidence, any device that is connected to a communication network is susceptible to security breaches and illegal acts. But these connected devices are directing human life, so device security turns out to be paramount (Abomhara & Køien, 2014). Particularly while the IoT is changing "heterogeneous networks" into "super-heterogeneous networks" of intelligent physical devices, its protection is one that is very multifaceted as well as complex. There exists a variety of methodologies which provide security to networks which are available today, though none is reliable (Swamy et al., 2017). A simple IoT application makes a huge difference in potential value by collecting insights about not only to whom the companies are speaking to, but when, where, how much, and in which way. Certainly, the IoT is the most incredible technological innovation that can be called a boon. But when it connects everything to the Internet, all these personal and business gadgets become vulnerable to several security attacks. Most proficient companies and cyber security experts are doing their best to safeguard things for customers, but there is still more to be done (Ali et al., 2018). These are some crucial things needed to secure IoT devices and networks:

- Changing the router's credentials. Never keep basic or default settings. Routers are usually named according to the identities of their manufacturers or networks; these provide hackers with a significant clue to enable them to obtain access. It's always recommended to avoid using your own name or address which may be advantageous and provide helpful traces to opponents who try to obtain admission into systems.
- Using powerful passwords is always advised, and one should aim at using random ones that contain a mix of characters, alphabets, and symbols.
- Never use open or public Wi-Fi whenever there is a need to access IoT networks through personal computers or smart phones. It is comparatively easier to crack Wi-Fi networks that are open in public places like cafes and restaurants. It is best to use a virtual private network (VPN).
- It is better to employ guest networks. It is a good idea to employ guest networks for customers who want to use wireless networks in their home; it never grants them admission into the network systems or to personal mail and other types of accounts. The guest network can be used for IoT devices. So even if an opponent hacks a device, they may get trapped within the guest network and can't obtain original Web access.
- Adopt robust encryption schemes like Wi-Fi protected access (WPA).
- Take extraordinary care in securing top-level control of an IoT network. It is a good suggestion to apply two-factor authentication (2FA) that uses biometrics like finger prints, a password, or a dongle to ensure that an attacker will not be able to produce two forms of evidence.

Other security measures might include:

- Cloud-based technologies are often vulnerable, so ensuring their security is advisable.
- Always test IoT physical devices.
- Avoid universal plug and play features.
- Authenticate IoT devices before connecting to them.
- User credentials should be changed frequently.
- Update mobile apps and IoT gadgets in a regular manner.
- Beware of IoT security breaches and risks.

Modern IoT ecosystems are complex. A security risk is present at every step along the IoT journey, and there is a huge number of hackers who exist to take advantage of a system's vulnerabilities. So IoT businesses should undergo a thorough security risk evaluation that inspects vulnerabilities in devices, networks, and customer back-end systems.

1.3 INDUSTRIAL INTERNET OF THINGS (IIoT)

The IIoT is the utilization of smart devices like sensors for enhancing manufacturing and industrial processes; it is also known as Industry 4.0. IIoT leverages the power of real-time analytics and smart machines and takes advantage of the data that machines have generated. The fascinating idea behind IIoT is that smart machines are not only better than human beings at analyzing and capturing data in real time, but they are better at communicating significant information that can be employed to make business decisions more accurately and quickly. This fourth industrial revolution is the most troublemaking factor in industrial automation and affects the industries of healthcare, energy, transportation, and manufacturing (Jeschke et al., 2017). The IIoT ecosystem contains assets that can sense, communicate, and store information about them, as well as public/private data communications infrastructure, analytics, and applications that generate business information from raw data and other entities such as people. IIoT is a division of IoT and focuses especially on industrial applications such as manufacturing, energy, transportation, and agriculture. Remarkably, IIoT has diverged technological requirements that give it an increased level of complexity, interoperability, and security.

1.3.1 Industry 4.0: Overview

Industry 4.0 and IIoT are different concepts that are too frequently used interchangeably. The term "Industry 4.0" conveys the advent of a new industrial rebellion through smart manufacturing. IIoT and Industry 4.0 consist of actuators, sensor networks, robots, appliances, business processes, machines, and personnel; hence, a huge amount of data of a diverse nature is generated. Simply put, IIoT is a subdivision of Industry 4.0, though it is not an identical term. IIoT permits further analysis and actions that are key drivers of Industry 4.0. Industry 4.0 cannot exist without IIoT, but IIoT would not be efficient without the bigger-picture support of Industry 4.0 (Conway, 2016). In many ways, IIoT is technology implementation using interconnected devices or retrofitted actuators and sensors, data transmitters, wireless tools, and so on. Industry 4.0 is more of an approach driven by emerging technology like IIoT, but sustains a broader range and has a greater vision. IIoT is the function of connected sensors, instrumentation, and other devices to machinery and processes in industrial settings. According to the Boston Consulting Group, 50% of IoT expenditure will be driven by distinct manufacturing, utilities, and transportation by 2020. IIoT applications are present in several industries: aerospace (planes, airports, drones, and automated vehicles); agriculture (smart farms); automotive (connected, autonomous, and semi-autonomous vehicles); healthcare (robotic surgery, connected healthcare, and medical imaging); industrialization (connected factories); military (military vehicles, simulations, military training and operations); oil and gas (refining and exploration); smart cities (parking and infrastructure, municipal services, etc.); transportation (trucks, buses, subways, trains, and Hyperloops). IIoT runs across many industrial applications that include autonomous systems, industrial robotics, and drones.

1.3.2 IIoT Applications

The IIoT has obviously proved to the world its own flexibility with operations that are available across many endeavors and including many case scenarios of IIoT use. It should be noted that implementation of IIoT needs to develop regarding forthcoming industrial productivity. A study made by analytics professionals proved that 79% perceive that the analytical maintenance of industrial goods and instruments will become a fundamental rationale of manufacturing industrial data analytics in forthcoming product developments and 15% expect industrial analytics to be a major success factor in today's businesses; 69% believe it may become even more important in the following decade. Typically, IIoT solutions include:

- Things: Internet facilitated devices such as sensors and actuators.
- Connectivity: communication among "things" through a process, software, a network, and a system.
- Communication: through Wi-Fi protocols such as 3G, 4G, 5G, or LoRaWAN.
- Data: how it is collected, stored, and processed is critical to the value of IIoT
- A cloud platform: a centralized and safe third-party platform service that can hold data.
- Analytics dashboard/software: for monitoring, gathering, and analyzing data and assets that are to be controlled.
- Intelligence and action: the gathered information should be analyzed using either humans or artificial intelligence. As a result of this analysis, action that is taken by humans is executed.

Smart meters have gained tremendous attractiveness in modern years. A smart meter is more of an Internet proficient tool, as compared to traditional meters, and they are even capable of reporting the amount of vital sources such as energy, irrigation, power, and natural gas that is consumed (Molina-Markham et al., 2010). In predictive maintenance, systems analyze the condition of the equipment beforehand and have an ability to lessen the operational industrial expenditure significantly, which results in considerable monetary benefits for manufacturers. Fleet management assists manufacturers to reduce and overcome the risks associated with productivity. The connected technology of IIoT has given power to the fleet business with improved efficiencies in managing the fleet and decreasing staff and transportation costs significantly. Connected vehicles in IIoT are furnished with the Web for automating normal driving tasks. They employ different network technologies that speak to the driver. Optimizing industrial robotic systems in manufacturing employ robots to put them into production: the autonomous machines carry out the same task repeatedly until they achieve sufficient accuracy. The underpinning training and learning technology are frequently used to teach robots and independent machinery. Using these techniques, a robot can moderately or quickly be trained to perform a task under the direction of a human being. The "brains" of such robots are usually neural networks.

1.3.3 IIoT Architecture

IIoT architecture, as shown in Figure 1.2, involves message queuing telemetry transport (MQTT) protocol for communicating with a huge number of devices. This protocol is a secure and lightweight one and employs the well-known publish/subscribe transportation method. The MQTT module is used to gather data from several edge of network devices. Data accumulated by the MQTT engine will be available to a client with the reporting module or vision module. IoT edge devices can operate as gateways, providing a connection between other devices on the network and the IoT hub. Supervisory control and data acquisition (SCADA) is a system of hardware and software elements that permits industrial organizations to manage industrial processes at remote locations and monitor, gather, and process real-time data to provide direct interaction with devices such as valves, pumps, sensors, motors, and more by using human–machine interface (HMI) software and recording incidents in a log file. Ignition software started off as a SCADA/HMI application designed to become a platform for industrial automation. Edge gateways play a critical role in IIoT architecture. The IoT edge hub module acts like an IoT hub, so it can handle connections

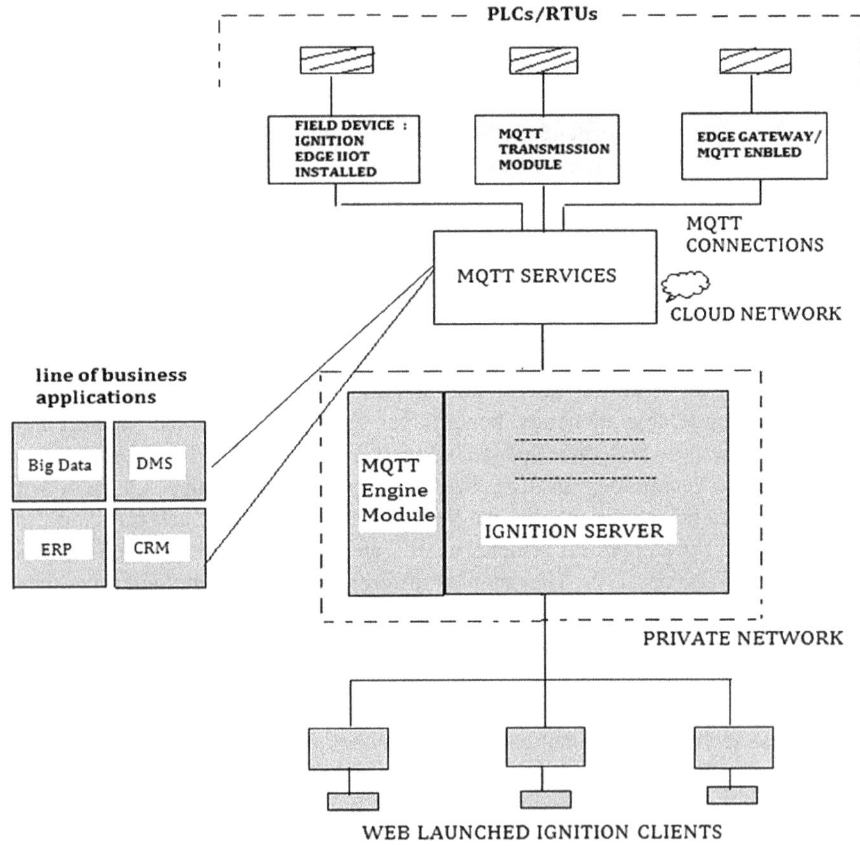

FIGURE 1.2 IIoT architecture.

from any devices that have an identity with an IoT hub, including other IoT edge devices. This type of gateway pattern is transparent because messages can pass from downstream devices to an IoT hub as though there were no gateway between them. For devices that can't connect to an IoT hub on their own, IoT edge gateways can provide that connection. This type of gateway pattern is called a "translation" because the IoT edge device has to perform processing on incoming downstream device messages before they can be forwarded to the IoT hub. These scenarios require additional modules on the IoT edge gateway to handle the processing steps.

Field devices such as Ignition Edge with the MQTT plug-in, Ignition Server with MQTT Transmitter Module, or Third-party MQTT enabled service can be used to connect to a programmable logic controller (PLC). PLC is a programmable microprocessor-based machine that is employed in manufacturing for controlling lines of assembly and machinery on the floor of the shop as well as many other types of electrical, mechanical, and electronic equipment in a plant (Schilling et al., 2018). A PLC is invented for real-time use in rocky industrial environments. PLCs are fixed to sensors and actuators, and are categorized by many types of I/O ports. This device should be capable of publishing tag values to an MQTT server. MQTT is a standard machine-to-machine data transfer protocol. MQTT Broker is a server system that communicates with field devices through MQTT which is an ignition gateway with an installed MQTT distributor module, or sometimes a third-party server, such as a Chariot MQTT Server. A subscriber is nothing but software that subscribes to topics in the Broker. With respect to ignition, this is the gateway with the MQTT engine module installed. A programmable ignition gateway is a service which runs in the environment of our server's operating system. When running, it is connected to a Web browser. Any computer connected to the same network can access the gateway, based on the gateway security settings, by using the host name or IP address of the computer where ignition is installed. An ignition-based IIoT is an end-to-end IIoT solution which unites the remarkable effectiveness of the MQTT data-transfer protocol with the development power and unrestrained data acquisition of the ignition industrial application platform. Using ignition, one can simply connect to plant floors and field devices at the edge of the network, and shove data from many devices across many places through a central MQTT infrastructure to both business and industrial purposes. Security and interoperability certainly are the biggest challenges while designing IIoT architecture. A major problem in designing IIoT architectures is the interoperability between devices and instruments that use diverged protocols and different architectures. Ignition is a brilliant resolution for interoperability because it is built on open-source and benchmark information technologies.

1.3.4 Impact of IIoT on the Economy

The business prospects of IIoT are enormous. In accordance with industry analytics, research, and consulting (ARC) done in June 2016, IIoT trade is set to be approximately $124 billion by 2021 at a higher compound annual growth rate (CAGR). It's difficult to estimate the economic impact of IIoT, as it affects GDP growth, growth distribution, regulation, and competition. IIoT undoubtedly boosts economic growth by means of some key metrics (e.g., GDP). IIoT technologies provide the opportunity to bring manufacturing towards a qualitatively new stage. Consequently, it makes it feasible to integrate agile production systems and also digital control systems into manufacturing, and that makes it easier to control production, accelerate it, and considerably increase its suppleness. Regarding these it is advisable to commence potential technologies from time to time and, owing to the higher costs, the effective use of funds happens to be a key issue. It is very important to provide enough interest on the capability of employees and promote their respective skills. Utilizing the maximum number of automated manufacturing tasks (advanced manufacturing) by means of implementation of IIoT technologies helps to link the activities of various divisions (including consumers, suppliers, logistics, and marketing in the production process by obtaining product details from them in real time) through the creation of only one information field. This helps in increasing the ease of production, the speed of a product's release to market, as well as the optimization cost. The development of IIoT technologies in Russia is based on the strong competencies of the specialists in the field of algorithms and software. This provides good opportunities for the equal participation of Russian organizations in international consortiums and projects related to the IoT. Plentiful cases of use exist in the implementation of the IIoT at the national level, as can be seen with smart metering in Germany at the enterprise level, a smart factory in Bernecker and Josef Rainer (B&R) industrial automation in Austria, and so on.

1.4 SECURITY CHALLENGES IN INDUSTRY 4.0

In Industry 4.0, organizations ought to avoid external as well as insider cyber attacks regarding the need to manage the complicated cyber security sphere. Smart industry, by virtue of its environment and scope, is connected to a huge number of devices and systems; therefore, if any system is expanded and diverged with complexity, there can be considerable increase of unforeseen security loopholes. Consider for example physical products, like bolts. If a hacker gains admittance into the network of a bolt manufacturing plant, and if the setting of the machine that controls the strength of the bolts is modified, this will cause the bolts to be unable to withstand stresses. Sometimes a person may die when a machine or a device fails because the risks are based on how we handle IIoT devices in the firm. Legacy security is equally a big concern. One of the major issues in IIoT deployment is that legacy or inheritance tools can be retrofitted with devices of IoT. This leads to longer life cycles of equipment and successful integration in IIoT, when building a smart factory that is not supplied with costly capital equipment purchases. The main motive for IIoT is to automate the operations of manufacturing industry without the want for human involvement. This intensive automation needs supply chains and the construction lines to be more refined, and needs self-optimizing instruments and artificial intelligence and self-configuration to finish complex jobs for delivering better cost efficiencies and excellent services and goods. However, at the beginning, these innovations brought critical and new threats, particularly in the field of security. In fact, each time there will be a need for extensive amendment, as new inborn challenges occur concerning information security. Integrating new methods and systems along with their access always raises fresh data security breaches. IIoT technologies face many security challenges, as these are divergent threats that affect both entities and organizations, and consequently damage financial growth. As manufacturing business organizations employ several IIoT solutions, they still need to be acquainted with security vulnerabilities, a lack of visibility, unknown risks linked with information technology/operational technology (IT/OT) convergence, and in-house threats (Tuptuk & Hailes, 2018). And when security competence is considered as a selection criterion, then obviously the key challenge is the lack of wide-ranging cyber security solutions. Adopting IIoT technologies may be resisted in the IT industry until and unless it has a solid security feature. So, industries are more vigilant with IIoT automation technologies.

1.4.1 How Secure Is IIoT?

In Industry 4.0, security threats are assumed to be thwarted by focusing on the importance of the collaboration between several partners in the network. Therefore, we observe that it is important to discover solutions that promise trust and transparency within the platform. Whenever a DoS attack is successfully achieved by attacking servers through an enormous number of requests for all existing sources, by means of sending to the server systems, a vast amount of malicious input data which can break down and collapse processes, by disabling or damaging a sensor or infiltrating a virus in the system, clearly prevents it from functioning customarily. Industry 4.0 depends upon a large number of interrelated processes as well as systems, so a DoS attack is a serious hazard in such environments. As cloud computing gains attractiveness, and is extensively employed in concept of the smart industrial unit, then more illicit actors will discover new-fangled ways to take advantage of a server's susceptibilities, like DoS (Pereira et al., 2017). The havoc created by this cyber attack may become very damaging for a company. Such attacks damage equipment such as sensors and servers and consequently servers and other devices have to be restored or returned to normal work processes. Sometimes networks have to be reprogrammed and hosts ought to be reinvented. This damage causes financial and operational losses, service interruption, and training for device or engine operators, and a need to design complex protocols for resuming operations. In the present scenario, DoS attacks are habitually unpredictable as well as extremely hard to manage for the industrial sector. Extended systems and supply chains are Industry 4.0 characteristics that make it possible to reconnect organizational physical environments and to carry the potential to construct the supply chain more resourcefully.

But supply chain elements have intrinsic security loopholes. Often, security threats originate with the suppliers, where they may attempt phishing assaults and the theft of privileged permissions that result in a massive data disclosure. The primary susceptibility is commonly in the top-most layer of the supply chain that reaches to other managerial developments from other reliant actors. Access control by means of authentication mechanisms, security awareness, behavioral analysis, and cryptographic processes are said to be vital security instruments which can assist in preventing supply chain hacks (Wang et al., 2004). The Industrial Revolution of the 18th and 19th centuries brought about immense societal and industrial changes, and Industry 4.0 is bringing diverse modifications to the way that industries work and how they collaborate and innovative. Highly interconnected industries are capable of presenting various improvements to all manufacturing companies and healthcare utilities. Big data generated in IoT gives practical information concerning decision making and provides insight into the improvisation of functions. On the other hand, innovative “things” generally bring unpredicted and unforeseen security risks and other threats exploited by enemies, as they are vigorously functioning to make sure that they also acquire the benefits of Industry 4.0. It is also essential to raise the consciousness of people who are engaged in production, starting from instrument and software developers, and planning and development engineers who work in the plant engineering side.

1.4.2 IIoT Security Challenges

As technology advances in the world with respect to manufacturing products by means of IIoT, understanding the threats and vulnerabilities associated with its deployment is very important. Organizations launching a manufacturing and IIoT initiative, or connecting legacy technology for remote monitoring and automated access, need to take into consideration all of the impending risks and vulnerabilities associated with those resolutions. As hackers turn out to be more sophisticated in employing the same AI technology and data tools as those used to build IIoT systems, the risks, vulnerabilities, and data breaches grow. There are a number of locations where a violation can happen, either within the company, the network of systems, or susceptible areas where there exist possibilities of security breaches (Ali et al., 2018):

- Insecure Web interfaces. The sites where users connect with IoT sensors and devices suffer from insufficient default secrets, session management issues, and password vulnerabilities within the scope of the network.
- Insecure network services. If the network is not secure, attackers may succeed in gaining admission to the communication network from open ports, DoS attacks, and buffer overflow.
- Weak encryption. Weak encryption can permit interlopers the facility to gain data in transit among devices.
- Vulnerable portable device interfaces. Many companies offer a field service as an addition to their industrialized operations for maintenance and repair. So mobile interfaces experience the same problem of authentication and encryption. There exist plenty of hazards associated with cyber security vulnerabilities in implementing IIoT. Some of them are described in detail below.

1.4.2.1 Device Hijacking

Device hijacking happens whenever a malevolent actor gains control of endpoint sensors or devices in IoT, and the owners are ignorant of how the breaches occurred. Based on how “smart” the endpoint devices are, this assault can vary according to the severity of the risk it creates. If an IoT sensor is hacked by ransom ware or spyware, an enemy can take control of the movements of the end-point device itself. This is especially important if the device has automated functionality and controls the function of an Internet-connected product in the field. This can happen if we fail to modernize and set IIoT devices appropriately.

1.4.2.2 Data Siphoning

Data siphoning concentrates on "eavesdropping" style attacks, which target the data that is being carried by an IIoT device quicker than the end user. In this situation, intruders eavesdrop on the communication traffic from the sensors or endpoint devices to the main network for collecting information to which they shouldn't have admission. This type of attack is most concerning when the data sent by an IIoT device is very sensitive or damage is caused when it gets into the wrong hands, which is more crucial for highly regulated industries like healthcare, defense, and aerospace. It is also critical to see whether a device is tapped to send information that allows a competitor to achieve access to an important IP address. In such a case, ensuring that all the information being broadcast is appropriately encrypted and protected is extremely important.

1.4.2.3 Denial of Service Attacks

Another common threat with IIoT devices is vulnerability to a distributed denial of service attack on all devices and the internal network itself. In this threat, attackers employ a device or network, which may be centralized, for flooding the endpoint devices with a huge amount of traffic so that they cannot complete the task they were on. Fundamentally, a DDoS attack only targets the endpoint devices of IIoT. It is important for organizations to be aware of DDoS assaults and damage, particularly when relying on the endpoint devices which are necessary for production to continue, and for products to work properly in the field. In this context, a security solution that completely protects the devices from the outside world, and other networks it speaks to, is especially attractive.

1.4.2.4 Data Breaches

This threat involves attacking the IIoT devices in a centralized network where sensitive and important information is placed. Since the scope of this attack concerns many devices, owing to the conventional technology alarms stated above, it uses them as a major target and exploits them as a "doorway" to critical business networks. Attackers use them as a way to attain entry point to the enterprise networks and gain admission to critical data that needs to be protected, including: clients' sensitive data or partners' business information, personally identifiable information, intellectual property or trade secrets, health data, and financial data. As observed earlier, the better way to save IIoT devices from being exploited as an entry point to networks is to secure the devices themselves. A hardware-based VPN solution is the best option to achieve this.

1.4.2.5 Device Theft

Another concern with endpoint equipment in the field is the stealing of physical devices. This risk is intense when endpoint equipment stores crucial information which may cause apprehension if that information ends up in the wrong hands. Often, IoT industries deploy their products protected from this risk by avoiding the storage of any sensitive data on the endpoint devices themselves, and trusting the cloud-based infrastructure to hold that information.

1.4.2.6 Man-in-the-Middle or Device "Spoofing"

This threat shows the talent of a hacker who places him or herself in the middle of the cloud or a centralized network and an IIoT endpoint device, and "pretends" the device is sending the data. This is most alarming if the traffic originating from the endpoint device is such that it might be used to alter production information or control information in this field. If we consider the bolt manufacturing example, in this case when an attacker pretends to be an IIoT sensor, he or she can send false information in return, which causes manufacturing machines and equipment to modify their processes or calibration, which might result in defective bolts being made. It is essential to sufficiently educate employees regarding security constraints and standards for the successful implementation of safety measures in an organization. Raising awareness about security requires the full involvement of all industrialized surroundings which might help to overcome present deficiencies in this environment. Apart from that, introducing mandatory research groups and expert classes in this area in advanced educational domains and institutions would aid in preparing personnel. Implementation of IIoT towards Industry 4.0 is presumed to be a critical job that influences a large area of today's mechanized industries, especially with respect to security concerns. Many manufacturing industries at present are not completely coping with their security vulnerabilities, though they have moved towards implementation of the Industry 4.0 paradigm. In general, they only attend to security matters when a sobering and unpleasant situation occurs. Therefore, it is critical that manufacturing industries strategically progress toward installing and running the security-assured conformity practices that Industry 4.0 always requires.

1.4.3 IIoT Security Solutions

The long-term gains from the enormous prospects offered by Industry 4.0 will persist if modern industries set up successful manufacturing systems along with competent security monitoring systems for smart industries. Conventional security schemes – like installing antimalware and antispyware, firewalls, and network monitoring and managing systems – can only shield explicit parts of the IT infrastructure from possible threats. Alternately, hackers may focus on identifying various mysterious security loopholes. The right approach for improving IIoT security is ensuring endpoint security and integrity methodologies which rely on secure-by-design approaches that are appropriate for endpoint instruments (Vermucht, 2020).

1.4.3.1 Endpoint Security-by-Design

An initial step in the direction of achieving secured IIoT endpoint "things" or devices is obviously conducting a vulnerability and risk analysis to decide whether they fulfill the relevant industrial standards. The following are associated specific safety and security measure requirements and descriptions of methods to lessen these problems.

1. Endpoint identity: Manufacturers of industrial instruments should maintain suitable validation methods for authenticating a user by means of their credentials so as to provide user control and access management functions for their services. As an example, IIoT machines should adopt a public key infrastructure (PKI) framework which identifies the cryptographic keys of users employing a certification authority (CA). Secure socket layer (SSL) connections are not verified or tested for authenticity by means of verifying server certificates; also clients may suspect that the system has become subjected to man-in-the-middle assaults as it may not recognize whether it is talking to a trustworthy server or malicious code that is imitating the server. Client services and applications should never depend on domain addresses because domain name system (DNS) services are also vulnerable. While validating the server certificates, it is very important that the certificate signatures and their validity periods are verified. Self-signed certificates by unfamiliar parties must not be acknowledged as root CA certificates. Server systems must contain certification support and provide a choice for legitimate, safe, authentic, and genuine client-side messages.
2. Secure boot: The primary computer system security function is a "secure boot" that ensures the authenticity as well as the integrity of the running code. For example, the U-boot integrity prerequisite is a very significant security-related Linux system obligation and recommends U-boot authentication of the OS kernel by signing the kernel image before loading it. The proposed cryptographic schemes for this rationale contain RSA-2048 with SHA-256 hash, and Elliptic Curve Digital Signatures using the P-256 or P-384 curves with SHA-256 hashes or higher.
3. Cryptographic services: Manufacturing and engineering instruments have to employ and implement cryptographic techniques across well-known transport protocols, applications, and storage. Feeble or substandard security mechanisms and policies bring communicating systems into the danger of identity theft, tampering, eavesdropping, and malware injection that may result in equipment captures or submission of erroneous information to networks and their services. For example, OPC-UA is a standard for information data exchange, manufacturing, and business communications, and is the most exercised protocol in the market for IIoT. One security constraint in favor of this practice is that an OPC-UA server must adopt powerful cryptographic practices and policies. This protocol presents numerous profiles for configuring security policy. Basic SHA-256 is the safest, and thus the recommended settings are: AES128-SHA256-RSAOAEP, AES256-SHA256-RSAPSS, PUBSUB-AES128-CTR, PUBSUB-AES256-CTR, BASIC128RSA-15, and BASIC256.
4. Secure communications: IIoT instruments should employ several protocols for communication as well as protection by means of cryptographic techniques and ensuring authentication by implementing a firewall barrier for network and user access control, because communicating online with no authentication, integrity, and encryption can cause data disclosure, spoofing, and corruption. This indicates clearly that the entire communication might be modified,

interpreted, and injected, which may cause device takeover as well as device damage. The security standards of VPN servers majorly depend on how the trader integrates them into its device, which is very hard to do properly. While using VPN servers, good practice commendations and well acknowledged security supports should be pursued to guarantee that the IIoT services are perfectly built.

5. Event management and monitoring: IIoT endpoint tools and devices should always perform real-time as well as continuous monitoring. The most crucial part in achieving this is employment of embedded runtime agents which are dedicated and designed for each device depending on the systematic investigation of its firmware executables. These spotlights on watching the vulnerabilities of devices they are prone to, and ensuring the needs of the agent such as runtime security, together with processor, memory, and database' never affect the device's functionality. For instance, the log information security requirement of auditing insists that these alerts as well as logs are to be produced for significant security-related incidents. So, it should be mandatory to record the subsequent features of events such as the component for which the event took place, the date and time of the event, the type of event, the outcome of the event, and the subject's identity.
6. Detection tools: For securing each state of the IT infrastructure, an "automated risk identification" module is employed by RADAR Cyber Security, which is a European market leader for proactive IT security monitoring and risk detection where both are managed services. The following modules are suggested by RADAR Services.
   - Advanced Cyber Intrusion Detection (ACID): Discovering hazardous malicious code or malware, abnormalities, and several types of vulnerability threats and risks within the traffic of a network by using digital signatures and behavioral intrusion detection systems.
   - Host-based Intrusion Detection System (HIDS): Using analysis, compilation, connecting servers, client log files, and random alerting whenever certain attacks have happened, or exploitation and errors are detected. The integrity of data in the local system is verified. Root kits, trojans, other types of attacks, and viruses are acknowledged.
   - Vulnerability Management and Assessment (VAS): An assessment of impending security errors in the software, operating systems, and monitoring of anomalies of all information is transmitted via the network.
   - Advanced Email Threat Detection (AETD): Employment of future generation sand box techniques in sensing email hacking and malware.
   - Software Compliance (SOCO): Automatic examination of the faithfulness of fulfilling rules and guidelines, and instant description of vulnerabilities to reduce risks.
7. Analysis and evaluation of experts: Security can be measured by ethical hackers and specialized experts. They measure, examine, and analyze outcomes using the most up-to-date technical knowledge, and they are also accountable for the progressive growth of industrial, automated, or computerized

controlling systems. These results can be observed as part of a big picture and an examination should be considered crucial in certain critical events, particularly in IT industry infrastructure and the newest developments inside or outside the company. Security authorities should respond faster to present perfect instructions regarding troubleshooting procedures and they must persistently acclimatize to all of the rules and policies followed by advanced correlation engines, risk identification modules to mark and eradicate vulnerabilities, and other kinds of attacks without hindrance.

1.5 CONCLUSIONS AND FUTURE WORK

1.5.1 Conclusions

The IoT assists in several day-to-day activities of life in the business sector, such as proficient resource utilization, minimizing individual effort, and saving time. IIoT enables intelligent engineering and industrial processes by employing data analytics for transformational business results by connecting machines and people. The main drivers of IoT implementation are increasing operational competence, production, creating innovative production and business opportunities, and maximizing asset consumption. The emergence of the IIoT movement has created both expectation and confusion among stakeholders that are accountable for operating industrial plants. There are many IIoT implementation challenges presently faced by organizations. This chapter has provided an overview of IIoT, implementation challenges, and applications. Industry 4.0 describes the fourth industrial revolution, which depends on the IoT. Moving in the direction of Industry 4.0 introduces different industrial, technological, and scientific challenges, together with higher impacts on today's manufacturing industry, especially in the security domain. While there are lots of benefits that IIoT brings to the world, this technological leap has generated numerous security concerns that need to be addressed in the case of critical connected devices. Industry 4.0 is further susceptible to cyber assaults due to connected and smart business practices. In this chapter we have learnt about IIoT security challenges in the Industry 4.0 paradigm and the recommended security solutions.

1.5.2 Limitations

The exponential growth of IIoT creates a fundamental need for vigorous cyber security policies, practices, and well-described standards which provide consumers with assurance that their connected devices are operated securely throughout their lifecycle. This chapter has focused on the endpoint security of IIoT devices where it offers an approach to the shielding of computer networks that are remotely connected to client devices. Still, we need to explore several security standards that may encompass all elements of security in the IIoT from product manufacturing and development to product and system features, operations, and delivery in the present scenario of Industry 4.0; these security aspects have not been focused on primarily in this chapter and still need to be investigated.

1.5.3 Future Work

The architecture of the IIoT is made of many elements (sensors, connectivity, and gateways) to manage devices and a variety of application platforms. IIoT architecture is a broad area of research and designing IIoT application is a multidisciplinary activity. This chapter has discussed the fundamental architecture of the IIoT, but further study could be extended to current advanced architectures of the IIoT that are intended to ensure security and the employment of agile methods. Therefore, several types of architecture of the IIoT could be studied in the future which would facilitate advanced security and protection approaches applied to different levels of architecture based on the exact means of a network assault.

REFERENCES

Abomhara, M., & Køien, G. M. (2014, May). Security and cyber in the Internet of Things: Current status and open issues. In *2014 International Conference on Privacy and Security in Mobile Systems (PRISMS)*, 1–8, IEEE.

Al-Fuqaha, A., Guizani, M., Mohammadi, M., Aledhari, M., & Ayyash, M. (2015). Internet of things: A survey on enabling technologies, protocols, and applications. *IEEE Communications Surveys & Tutorials, 17*(4), 2347–2376.

Ali, S., Bosche, A., & Ford, F. (2018). *Cybersecurity Is the Key to Unlocking Demand in the Internet of Things*. Boston, MA, USA.

Berrehili, F. Z., & Belmekki, A. (2016, May). Privacy preservation in the Internet of Things. In *International symposium on ubiquitous networking* (163–175). Springer.

Boyes, H., Hallaq, B., Cunningham, J., & Watson, T. (2018). The industrial internet of things (IIoT): An analysis framework. *Computers in Industry, 101*, 1–12.

Conway, J. (2016). *The Industrial Internet of Things: an evolution to a smart manufacturing enterprise*. Schneider Electric.

Dean, A., & Agyeman, M. O. (2018, September). A study of the advances in iot security. In *Proceedings of the 2nd International Symposium on Computer Science and Intelligent Control* (1–5).

Jeschke, S., Brecher, C., Meisen, T., Özdemir, D., & Eschert, T. (2017). Industrial internet of things and cyber manufacturing systems. In *Industrial internet of things* (3–19). Springer.

Jia, X., Feng, Q., & Ma, C. (2010). An efficient anti-collision protocol for RFID tag identification. *IEEE Communications Letters, 14*(11), 1014–1016.

Jindal, F., Jamar, R., & Churi, P. (2018). Future and challenges of internet of things. *International Journal of Computer Science and Information Technologies, 10*(2), 13–25.

Molina-Markham, A., Shenoy, P., Fu, K., Cecchet, E., & Irwin, D. (2010, November). Private memoirs of a smart meter. In Proceedings of the 2nd ACM workshop on embedded sensing systems for energy-efficiency in building (61–66).

Pereira, T., Barreto, L., & Amaral, A. (2017). Network and information security challenges within Industry 4.0 paradigm. *Procedia manufacturing, 13*, 1253–1260.

Schilling, K., Storms, S., & Herfs, W. (2018, July). Environment-integrated human machine interface framework for multimodal system interaction on the shopfloor. In Ahram, Tareq, Karwowski, Waldemar (Eds.), *International conference on applied human factors and ergonomics* (374–383). Springer.

Sicari, S., Cappiello, C., De Pellegrini, F., Miorandi, D., & Coen-Porisini, A. (2016). A security-and quality-aware system architecture for Internet of Things. *Information Systems Frontiers, 18*(4), 665–677.

Suo, H., Wan, J., Zou, C., & Liu, J. (2012, March). Security in the internet of things: a review. In *2012 International Conference on Computer Science and Electronics Engineering* (Vol. 3, 648–651). IEEE.

Swamy, S. N., Jadhav, D., & Kulkarni, N. (2017, February). Security threats in the application layer in IOT applications. In *2017 International Conference on I-SMAC (IoT in Social, Mobile, Analytics and Cloud) (I-SMAC)* (477–480). IEEE.

Thakur, K., Qiu, M., Gai, K., & Ali, M. L. (2015, November). An investigation on cyber security threats and security models. In *2015 IEEE 2nd International Conference on Cyber Security and Cloud Computing* (307–311). IEEE.

Tuptuk, N., & Hailes, S. (2018). Security of smart manufacturing systems. *Journal of manufacturing systems, 47*, 93–106.

Vermucht, M. (2020). The Keys to Securing Industrial IoT (IIoT) Environments. https://www.vdoo.com/blog/industrial-IoT-security.

Vignesh, R., & Samydurai, A. (2017). Security on Internet of Things (IoT) with challenges and countermeasures. *International Journal of Engineering Development and Research, IJEDR, 5*(1), 417–423.

Wang, L., Wijesekera, D., & Jajodia, S. (2004, October). A logic-based framework for attribute based access control. In *Proceedings of the 2004 ACM workshop on Formal methods in security engineering* (45–55).

Wunck C., & Baumann, S. (2017). Towards a process reference model for the information value chain in IoT applications. *IEEE European Technology and Engineering Management Summit (E-TEMS)*, Munich, 1–6. doi:10.1109/E-TEMS.2017.8244228.

Zeadally, B. H. R. K. S., & Khoukhi, A. F. L. (2018). Internet of Things (IoT) Technologies for Smart Cities.

2 A Demand Response Program for Social Welfare Maximization in the Context of the Indian Smart Grid

A Review

Sachin Ramnath Gaikwad and R. Harikrishnan

Symbiosis International Deemed University (SIDU), Pune, India

CONTENTS

2.1 Introduction 25
2.2 A Smart Grid for India: A Dire Need 27
2.3 Progress of the Indian Smart Grid 29
    2.3.1 SG Initiatives in India 29
    2.3.2 Indian Governmental Energy Organization 30
    2.3.3 Smart Grid Projects and Their Experience in India 31
2.4 DRP-based Social Welfare Maximization Model 32
2.5 Literature Survey, Research Findings, and the Framework of SWMM 33
2.6 Challenges and Discussion 35
2.7 Conclusion and Future Scope 36
References 36

2.1 INTRODUCTION

In the 21st century, the complexity of demands and challenges on the electricity grid are rising continuously. In the 1990s power generation was centralized and mostly away from the load side.

Power is transmitted from the generation plant to users without exchanging information. Despite this, throughout the world, traditional electricity systems face many challenges like power blackouts, transmission-distribution losses, increasing emissions of greenhouse gases (GHGs), reducing non-renewable resources,

DOI: 10.1201/9781003146711-2

environmental imbalance, and climate change (ISGM, 2017). On 13 August 2003, North America faced a major power blackout that lasted for around four to seven days, and nearly 50 million people were affected. In the subsequent analysis, it was found that data communication was inadequate. On 30 and 31 July 2012, the eastern and northern states of India also faced a major blackout. Some other instances of major blackouts worldwide occurred in: the Greek Island of Kefallinia (January 2006), San Diego (September 2011), Vietnam (May 2013), Bangladesh (November 2014), and Ukraine (December 2015) (BDASG, 2020). Despite these lapses, there has been a continuous increase in demand for electricity among residential, commercial, and industrial users in recent years primarily due to the development of new electronic devices and technologies like electric vehicles (EVs). On the other hand, the complexity of the electric grid is also increasing with less predictability of the loads. With this, it is also important to take into consideration global warming issues and the reduction of carbon emissions to achieve sustainable development goals. Considering all these aspects, every country is trying to upgrade its existing grid with advanced technologies. The U.S. Department of Energy (DoE) has defined a smart grid (SG) thus: "It is a modernized grid having automation technologies, integrates distributed generation stations, and allows the flow of bi-directional energy and information with on-demand reliability" (DoE, 2019).

There is an urgent need to revise and modify our existing electricity system using advanced technology. In this, implementations of energy efficiency and energy conservation-related regulations, code, standards, and programs play a significant role in demand-side management. The improvements in the overall energy efficiency of the power system depend on several major initiatives. The initial need is to focus on spreading awareness about the smart grid to prosumers (producer + consumer), utility providers, and other stakeholders. Thereafter, the need is to focus on micro-grid regions to achieve the decentralization of power generation using distributed generation (DG). DG is a major transformational change towards energy independence and energy islanding. A micro-grid has the potential to integrate renewable energy sources (RESs) for DG along with the main grid system (ISGM, 2017; NSGM, SGPP, 2019).

Figure 2.1 shows the integration of RESs with the main grid at different user regions, such as residential, commercial, and industrial regions. The RES-based DG helps to develop the energy island system; as a result traditional boundaries such as grid dependency on centralized generation are now drastically disappearing at generation, transmission, and distribution levels. RESs provide an opportunity for end-users to become producers. "Prosumer" is the new term coined to represent this (ISGM, 2017).

Section 2.2 focuses on the need for SG in India. Section 2.3 covers the progress of the Indian smart grid. Section 2.4 proposes the idea of a social welfare maximization model (SWMM) based on a demand response program (DRP). Section 2.5 is a literature survey and discusses research findings and a framework for an SWMM. Section 2.6 concerns challenges to an SWMM in an Indian context. The last section 2.7 concludes.

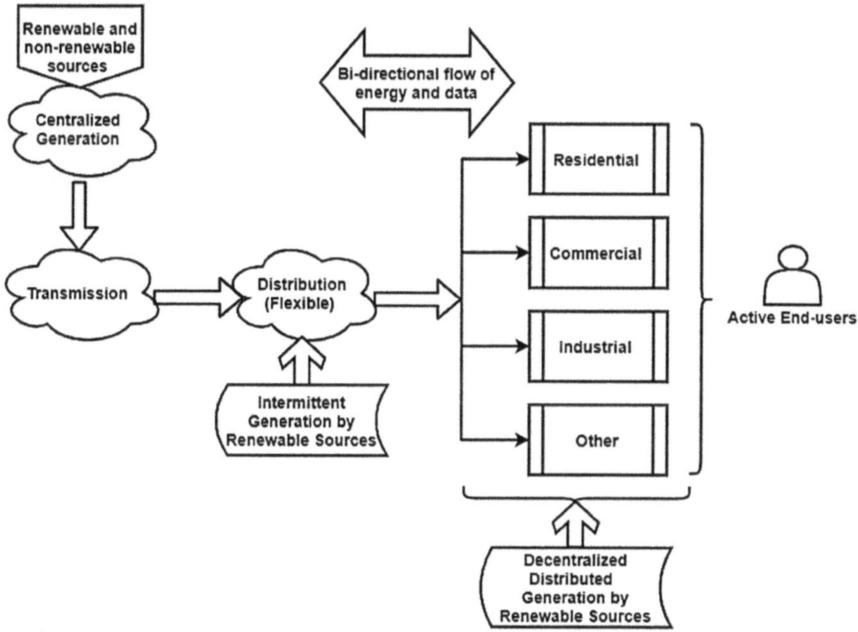

FIGURE 2.1 Framework of SG model (ISGM, 2017).

2.2 A SMART GRID FOR INDIA: A DIRE NEED

India has two main challenges: one is the national-level challenge to balance the country's speedy and sustainable power growth. The second is to fulfill the climate change bindings of international organizations. India is a developing country with diverse problems in the power sector and is also committed to reducing carbon emissions in response to Conference of Parties (COP) decisions 1/CP.19 and 1/CP.20 for the period 2021 to 2030 (ESSRI, 2019). In India, the installed capacity of electricity was 364,169.59 MW on 31 October 2019 out of which the central government contribution was around 25%, the state government contribution was around 29%, and the contribution of the private sector was around 46% (GoI, MoP, ISGF, 2017). Among all the major contributing energy sources, coal power was the highest (around 86%), followed by thermal energy (around 63%). So, there is a dire need to shift towards eco-friendly, clean-green energy sources like RESs, namely wind, solar, biomass, and hydropower. In India, the RES sector installed capacity is around 22% of the total installed capacity as of 30 April 2019. So, the few solutions are likely to integrate more RESs with the main grid to spread awareness of the benefits of RESs and achieve the active participation of consumers through various price and incentive-based DRPs (ISGM, 2017). India had faced a major power blackout on 30 and 31 July 2012 in which 700 million people from eastern and northern Indian states were affected. The consequent analysis found that the 400 kV Gwalior-Bina double transmission line in UP was overloaded (BDASG, 2020). India is also facing higher aggregate technical and commercial losses, which is above 25% in the electricity transmission-distribution network in towns. The Government of India (GoI) has taken initiatives through various schemes to tackle all the diverse challenges to achieve the maximization of social welfare (ISGM, 2017).

According to the Indian power blackout analysis in 2012, it was found that the transmission system operator failed to have real-time information, monitoring, and control. In its aftermath, GoI focused more on real-time monitoring systems using wide area monitoring system technology based on a phasor measurement unit (PMU). Hence, India had to modernize the conventional power grid to satisfy future perspectives (BDASG, 2020).

Consumers, utilities, and regulators are the key drivers of the SG system, as shown in Figure 2.2. The effective coordination between these key drivers is important to obtain social welfare maximization (NSGM, SGPP, 2019). There are three major advantages of SG in the context of the maximization of social welfare, namely: social development, economic development, and environmental and sustainable power development, which are discussed here. The advantage includes 24/7 electricity for all, the active participation of consumers, real-time demand side management (DSM), power system security, power quality and reliability, and the scope to create a new business model, which creates better competition, reduces the overall electricity cost, and maximizes the benefits of utility. Moreover, India is committed to reducing carbon emissions regarding COP for the period 2021 to 2030, and to reduce its GHG emission and carbon footprint (ESSRI, 2019). GoI has instigated a few initiatives for attaining sustainable power development, such as the ash utilization action plan, replacing coal-based power stations with environmentally friendly ones, and adopting a clean development mechanism (CDM) and ISO 14001 (GoI, MoP, 2019; Figure 2.2).

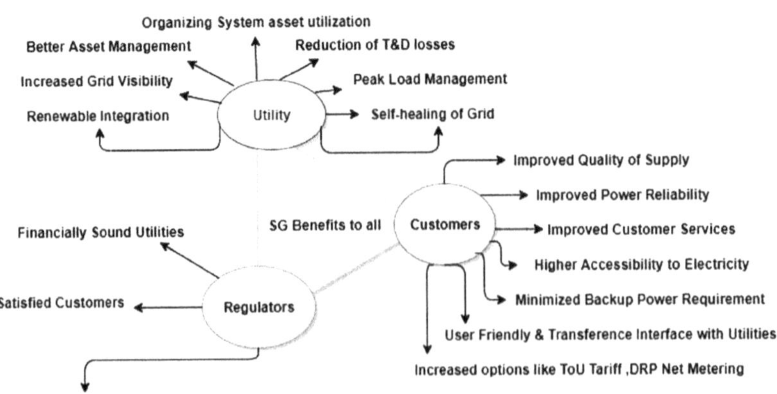

FIGURE 2.2 Key drivers of the SWMM (NSGM, SGPP, 2019; ISGM, 2017).

2.3 PROGRESS OF THE INDIAN SMART GRID

The power system is the pillar and engine of the nation's overall growth. After the 2012 blackout, GoI realized the limitations of the existing grid and then actively started working on its modernization. GoI has taken the initiative to develop SG for social welfare maximization as given below (BDASG, 2020). GoI established the Restructured Accelerated Power Development and Reforms Program (R-APDRP) in 2008. The purpose of R-APDRP is to reduce various power losses in the electric grid system. It collects truthful baseline data using the adoption of information and technology in the area of energy accounting and decision making. It also strengthens the regulations on distributed projects. So, before launching the SG vision and mission plan it is essential to build a foundation for SG development. The vision and mission statement is a preamble and illustrates a road map with guidelines for SG development. In 2013 GoI's Ministry of Power (MoP) launched the "Smart Grid Vision-Mission and Roadmap." The outline of the Vision Statement was: "Transforming the Indian electricity grid in terms of security, sustainability, adaptability, and digitally-enabled ecosystem which will provide stable, quality, low cost and stable power to all by the active participation of stakeholders." The Mission Statement outline was: "To enable real-time connectivity and reliability for green power to all citizens by integrating renewable energy sources by distributed generation in a micro-grid" (NSGM, SGPP, 2019; NSGM, Framework 2018).

2.3.1 SG Initiatives in India

The development of SG falls under MoP, GoI. SG deployments, observation, and results analysis are conducted by a separate advisory body called the India Smart Grid Forum (ISGF) and the India Smart Grid Task Force (ISGTF), which was launched in 2015. Figure 2.3 shows the important timeline of activities of the National Smart Grid Mission (NSGM), spanning the years 2010 to 2018. ISGF and ISGTF were launched in 2010 by MoP. They are important bodies and both support each other. In 2011, NSGM defined the functional specifications of smart meters, which are essential in order to achieve real-time communication within utilities and for consumers to know their energy consumption curve (ECC), data analytics, monitoring, control operations, and so on. After that MoP sanctioned 14 pilot projects in the year 2012. MoP developed the foundation to launch big investment into SG. GoI then launched its vision, mission, and roadmap for Indian SG. This initiative provided a constructive approach, direction, and shape to SG development across India. In 2014, IIT Kanpur took the initiative to work on a smart city pilot project.

The project included 11 substations with one 33 kV/11 kV main station and the others of 11 kV/0.415 kV. They also completed 20 automation projects which involved the installation of a rooftop photovoltaic system. In 2015, NSGM established IS16444 and released an SG regulated model. In 2016, four pilot projects were sanctioned based on advanced metering infrastructure (AMI) functional requirements. In 2017, an important amendment to IS16444 was made and IS15959 was published. Four more projects were sanctioned in 2018 (NSGM, IITK, 2019).

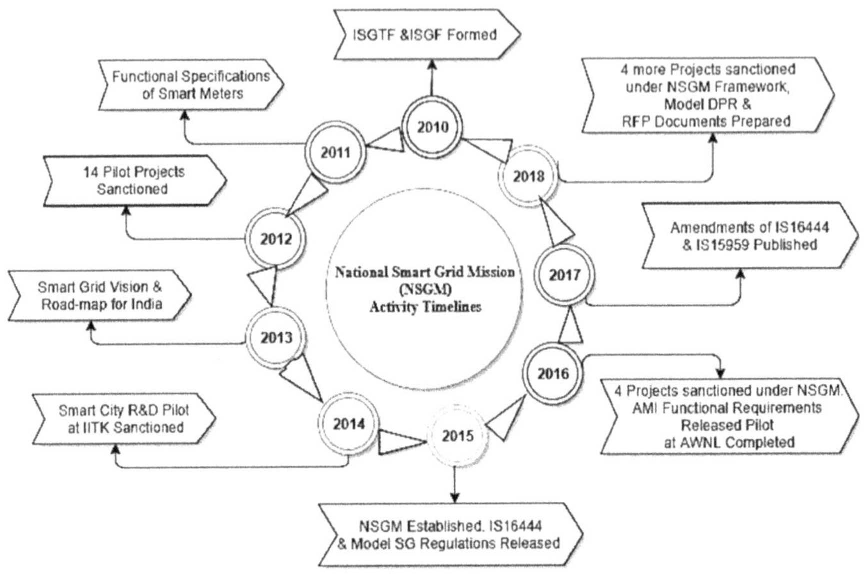

FIGURE 2.3 GoI's National Smart Grid Mission activity timelines between 2010 to 2018 (NSGM, SGPP, 2019).

The important initiates in the year 2015, in terms of launching schemes, were: the Integrated Power Development Scheme (IPDS), Ujwal Discom Assurance Yojana (UDAY), Deendayal Upadhyaya Gram Jyoti Yojana (DDUGJY), and Energy Efficiency Services Limited (EESL) (ISGM, 2017). The IPDS is a new form of the R-APDRP scheme, in which funds are provided for reducing various losses, development of infrastructure, real-time billing, effective collection, monitoring systems, and so on. Moreover, the DDUGJY scheme is focused on rural households and agricultural purposes. It aims to provide 24/7 electricity to all if household lines and agricultural lines are separated in villages. UDAY is providing a liability restructuring plan for distribution companies. This is an important scheme and has the potential to change the face of the power sector in India. EESL is a MoP, GoI initiative. Its objective is to support the activities of CDM projects, an energy efficiency (EE) program, and reducing the carbon footprint (Desi Smart Grid, 2013; EESL, 2019; GoI, MoP, DDUGJY, 2019; PIB, 2019).

2.3.2 Indian Governmental Energy Organization

In India, MoP is the supreme body that takes centralized decisions on SG development. Additionally, GoI promotes the private sector via a public–private partnership (PPP) model by providing various funds, infrastructure facilities, and so on. The ISGTF organization works along with MoP, the Department of Science and Technology (DST), the Ministry of New and Renewable Energy (MNRE), and the Ministry of Communications and Information Technology (MCIT). Other governmental organizations include the Central Power Research Institute (CPRI), the

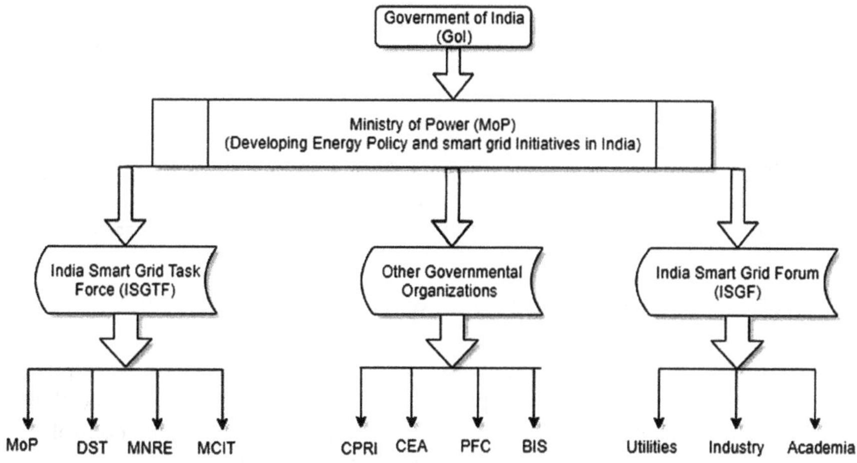

FIGURE 2.4 GoI's energy organizations for SG implementation (Samantaray, 2014).

Central Electricity Authority of India (CEA), the Power Finance Corporation (PFC), and the Bureau of Indian Standards (BIS). The ISGF works directly with utilities, industry, and academic organizations as shown in Figure 2.4 (Samantaray, 2014).

The following are the three main advisory bodies which are supporting Indian SG development. First, ISGF is a PPP initiative to speed up SG development. It is an advisory body on government initiatives for SG and works with different national and international agencies. Second, ISGTF plays a key role in raising awareness, coordination, and integration on SG-related initiatives. Moreover, ISGTF does the effective coordination and integration between other inter-ministerial bodies like MoP, DST, MNRE, and MCIT as shown in Figure 2.4. ISGTF comprises five working groups. Third, Shakti Sustainable Energy Foundation was established in 2009 with the aim of providing a sustainable energy future in India. It works on policies and programs for encouraging EE and integrating RE with the main grid (GoI, MoP, ISGF, 2019; ISGM, 2017; NSGM, SGPP, 2019).

2.3.3 Smart Grid Projects and Their Experience in India

Any new development in technology is done through pilot projects in different parts of the country. Based on the outcome of the pilot projects, awareness programs are implemented with the stakeholders which creates a positive business environment in the market. The key objective of an SG project is to enrich the overall efficiency, connectivity, and communication within the generation stations, transmission lines, and distribution stations. At the same time network security, visibility, and access has to be taken care of. Due to global climate changes and CO_2 emissions issues, the priority is to integrate RESs with the main grid (Desi Smart Grid, 2013; ISGM, 2017).

Figure 2.5 shows the general phases of an SG pilot project development, from the awareness step to the final full-scale project implementation (NSGM, SGPP, 2019). Moreover, the outcomes of pilot projects are important to plan future

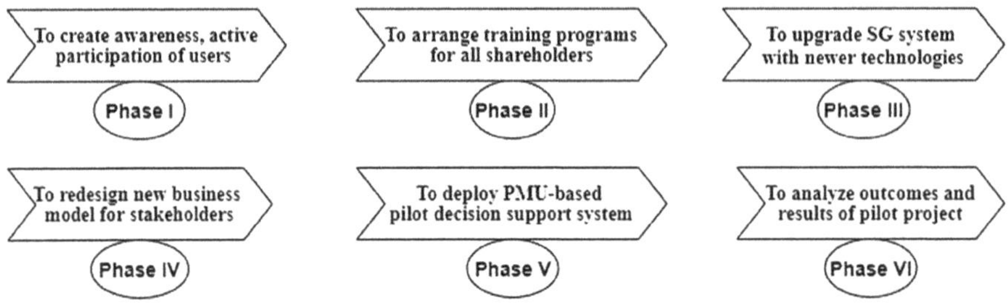

FIGURE 2.5 Phases of an SG pilot project development (NSGM, SGPP, 2019).

full-scale project policies and road maps. Here, we discuss a few of the important technical experiences of small-scale project works. The pilot projects used different technologies such as AMI, PMU devices for real-time monitoring and control, and analysis from a remote location in an easy way. The PMU data helps us to know the nature of the energy consumption of users. It is important in demand forecasting, peak load management, the management of dynamic tariffs, and so on. It is also possible to remotely connect or disconnect the system. At the micro-grid level, an outage management system is implemented and also integrated with RESs. There are also improvements in the power quality and the reliability of the power supply when using network control management. Moreover, the IPDS of MoP has implemented initiatives for SG development. Their primary focus was on technology demonstration and then testing of the pilot base across India. Moreover, for the implementation of SG projects, GoI provides a 50% fund. The NSGM website shows the funded project list under NSGM mission innovation by the National Smart Grid Mission Project Management Unit (NPMU). Some of the listed projects have international collaboration. Most of the projects listed are granted funds by GoI and are RES based (GoI, MoP, ISGF, 2019; ISGM, 2017; NSGM, NPMU, 2019; Figure 2.5).

2.4 DRP-BASED SOCIAL WELFARE MAXIMIZATION MODEL

There are five main types of optimization models: minimization of electricity cost, maximization of social welfare, minimization of aggregated power consumption, minimization of both electricity cost and aggregated power consumption, and both the maximization of social welfare and the minimization of aggregated power consumption (Kimata et al., 2020; Vardakas et al., 2015). This section proposes the thematic idea of a DRP-based SWMM (DRP-SWMM). The DRP-SWMM takes care of the welfare of different stakeholders, including consumers, prosumers, utilities, and the environment. The few important characteristics of DRP-SWMM are that is saves non-renewable resources, focuses more on EE, EC, DSM, and DR program techniques, reduces utility investment and maximizes profits, minimizes consumer energy consumption and cost, integrates green energy sources, and reduces energy theft and energy losses. The increase in fuel consumption is directly related to carbon emissions. Electricity demand and its complexity is unstoppable. So, we have to balance all circumstances while designing and developing the social welfare maximization model. Along with this, it is also vital to improve the EE of systems, and reduce the energy demand, energy consumption, and power losses, through different DSM techniques and DRP (Gaikwad & Harikrishnan, 2021). Furthermore, ISGM (2017) discussed a "three-legged stool" approach for energy optimization. This approach consists of three parts: the integration of RESs with the main grid, using energy-efficient appliances and adopting energy-saving habits. Overall, this approach supports effective load management and also helps to reduce the load shedding effects (ISGM, 2017).

2.5 LITERATURE SURVEY, RESEARCH FINDINGS, AND THE FRAMEWORK OF SWMM

Jamil et al. (2019) applied optimization techniques to household appliances to reduce electricity cost. The author narrates the application of a load shifting technique using optimization algorithms, namely: the Cuckoo Search Algorithm, Earthworm Optimization, and a Hybrid Technique Cuckoo-Earthworm Optimization. The hybrid optimization technique gives better results by reducing the energy cost by 49% which is due to the shifting of appliances in low price hours. If consumers are not following the load scheduling and want to reduce the waiting time of appliances then the higher energy charge may be applicable. This paper also highlights the future scope for selling the excess electricity produced by one household to other households. The paper used a simulated dataset along with a few case studies of ground level problems which were evaluated by using a real-time dataset, and which needs to integrate the energy storage system (ESS) (Jamil et al., 2019). Moreover, ESS helps to optimize electricity bill cost and increase power reliability (Gaikwad & Harikrishnan, 2021).

Nakabi and Toivanen (2019) used two models: a fully connected neural network (NN) and a long short term memory (LSTM) to learn the impact of electricity price on household electricity consumption. These models consider electricity usage patterns of shiftable appliances and thermostatically controlled loads in a house. The NN-based first model is used to forecast the load a day ahead by considering two types of loads – shiftable appliances and basic load without thermostatically controlled loads (TCLs) – by using a multilayer perceptron. The LSTM-based second model forecasts the hourly consumption from TCLs using time-series data on energy cost and external temperature using a recurrent technique. A combination of both these models was used to forecast the total household consumption. The load prediction is obtained by considering parameters like external temperatures, electricity prices, and history loads. The LSTM model shows good accuracy in forecasting the electricity consumption pattern with incomplete data of the inside temperatures of households. It is also used to understand the electricity consumption of TCLs by including energy cost and outside temperature datasets. The main challenge is to work with uncertainties in dwelling devices and inside temperature. This study supports achieving an SWMM through different parameters, such as cost optimization, electricity consumption reduction by using price-based DRP schemes, demand flexibility with balancing load demand and energy supply, and integration of RESs to obtain GHG emission reduction. Also, artificial-intelligence-based models have not considered users' data safety and privacy issues (Nakabi & Toivanen, 2019).

Energy optimization provides an effective solution to improve the efficiency of an electricity grid system. Monfared et al. (2019) proposed the active participation of consumers using cost-based and incentive-based DRP to achieve an SWMM. The authors discussed a retail electricity pricing mechanism and a hybrid price-based demand response (HPDR). The day-ahead scheduling model considered the different uncertainties of decision-making variables, parameters, energy production units, and load dispatch at the micro-grid level to improve the accuracy of the DRP-based model. Analysis shown that the peak-to-valley index and the coefficient of variation percentage is reduced by 12 and 25% respectively, and that the social welfare indicator was improved by 18%. According to the authors, day-ahead real-time pricing (RTP) is not a simplified tariff plan, as compared to other methods. The HPDR program has a better social welfare index than other methods (Monfared et al., 2019).

Barabadi and Yaghmaee (2019) mainly focus on the active participation of consumers and also achieve a social pricing/billing mechanism using day-ahead DRP. The role of DRP is to minimize the energy cost through load shifting techniques. Moreover, the authors use the quasiconvex cost function for ascertaining the base-load price. They applied game theory, utility theory, and prospect theory (PT) to find out the inefficiencies in used DRP. The result shows that the expected social pricing/billing mechanism is capable of controlling electricity consumption patterns through suitable norms applied to the majority population. The authors state that there is a need to focus on designing a robust algorithm, which will automatically adjust the consumer's cluster – based on electricity consumption – and also accordingly provide a penalty and incentive within a reasonable range. The integration of RESs with the main grid was not discussed by the authors.

Huang et al. (2019) applied dynamic pricing DRP to achieve many objectives such as the active participation of end-users, reducing grid power imbalance, and improving the security and efficiency of the electricity supply. The game-theory approach is used to obtain the bi-directional communication of energy and information between buildings and energy service providers. The task is challenging due to the involvement of various issues like secrecy, communication complexity, and high computation load. These authors applied a genetic algorithm (GA) on dynamic DRP to deal with the mentioned issues. A GA optimizer was applied to obtain a better dynamic price as per the total demand response results. Moreover, the proposed model is validated through two different energy sources: thermal power and renewable energy. According to the results, the dynamic price-based DRP reduced the power imbalance by 59% through thermal power plants and 81% through renewable energy systems. Consideration of different uncertainties and diverse energy characteristics in demand/supply prediction is lacking. The weather conditioning parameters in load prediction and load management were also lacking. The optimization of energy in cost, resources, and CO_2 emissions were also missed.

Figure 2.6 is the proposed framework for SWMM, which includes load forecasting for end-users, utilities, and weather conditioning. This is the SG that exchanges energy and information in a bi-directional way. This implementation needs AMI and active participation of consumers. However, integration of RESs with the main grid at the utility and end-user levels will lead to clean energy generation. This proposed work addresses the benefits of end-users, utilities, and the environment. So, for

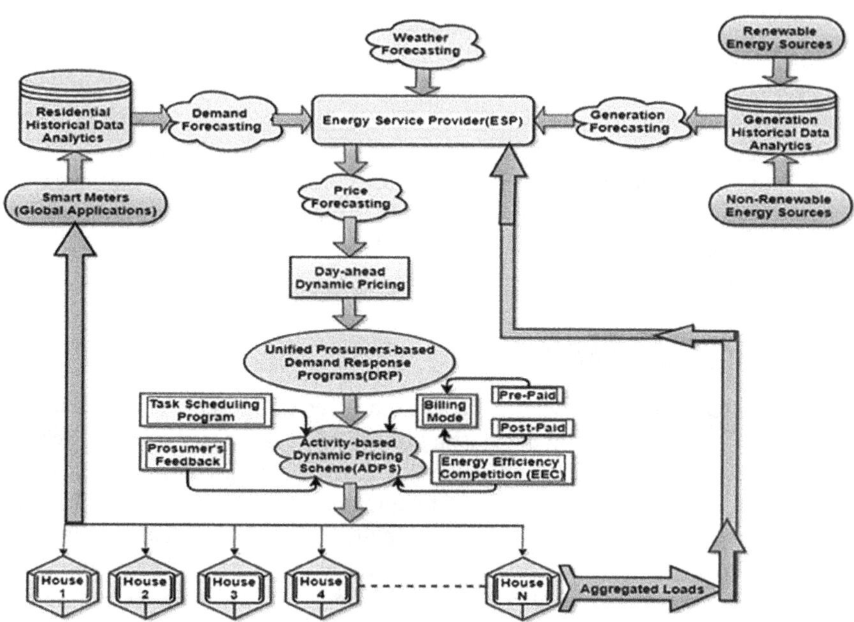

FIGURE 2.6 Framework of SWM model in SG.

effective implementation and DSM, there is a need to focus on DRP schemes. Figure 2.6 shows the Activity-based Dynamic Pricing Scheme, which stimulates the active participation of consumers through task scheduling programs, prosumers feedback, billing mode (pre-paid and post-paid), and EEC.

2.6 CHALLENGES AND DISCUSSION

Developing countries like India face a big challenge to implement new technology from the start because it requires huge capital and advances in technological investment. Moreover, GoI has the moral responsibility to take the initiative to deploy SG technologies across the country. The NSGM institutional organization is working for deployment of SG characteristics and brings them in line with the vision and mission plan. Some of the challenges and principles in its implementation are discussed here.

There is a dire need to create a healthy business model to attract new investment in SG development. It is essential to have strong technical development and open consultation, association, participation with different stakeholders, creation of various opportunities through pilot project implementation, and to allow the introduction of new concepts. There is also a need to provide facilities to participate in SG development, by supporting the technologies and business models. It is essential to support automated technologies and digital-based structural design for energy management and communication. It is also required to integrate skilled and people experienced in the field, and to make resources available in several institutional programs across India. There is a need to develop suitable standards, policies, regulations, capacity building initiatives, research and development laboratories, and so on. So, the NSGM is an official platform of MoP for working on the deployment of technologies and standards for the development of utilities and consumers. In the primary phase, pilot-based demonstrations and testing of technologies are initiated in different parts of the country. Based on the outcomes of small-scale projects, NSGM will plan the implementation of a full-scale project to the public and private capital investors through a new business model. GoI is trying to develop a social welfare maximization environment in India (ISGM, 2017; NSGM, SGPP, 2019; NSGM, IITK, 2019).

2.7 CONCLUSION AND FUTURE SCOPE

Electricity demand is continuously growing with adverse effects on resources and the environment. Modernization in the existing power grid system is required to achieve sustainable development and power reliability. This chapter has discussed the major initiatives of GoI on SG and pilot project deployment across the country. The outcomes of these pilot projects helped to develop the basic infrastructure, research and development labs, tools, standards, regulation, and schemes. This chapter has mainly focused on SG development with the SWMM using a DRP. This approach takes care of the maximization of the benefits of different stakeholders along with demand-side management and energy optimization. The experiences and outcomes of SG pilot and large-scale projects are helpful for other developing countries like India to develop their SG infrastructure.

This thematic study using grey literature in the Indian context can be extended to develop SWMM using hybrid incentive and RTP-based DRP. Moreover, there is a need to design and develop an algorithm on a data-driven model for dynamic pricing schemes using the SWM approach in a micro-grid.

REFERENCES

Barabadi, B., & Yaghmaee, M. H. (2019). A New Pricing Mechanism for Optimal Load Scheduling in Smart Grid. *IEEE Systems Journal, 13*(2), 1737–1746. https://doi.org/10.1109/JSYST.2019.2901426.

Big Data Analytics for Smart Grid (BDASG). (2020). Available online on, https://onlinecourses.swayam2.ac.in (accessed on February 15 2020).

Department of Energy (DoE). (2019). Available online on, https://www.energy.gov/ (accessed on December 17 2019).

Desi Smart Grid News Item Smart Grid Pilot Projects Under Execution in India (2013–2016) (Desi Smart Grid) (2013). Available online on, https://desismartgrid.com/?p=1565. (accessed on December 17 2019).

Energy Efficiency Services Limited (EESL). (2019) Dashboard, Ministry of Power, Govt. of India. Available online on, https://www.eeslindia.org (accessed on December 15 2019).

Energy Storage System Roadmap for India (ESSRI:2019-2032) (2019). Available online on, https://niti.gov.in/sites/default/files/2019-11/ISGF.pdf (accessed on December 15 2019).

Gaikwad, S. R., & Harikrishnan, R. (2021). Social Welfare Maximization in Smart Grid : Review. *IOP Conference Series: Materials Science and Engineering, ASCI 2020*. https://doi.org/10.1088/1757-899X/1099/1/012023

Government of India, Ministry of Power (GoI, MoP, ISGF). (2019). Available online on, http://www.powermin.nic.in, ISGF Bulletin Oct 2019.doc. (accessed on December 16 2019).

Government of India, Ministry of Power (GoI, MoP). (2019). Available online on, https://powermin.gov.in/content/protection-environment (accessed on December 16 2019).

Government of India, Ministry of Power, Deen Dayal Upadhyaya Gram Jyoti Yojana (GoI, MoP, DDUGJY). (2019). Available online on, http://www.ddugjy.gov.in (accessed on December 14 2019).

Huang, P., Xu, T., & Sun, Y. (2019). A genetic algorithm based dynamic pricing for improving bi-directional interactions with reduced power imbalance. *Energy and Buildings*, *199*, 275–286. https://doi.org/10.1016/j.enbuild.2019.07.003.

India Smart Grid Forum (ISGM). (2017). Smart Grid Handbook for Regulators and Policy Makers, November 2017. Available online on https://indiasmartgrid.org/reports/Smart%20Grid%20Handbook%20for%20Regulators%20and%20Policy%20Makers_20Dec.pdf (accessed on December 15 2019).

Jamil, A., Javaid, N., Khalid, M. U., Iqbal, M. N., Rashid, S., & Anwar, N. (2019). An energy efficient scheduling of a smart home based on optimization techniques. *Advances in Intelligent Systems and Computing*, *773*, 3–14. https://doi.org/10.1007/978-3-319-93554-6_1

Kimata, S., Shiina, T., Sato, T., & Tokoro, K. I. (2020). Operation planning for heat pump in a residential building. *Journal of Advanced Mechanical Design, Systems and Manufacturing*, *14*(5), 1–11. https://doi.org/10.1299/jamdsm.2020jamdsm0076

Monfared, H. J., Ghasemi, A., Loni, A., & Marzband, M. (2019). A hybrid price-based demand response program for the residential micro-grid. *Energy*, *185*, 274–285. https://doi.org/10.1016/j.energy.2019.07.045.

Nakabi, T. A., & Toivanen, P. (2019). An ANN-based model for learning individual customer behavior in response to electricity prices. *Sustainable Energy, Grids and Networks*, *18*. https://doi.org/10.1016/j.segan.2019.100212.

National Smart Grid Mission (NSGM Framework) (2018). Available online on https://www.nsgm.gov.in/sites/default/files/NSGM-Framework-Final.pdf (accessed on December 14 2019).

National Smart Grid Mission (NSGM, IITK). (2019). Available online on, https://www.iitk.ac.in/smartcity (accessed on December 17 2019).

National Smart Grid Mission (NSGM, NPMU). (2019). Mission Innovation Brief by NPMU. Available online on, https://www.nsgm.gov.in/sites/default/files/Mission_Innovation_Brief_by_NPMU.pdf (accessed on December 14 2019).

National Smart Grid Mission (NSGM, SGPP). (2019). Available online on https://www.nsgm.gov.in/sites/default/files/SGPP_Impact_Assessment_Findings_by_QCI.pdf (accessed on December 19 2019).

Press Information Bureau (PIB). (2019). Government of India, UDAY (Ujwal DISCOM Assurance Yojana) for financial turnaround of Power Distribution Companies. Available online on http://pib.nic.in/newsite/ (accessed on December 15 2019).

Samantaray, S. R. (2014). Letter to the editor: Smart grid initiatives in India. *Electric Power Components and Systems*, *42*(3–4), 262–266. https://doi.org/10.1080/15325008.2013.867555.

Vardakas, J. S., Zorba, N., & Verikoukis, C. V. (2015). A Survey on Demand Response Programs in Smart Grids: Pricing Methods and Optimization Algorithms. *IEEE Communications Surveys and Tutorials*, *17*(1), 152–178. https://doi.org/10.1109/COMST.2014.2341586.

3 Cloud Computing Security Framework Based on Shared Responsibility Models

Cloud Computing

Umesh Kumar Singh and Abhishek Sharma
Institute of Computer Science, Vikram University, Ujjain

CONTENTS

3.1 Introduction 39
3.1.1 Cloud Computing Reference Models 40
3.1.2 Cloud Computing Architecture 41
3.2 Related Work 44
3.2.1 Cloud Computing SRM for IaaS, PaaS, and SaaS 45
3.2.2 Analysis of SRM Implementation by Leading Cloud Service Providers 47
3.3 Proposed Cloud Computing Security Framework Based on an SRM and Its Application 51
3.4 Results and Conclusion 53
3.5 Future Work 54
References 54

3.1 INTRODUCTION

Cloud computing (CC) is a new technological development that will continue to construct e-governance products in IT industries. The degree to which industries are shifting towards CC is growing quickly because of its massive benefits, such as high scalability, elasticity and flexibility of industry assets to achieve sudden variation in requirements, exceptional reliability, and constant availability in that assets can gain access to and from any place and time without any extra price for the installation and management of software and hardware infrastructure.

The basic requirement of administration and commerce is to implement e-governance, which requires constant infrastructure availability to optimize

DOI: 10.1201/9781003146711-3

downtime. E-governance products are able to adopt a limitless supply of central processing power, data storage, and internet speed during processes; cloud computing will be the perfect solution for them (MeitY 2014). CC denotes the entire infrastructure, platform, and applications delivered as services over the Internet, hardware and software systems at the level of the cloud data center provider. The reference architecture of CC by the National Institute of Standards and Technology (NIST) recognizes the key actors, their actions, and roles with CC for the execution model (IIIT Hyderabad 2010). The biggest challenge due to the introduction of cloud based implementation in industries is in the area of computing security. Handling the security related to an enterprise for their own privately deployed cloud as well as monitoring the actions of the cloud services provider (CSP) can be a significant assignment. Security is a major concern when sharing data critical to an organization with geographically distributed cloud platforms.

If a society transfers to consuming public cloud services, lots of the computing system infrastructure will be controlled by the third-party CSP. That's why many organizations are failing to maintain control over their computing areas and have issues related to security and privacy regarding the novel technology which is a big source of unique vulnerabilities.

Therefore, it is necessary to build various controls to reduce the dangers and offer a base layer of protection, raise trust levels, or the capabilities of cloud services, to avoid panic situations due to cloud usage. So, cloud computing security can be enhanced, if the CSP and cloud users understand the shared responsibility model and either follow the security standards or model their own security controls and standards.

In order to provide a secure environment for cloud users, risk assessment and analysis with mitigation policies are required. Through mitigation of vulnerabilities and analysis of risk the effect of attacks can be reduced. Risk management plays a very important role for computing controls in order to provide the assessment and management of the risk related issues for the cloud environment. It also provides a prevention mechanism and mitigation policies for the risk, which may be dangerous and affect business objectives.

The main motive of this chapter is to propose a CC-security framework based on the shared responsibility model (SRM), but before that we need to understand the stakeholders and their interaction with the CC environment. The chapter first introduces the CC reference model, then its architecture, followed by the various cloud actors involved. Then readers are introduced to the latest data breaches and then SRM for an SPI model. It is better to understand more about the top CSPs, what they offer, and compare them with service models.

3.1.1 Cloud Computing Reference Models

As per NIST (U.S. Department of Commerce 2010), cloud models include three service models, four deployment models, and five essential characteristics, or we can say they are 3–4–5 reference models. CC is nothing but virtual management related to data center assets placed centrally and stored within software defined pools. This explanation just scratches the upper layer of cloud-based services capabilities. Cloud

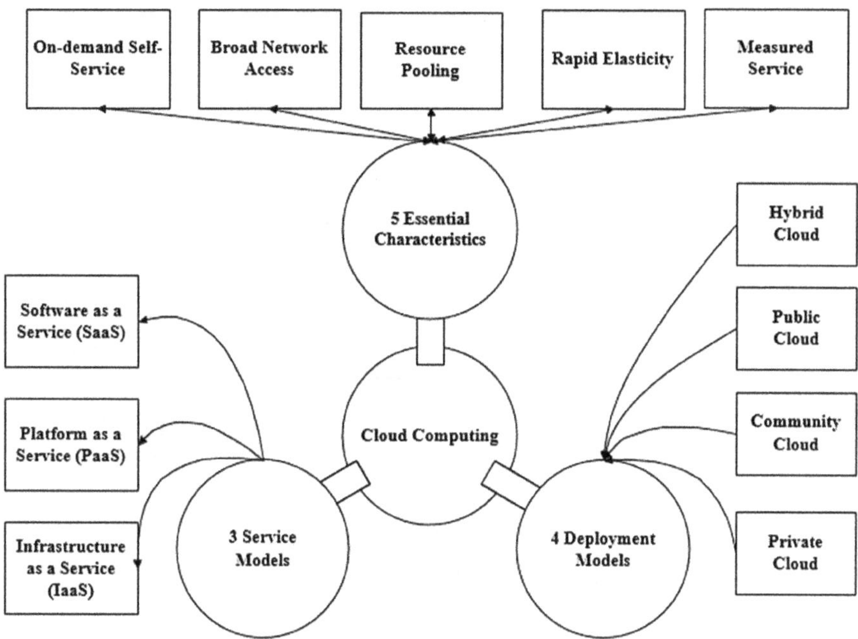

FIGURE 3.1 3–4–5 cloud computing reference model.
Source: National Institute of Standards.

solutions can provide on demand computing services to collaborating entities on the Internet from apps to keep data and execution power normal by payment for each activity. Now, it is understood from the 3–4–5 reference model of CC that the collaboration of hardware and software, which triggers the mandatory properties of CC, is actually a cloud setup. It contains both an abstraction layer (AL) and a physical layer (PL). The AL includes the software setup through the PL, which establishes mandatory characteristics. Theoretically the AL stands above the PL. The 3–4–5 reference model of the cloud can be represented as shown in Figure 3.1.

The PL involves the hardware assets required to provide the cloud services delivered, usually involving network components, workstation machines, and storing devices. This proficiency does not essentially preclude the usage of well-matched services, libraries, program writing languages, and tools from additional sources.

3.1.2 Cloud Computing Architecture

Cloud architecture is introduced as a recommendation for the recognition of the complete procedure containing consumers within a cloud. Currently, some CC architecture is used to build a cloud deployment setup. As the CC technique is actually used to reduce the conventional budget of computing assets and its setup, numerous organizations have gained interest in migrating towards CC systems. Cloud computing reference architecture (CCRA) describe blueprints which are useful to establish a CC environment. The relationships are also represented, and explain the roles and

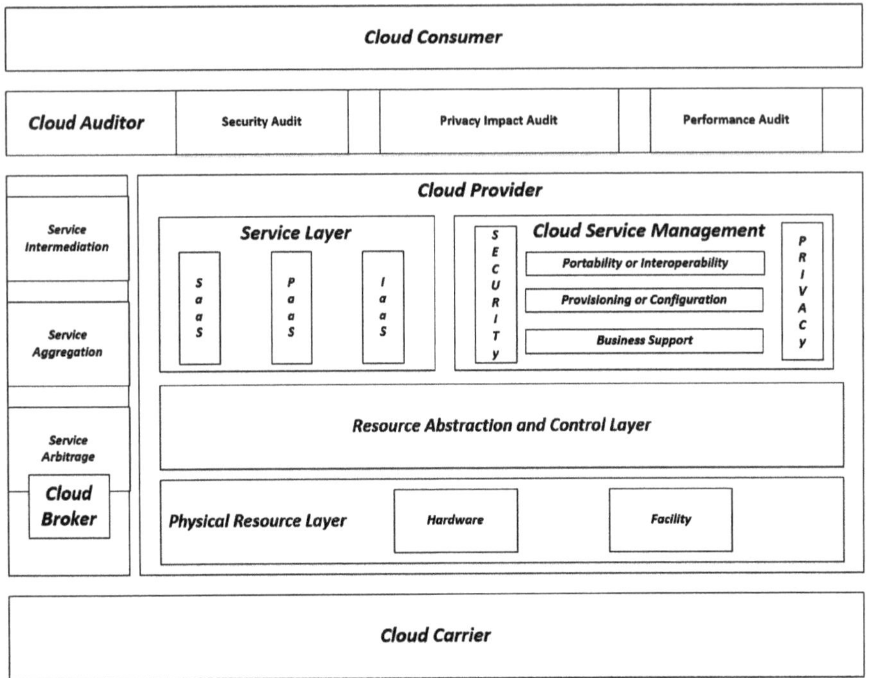

FIGURE 3.2 Cloud computing reference architecture.
Source: NIST.

responsibilities of various entities and how they collaborate and communicate with each other. The NIST CCRA (Amanatullah et al. 2013; Liu et al. 2011) recognizes the consumers, actions, and tasks in CC as shown in Figure 3.2.

For the most part there are five actors explained in CCRA: the service provider, service broker, auditor, consumer, and carrier. Every performer is an entity who may be an individual or an enterprise which contributes toward the operation, procedure, and execution of jobs in CC.

The **cloud provider** is a person, an enterprise, or an entity which owns a cloud infrastructure and is responsible for building stacks of services available to various parties and users. They are also named CSPs. In the case of SaaS, the cloud provider organizes, configures, updates, and maintains the task of the application software to the infrastructure of the cloud, providing services matched with the desired thresholds. The SaaS supplier accepts the duties for handling and monitoring the applications with a physical setup, whereas the users are restricted to a managerial mechanism for applications. In the case of PaaS, the CSPs manage the setup of the cloud platform and execute the cloud software. PaaS also delivers mechanisms like the runtime software execution stack, database, and supplementary middleware modules. In the case of IaaS, the CSPs obtain the computing assets' essential service, networks, storing devices, servers, and hosting setup. The CSPs execute the software required to make the computing assets accessible to the users over a collection of interfaces and computing asset perceptions, like virtual network interfaces and virtual machines (VMs).

The **cloud broker** maintains the usage, efficiency, and deployment of the cloud. Cloud brokers behave as a mediator who negotiates within the group of all consumers. If cloud users are uninterested in contracting or looking straight then they will play their role and take the place for them. They are also responsible for providing service intermediation, service aggregation, and service arbitrage in context with the cloud.

The **cloud auditor** is an entity or party which can execute independent evaluation as well as assessment of performance, cloud service controls, security and privacy, and information system operations for the implementation of the cloud environment. Auditors are supposed to verify agreed standards which is done by a review of objective evidence.

The **cloud consumers** are very important actors for any professional-use case, mainly in a cloud marketplace. So, consumers will regularly support the construction of cloud based services through consideration of their expectation. Based on the CC service model, users are categorized into four types: cloud provider, cloud broker, cloud auditor, and cloud provider in the SPI model. The users who required a CC physical setup with service computing capabilities prefer central computing assets. PaaS users envisage a platform that can build applications, software without installation worries related to issues of the building environment, languages, and security based tools. SaaS users correspond to those users who normally avail themselves of online apps for day-to-day needs.

A **cloud carrier** is an intermediary which ensures the availability of interconnectivity and the transportation of services from CSPs to users. They also deliver access for consumers to network, telecommunication, and additional access devices. CSPs are supposed to establish service level agreements (SLAs) for cloud consumers and carriers to enable services, as per the SLAs, and guarantee the delivery of dedicated and secure connections between them. Now, in order to meet the requirements and expectations of consumers, proper interaction and collaboration between all the actors involved is needed. This interaction can be represented as in Figure 3.3.

Apart from the above five actors, one more highly important actor called a cloud developer is involved, though it is hidden from the collaboration among the actors. A **cloud developer** is an individual or it may be an enterprise which builds cloud services. Several CC architectures or physical asset setups are described with their detailed surroundings for the apps in Azua and Goodman (2011), Buyya et al. (2009), Kirschnick et al. (2010), Peterson (2010), and Samimi et al. (2007). To successfully construct a CC application in an organization, it is required to redefine the methodical architecture of the software systems and implement CC. Additionally, it involves all the mandatory properties of CC.

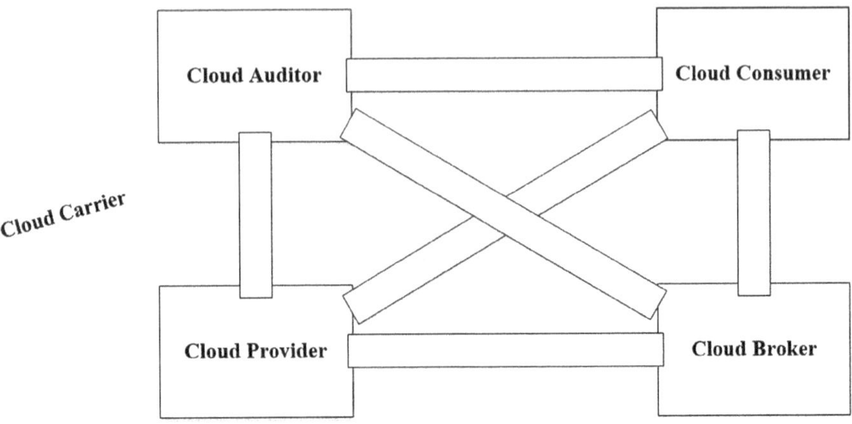

FIGURE 3.3 Actors in cloud computing and their interactions.
Source: NIST.

3.2 RELATED WORK

CC has been expanding in its acceptance quickly over recent years, because business requires additional efficiency and effective techniques at acquiring physical assets. Hence, a consumer site does not require any direct installation of computing assets and resources for delivering services to cloud consumers. Although every industry defines CC in its own manner, no detailed designation exists with CC standardization. On the other hand, the definition of CC can be described on the basis of whatever users want from it. Additionally, in present CC architecture, the actors are able to construct and plan for individual CC environments. So, for the best delivery of cloud services, the interaction of cloud actors needs to be established by knowing their participation, roles, and responsibility. CSPs are supposed to recognize the fundamental needs of a CC service, service model, CC service deployment, and security concerns. After service delivery, cloud consumers are able to use the delivered service, and CSPs are required to maintain, monitor, operate, and manage the main actions. CC delivers data access and checks which bodies can gain legal access to it. When the cloud is used, an external agency is entrusted to make conclusions regarding its platforms and data, in such a way not seen previously in CC. There should be in place a suitable technique to protect CSPs from consuming data for which they are not approved. It appears doubtful that any mechanical issues might entirely protect CSPs from harming customers' data in every case. That's why it requires a grouping of all the challenges related to nontechnical and technical means. Customers are required to have a lot of faith in the capability and financial strength of CSPs.

As per the 2020 Verizon Data Breach Investigations Report (www.verizon.com), an analysis of almost 4000 data breaches that occurred in the USA, Figure 3.4 shows the major causes of and reasons for these data breaches.

As per this report, the pattern can be for the most part connected with uncovered Web capacity found by security analysts and inconsequential outsiders. The organization found those misconfigurations are an essential driver of cloud security issues.

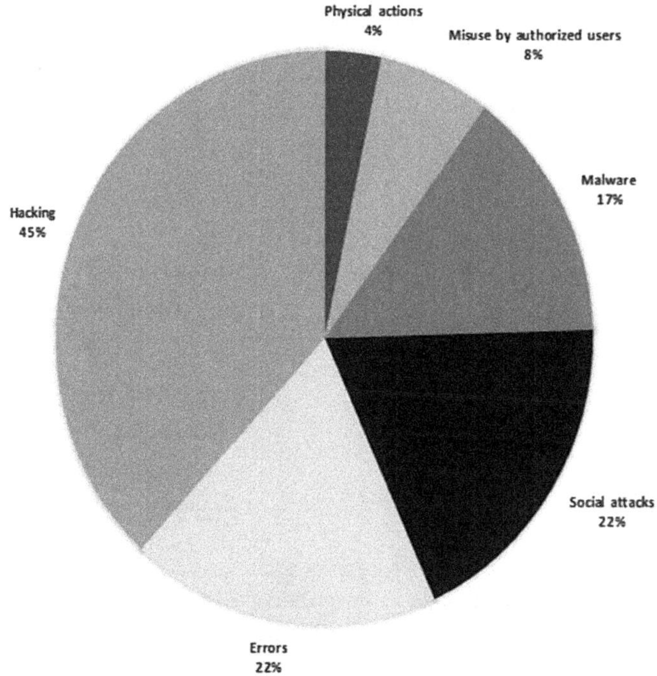

FIGURE 3.4 Causes of and reasons for data breaches.
Source: 2020 Verizon Data Breach Investigations Report.

A misconfiguration is the point at which a framework manager doesn't acquire a distributed storage framework or an information base effectively regarding cloud administration.

3.2.1 Cloud Computing SRM for IaaS, PaaS, and SaaS

Cloud customers and providers are required to share privacy and security responsibility for CC environments, though sharing levels must be managed in such a way that it will be maintained for all delivery models which control or affect cloud extensibility. As per Lanfear (2019), IBM (2019), and Amazon Web Services (2020a, 2020b), for best delivery of services to customers by CSPs, the roles and responsibilities in context with various models including an on-premises cloud data center is compared in Table 3.1.

Table 3.1 clearly shows that in the case of on-premises cloud data centers the whole responsibility of managing the assets falls to the cloud customer, whereas in case of the SPI model the responsibility is shared or distributed at various levels of abstraction. But infrastructure as a service is a supremely stretchable delivery model and delivers some of the application with example features. It is expected that consumers will protect the content, applications, and operating system. CSPs will still deliver some fundamental functionality, like low-level data shielding capabilities. Table 3.1 demonstrates the regions of responsibility shared between CSPs and CCs

TABLE 3.1
Cloud Responsibilities Stack Based on SaaS, PaaS, IaaS, and an On-Premises Cloud Data Center

	Cloud Service Models			
Cloud Responsibilities Stack	**SaaS**	**IaaS**	**PaaS**	**On-premises cloud data center**
User access	√	√	√	√
Data	√	√	√	√
Application		√	√	√
Runtime environment		√		√
Middleware		√		√
Operating system		√		√
Virtualization and hypervisor				√
Servers				√
Storage				√
Networking				√

according to an IaaS stack. The objective of PaaS is to allow programmers to construct individual customized apps using provided cloud platforms. It also demonstrates the regions of responsibility between the CSPs and CCs according to a PaaS stack.

In SaaS, CSPs usually facilitate a huge number of functionalities, which results in low extensibility for CC users. CSPs are additionally answerable for the privacy and security of application level services, especially within the public domain, whereas consumer groups must maintain their robust and secure requirements to ensure the availability of services. Private clouds can mandate additional flexibility to personal house requirements. Table 3.1 shows the regions of responsibility between CSPs and CCs according to a SaaS stack. So, it is the responsibility of consumers to protect, execute, and avail themselves of the applications they construct, the operating system, and the platform. CSPs are responsible for the isolation of independent working areas and customer applications among consumers.

Multi-tenancy within a cloud virtualized environment is one more key feature related to a public domain cloud. Basically, it permits CSPs to achieve resource consumption proficiently through splitting virtualized and shared assets among several consumers. As per the consumer's view, the conception of shared assets and infrastructure can be a vast issue. Consider the case where it is required to separate multiple customers and their related information. One enterprise works on rewriting the query at the level of the database, while another customer is interacting with hypervisors of hardware. CSPs must note all challenges like access policy, apps building or setup, data protection for ensuring secure, and multi-tenancy (Takabi & Joshi, 2010). Security with privacy must be designed correctly for clouds, because in the absence of them, a disaster may affect this transforming computing paradigm.

Table 3.1 shows the theoretical separation of cloud services under the umbrella of the CC shared responsibility model and maps them with respect to the IaaS, PaaS, and SaaS service models. But the practical implementation and separation of cloud services may be different among CSPs. It may also depend and very much so on the different types of deployment model, such as the public, private, community, and hybrid models. So, sometimes the CC SRM could be visualized as a shared confusion because it is practically impossible to separate the responsibility by drawing a line among the responsibility stack. That's why in IT industries, CC responsibilities are shared on the basis of the requirements gathered from the customer, use case, SLA, architecture, and implementation aspects.

3.2.2 Analysis of SRM Implementation by Leading Cloud Service Providers

Before focusing on the implementation of a CC shared responsibility model from an IT industries perspective, we need to know about the top CSP and its services in the present cloud market. As per the TechTarget & Insight Report (2021, https://www.statista.com/chart/18819/worldwide-market-share-of-leading-cloud-infrastructure-service-providers/), more than 60% of the cloud market is occupied by the top ten cloud providers. That's why, in order to supply the best services, it is required to visualize the CC-SRM of the leaders of the present cloud market. There are various CSPs, for example, Amazon, Microsoft, IBM, Google, Salesforce.com (CRM Software Salesforce 2020), Facebook, Twitter, Yahoo, and Rackspace open cloud. The comparative study of various service models related to top CSPs are listed in Table 3.2.

When selecting a cloud the consumer is required to consider two significant concerns. The first is whether the cloud satisfies the performance requirement of the company or not. The second is how much needs to be invested for the service usage. CSPs charge cloud customers according to the demand for services and usage of resources and instances, or as a pay-as-you-go model when there is no ongoing need for them. Every CSP consumes various products like security tools, networking, and data storage. Cloud users are able to select the application according to their requirements.

A number of CSPs (Karimunnisa et al., 2019) like IBM cloud, Microsoft Azure, AWS, and GCP offer free tiers for which the customer can use any cloud services with definite boundaries. After the usage defined reaches the boundary, it will continue on a chargeable basis and be converted to a pay-per-use method. This free use is very convenient for individuals, minor organizations, and those who want to gain exposure to the cloud environment before acquiring it commercially. Table 3.3 shows a comparison of various cloud services and products of top CSPs.

Tables 3.1, 3.2, and 3.3 represent the recent trends of CSPs, their offerings, and responsibilities based on the SPI model. In the case of Amazon Web Services (AWS) SRM, the customer and AWS both share the responsibility for compliance and security. The SRM is able to support the release of the client's working liability, as AWS drives, accomplishes, and pedals the host modules and VM layer below the security setup. The client undertakes the accountability and controlling of the guest OS which involves updating safety coverings and other related application software with the configuration of the AWS-delivered security-based firewall. The landscape of this SRM also delivers the elasticity with the consumer regulator to authorize the deployment. AWS is accountable for securing the physical setup that executes every service

TABLE 3.2
Comparison of Top CSPs based on Their Service Models

Service Models	Service Provider							
	VMware	**Rack space**	**Amazon**	**Google**	**Micro soft**	**IBM**	**HP**	**Sales force.com**
IaaS	VMware	Cloud Servers	EC2 (Elastic Cloud Compute)	Google Compute Engine	Micro soft Private Cloud	Smart Cloud Enterprise	Enterprise Services Cloud Compute	NA
PaaS	VMware VS Fabric	Cloud Sites	AWS Elastic Beans-talk	Google App-Engine	Windows-Azure	Smart-Cloud Application Services	Cloud Application Delivery	Force. Com
SaaS	Slide Rocket	Email & Apps	Amazon Web Services	Google Apps	MS Office 365	SaaS products	HP Software as a Service	Salesforce Com

TABLE 3.3
Comparison of Top CSPs based on Their Products or Services Offered

CSP	Cloud-based Product or Services Offered
AWS	Analytics, Application Integration, AWS Cost Management, Blockchain, Business Applications, Compute, Containers, Customer Engagement, Database, Developer Tools, End User Computing, Front-End Web & Mobile, Game Tech, Internet of Things, Machine Learning, Management & Governance, Media Services, Migration & Transfer, Networking & Content Delivery, Quantum Technologies, Robotics, Satellite, Security, Identity, Compliance, Serverless, Storage, VR & AR
IBM	Compute, Network, Storage, IBM Cloud Pak solutions, Management, Security, Databases, Analytics, AI, IOT, Mobile, Developer tools, Blockchain, Integration, Migration, Logging and Monitoring.
Microsoft Azure	AI + Machine Learning, Analytics, Blockchain, Compute, Containers, Databases, Developer Tools, DevOps, Hybrid + Multicloud, Identity, Integration, Internet of Things, Management and Governance, Media, Migration, Mixed Reality, Mobile, Networking, Security, Storage, Web, Windows Virtual Desktop.
Google Cloud	AI and Machine Learning, API Management, Compute, Containers, Data Analytics, Databases, Developer Tools, Healthcare and Life Sciences, Hybrid and Multi-cloud, Internet of Things (IoT), Management Tools, Media and Gaming, Migration, Networking, Operations, Security and Identity, Serverless Computing, Storage, Google Work Space, Meet, Chrome Enterprise, Google Map Platform.

offered. This physical asset is built with the hardware, software, networking, and amenities that execute AWS cloud services. "The diversity of responsibility is usually stated as Security 'of' the Cloud versus Security 'in' the Cloud" (Amazon Web Services (2020a, 2020b)). In the case of AWS the security responsibility is divided into two separated layers, the customer managed layer and the AWS managed layer. The customer managed layer is divided into a sequence of four parts. The first part includes customer data security; the second part consists of the security issues related with platforms, applications, identity management, and access management. Similarly the third part is responsible for managing the OS, network, and firewall configuration security issues. In the fourth part, security responsibility concerns managing client-side data, data integrity authentication, server side encryption, file system data, networking traffic, protection based on encryption, integrity, and identity,

whereas AWS is responsible for security issues related to software and hardware-based global infrastructure, like regions, availability zones, and age locations. A CSP is also responsible for providing protection for computer database storage as well as infrastructural networking. Cloud customer responsibility is based on the AWS services chosen by the consumer. This defines the volume of specification tasks the consumer needs to perform as part of their security responsibilities. Consider a case where a service like as Amazon EC2 is classified as IaaS and needs the consumer to implement the required security configuration and controlling tasks. The consumer, who set up an Amazon EC2 instance, is responsible for the supervision of the guest OS which involves patching a security and system update, any software application or utility set up by the consumer, and the AWS firewall configuration or security group at each instance.

Consumers are also accountable for maintaining relevant information, its cryptographical policy, categorizing resources, and identity or access management techniques to implement suitable authorizations. In case of abstracted services, like Amazon S3 and Amazon Dynamo-DB, AWS drives the OS, infra-layer, or platform, and consumers have the right to use the endpoints to access and save data. The current AWS-consumer SRM also spreads out to controls, as the accountability to control the IT background is handled by the AWS and consumers together. That's why it controls, actions and provides confirmation for jointly handled system controls. Functional control in the cloud is a responsibility of customers; CSP is able to help them. Though all consumers set up their own AWS environment, cloud consumers are able to migrate only a few of their IT controls to an AWS environment. This leads to a novel diversified control setup. Then consumers can avail themselves of cloud controls and access compliance records in order to implement verification actions and control estimation as needed. The controls that are managed by AWS, AWS-consumers, or both (Microsoft Azure 2020) are:

- Inherited controls: consumers are able to inherit from AWS (e.g., environmental and physical controls);
- Shared controls: AWS makes available the necessities for infrastructure and consumers are required to prepare control execution for themselves according to the usage of AWS services (e.g., configuration management, patch management, awareness and training).

The Google Cloud Platform (GCP) SRM is precise and flawless regarding where GCP security responsibilities finish and where the consumer's security responsibilities start. GCP Virtual Machine (VM) series, set up to defend assignments inside a Google project, supports consumers' tasks and roles within the shared responsibility model, although it is the responsibility of consumers to shield content, access policies, usage, deployment, Web application security, identity, operations, access and authentication, network security, guest OS, data, and packages which are deployed (Google Cloud 2020). Security issues related to audit logging, network, storage and its encryption, hardening kernels for inter-process communication, BOOT, and hardware security are the responsibility of Google CSP.

In the case of Microsoft Azure Cloud, the CSP realizes in what way various cloud service models touch the techniques and responsibilities which are jointly handled between consumers and CSPs. In the case of Azure SRM, seven responsibilities of the enterprise and the contribution of cloud customer and cloud provider are represented. It also represents the responsibility to attain compliance and protection of the computing environment. The consumers are responsible for the data, and its categorization is performed appropriately – its solution must be compliant with regulatory obligations. The responsibility for physical security is owned by CSPs only. The remaining responsibilities are on a shared basis for cloud consumers and CSP, such as auditing domains, IAM when using Azure Active Directory Services, and configuration of services like MFA set up for the cloud consumer (Microsoft Azure 2020).

3.3 PROPOSED CLOUD COMPUTING SECURITY FRAMEWORK BASED ON AN SRM AND ITS APPLICATION

The legal and technical strategy associated with security problems results in a reduction in the confidence of CC adaptation. Sometimes cloud safety subjects associated with the geographical location of the data center, network, and other general issues also obstruct the progression of CC adoption. In CC, all responsibility for appointing and maintaining effective security techniques are in the hands of the CSPs. To decrease consumers' fears of CC, CSPs are supposed to assure consumers that their applications and related data will be definitely protected. Security is said to be a risky hurdle for CC. That's why a shared responsibility model with effective participation of all the actors of security use cases are required to contribute to it. If it retains security for a long time in a competitive environment, this leads to success and a large profit in the future. So the top challenges and concerns are to work out a security framework for the cloud – a consumer model. In the present section, a CC security framework based on an SRM will be proposed which represents the roles and responsibilities of CSPs and/or consumers.

The proposed security framework shown in Figure 3.5 consists of four layers: (i) an infrastructure layer; (ii) a network layer; (iii) a virtualization layer; and (iv) an application layer with the provision of following and maintaining all the major objectives of security like availability, authenticity, integrity, accountability, and confidentiality.

An **infrastructure layer** is placed in the central part of the proposed CC security framework represented, which consists of CC power, cloud storage, cloud database, a cloud network, and various supporting cloud infrastructural resources. The CSP is responsible for management and security (including physical security) of the infrastructure layer's assets and resources as shown in Figure 3.5.

Above the infrastructure layer, the second layer is placed: the **networking layer**. Its responsibilities are categorized into three sections. The first section is represented by use cases numbered as #, for which a CSP is fully responsible, such as network scanning and monitoring services, network edge security services, external DDOS, as well as spoofing and segmentation of the logical network. The second section is represented by use cases numbered as **, for which the cloud consumer is fully responsible, such as data encryption using SSL/TLS, data segregation, data destruction, and configuration of network security.

The third section is represented with the help of use-cases marked with $, which represents the shared responsibility of a CSP and cloud consumer, such as security monitoring, network auditing, network controlling, and threat detection.

The **virtualization layer** is placed above the network layer as shown in Figure 3.5. This layer also represents the variety of security responsibilities with the help of three sections. The first section, represented by the use cases numbered as #, is the responsibility of CSP and is related to hypervisor and cloud virtualization management. It also includes the implementation and management of environmental security, component-based security, PaaS-based patching management, and consumer

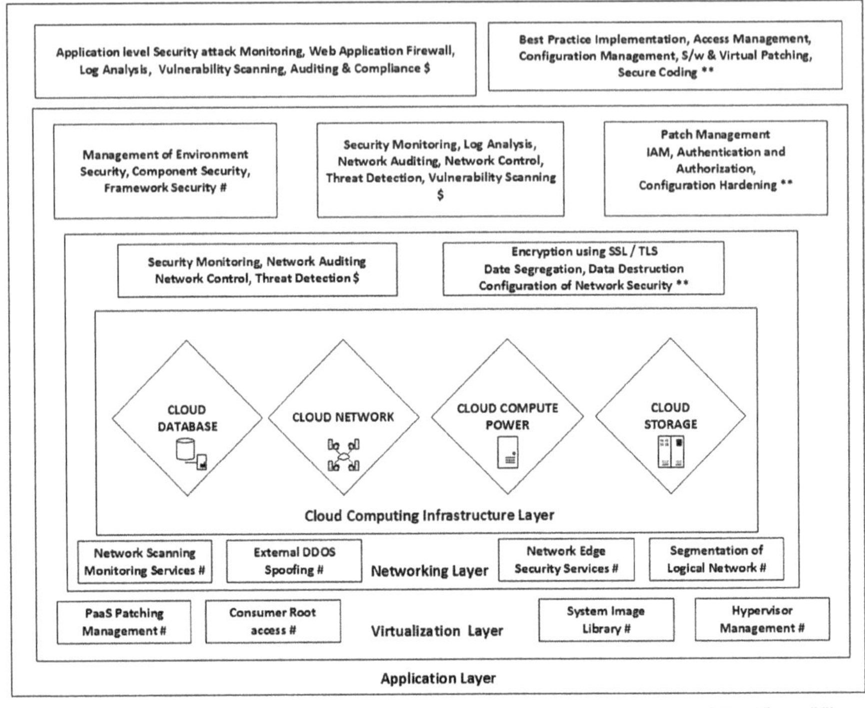

FIGURE 3.5 Proposed cloud computing security framework based on an SRM.

root access. CSP is also responsible for the system image library required for virtualization and hypervisor. The use cases numbered as ** represent the responsibility of cloud consumers, such as patch management, IAM, authentication and authorization, and configuration hardening at the level of hypervisor.

Similarly, in the third section, which is represented by use cases marked with $, both CSP and cloud customer are responsible and represented as shared responsibilities, such as security monitoring, log analysis, network control, network auditing, threat detection and vulnerability scanning, and/or assessment, especially for PaaS.

The **application layer** is placed at the top of virtualization layer. It plays an important role in the security framework because in the present situation the SaaS model is widely used in the field of business automation, engineering, technology, healthcare, e-governance, education, IT communication, and so on, especially during the COVID-19 pandemic.

The cloud customer can interact with the application layer, where the application-based security responsibilities are categorized in two sections. The use cases numbered as ** represent the responsibilities of a cloud customer, such as access management, configuration management, software, as well as virtual patching. They are also responsible for following best practices for implementing and coding secure applications.

The cloud customer and CSP both are responsible for the provisioning of application level security, attack monitoring, Web application, firewall management, log analysis, vulnerability scanning, auditing, and compliance, which is represented by use cases marked with $.

Within the multilayer security framework, every layer is enabled with a two-way communication interface with the help of which the layers can communicate with each other. The proposed SRM based cloud security framework is also capable of delivering the cloud services for various deployment models, such as private, public, hybrid, and community clouds, as well as service models like SaaS, PaaS, and IaaS. It is also capable of adopting and implementing cloud security standards, security guidelines, compliance, and law regulation, defined at each layered structure.

The proposed system is capable of applying in the various CC services: e-governance, e-commerce, data storage and backup services, communication, big data analysis, education, entertainment, backup management, social networking, cloud application development, healthcare, and many more. Consider an application of a proposed security framework in the area of healthcare or medicine. Frail cloud security is one of the trivial issues that block the full dissemination of the cloud in the medical services industry. Medical care experts have numerous motivations not to confide in the cloud, for instance, they can't part with their authority over clinical records. Cloud suppliers generally store their information in various server farms situated in various geographic areas. This addresses an unmistakable and favorable position, since information stockpiling within the cloud will be repetitive, and if there should arise a *force majeure*, diverse server farms will help recover the situation. Then again, this equivalent favorable position can represent a security challenge since information put away in various areas will be more inclined to burglary and misfortune. When all is said and done, there are numerous security risks related to the utilization of the cloud: the inability to isolate virtual clients, wholesale fraud, advantage misuse, and useless encryption are among the security concerns. The proposed SRM security framework provides help to move information into the cloud, where security difficulties ought to be relieved. It will screen the secured wellbeing information life cycle. Prior to choosing a cloud specialist organization, the framework provides various inquiries: Is supplier compliance confirmed, such as ISO/IEC 72001? Are the CSPs prepared for any danger or emergency situation, regardless of whether the supplier performs occasional security checks? Are the CSPs following security and protection regulatory guidelines? Is the specialist cooperative willing to have SLAs based on HIPAA, which imposes severe penalties for infringing activities? Diverse safety efforts like firewalls, interruption discovery, and cryptographic and confirmation strategies ought to be likewise checked. The proposed security framework illustrated in Figure 3.5 will provide guidelines about the responsibilities to be taken by the CSP and cloud customer. The proposed cloud computing security framework (CCSF) based on an SRM will be further helpful to cloud actors and stakeholders for mapping security standards, compliances, and regulations at various layers.

3.4 RESULTS AND CONCLUSION

The system presented here is able to build and introduce the security components required from the framework. In any security-based framework five activity components must be provided for validation: identification, protection, detection, response, and recovery. The recommended framework shown in Figure 3.5 is a layered approach which identifies the cloud assets and also distributes the security responsibility in three categories: customer responsibility, CSP responsibility, and joint responsibility. At each layer the mechanism of detection and recovery is also represented. So, this results in continuous monitoring capabilities which results in better and enhanced security levels. The presented framework is helpful to identify a baseline collection of controls for the assessment of technological capabilities. It also provides guidelines for the prioritization and implementation of security controls. It also states the primary road map for building the application and helping the security team. The proposed SRM-based CC security framework will be helpful for computing migrants and newcomers to understand more deeply the roles and responsibilities represented at each layer of the CC background. It also explains the duties of consumers and CSPs, both effectively and transparently, which results in extended security in the area of cloud application deployment.

We have also discussed the CC reference architecture and different kinds of actors involved as users of use cases with their interaction in order to implement the cloud and deliver the best services to cloud consumers. We have listed the various roles and responsibilities of cloud actors and focused especially on CSPs and cloud consumers. We have outlined CC-SRM implementation in the various world-leading IT industries, such as Amazon AWS, Microsoft Azure, and Google Cloud, with the products and services they offer. A comparison of top CSPs based on their service models and their products or services was also presented.

3.5 FUTURE WORK

As security is the primary concern of the CC environment, the security framework will guide both the cloud consumer as well as the CSP regarding the clear boundaries of individuals and their shared responsibilities at each level. Cloud actors are able to simulate the proposed SRM-based CC security framework, either in their in-house or external cloud environments for evaluation of security parameters and compliance.

Therefore, as future work, we need to perform research on the comparative analysis of the performance of the proposed framework by simulation and integration of security standards and guidelines for service and delivery models, which would be helpful for bench-marking the proposed framework. Additional research scope is in the field related to the integration of a cloud SLA with the proposed security framework which would support lots of CSPs to assure a service level that consumers require.

REFERENCES

Amanatullah, Y., Lim, C., Ipung, H. P., & Juliandri, A. (2013). Toward cloud computing reference architecture: Cloud service management perspective. *International Conference on ICT for Smart Society (ICISS), Jakarta* 1–4. https://doi.org/10.1109/ICTSS.2013.6588059.

Amazon Web Services. (2020a). *Amazon Elastic Compute Cloud*. https://aws.amazon.com/ec2/. Retrieved on December 5 2020.

Amazon Web Services. (2020b). *AWS Compliance*. https://aws.amazon.com/compliance/. Retrieved on November 28 2020.

Azua, M., & Goodman, B. (2011). Enabling smarter compliance architecture using social networks and cognitive agents. *IBM Journal of Research and Development, 55*(6), 445–455, https://doi.org/10.1147/JRD.2011.2165680.

Buyya, R., Yeo, C.S., Venugopal, S., Broberg, J., Brandic, I. (2009). Cloud computing and emerging IT platforms: Vision, hype, and reality for delivering computing as the 5th utility. *Future Generation Computer Systems, 25*(6), 599–616, https://doi.org/10.1016/j.future.2008.12.001.

CRM Software Salesforce. (2020). *Learning Center-CRM.* https://www.salesforce.com. Retrieved on December 6 2020.

Google cloud. (2020). *Google Cloud solutions.* https://cloud.google.com. Retrieved on December 2 2020.

IBM Corporation. (2019). Cloud-native security practices in IBM Cloud. *IBM Cloud.* https://www.ibm.com/in-en/it-infrastructure/z/capabilities/cloud-native-development. Retrieved on November 6 2020.

IIIT Hyderabad. (2010). *Cloud Computing for E-Governance.* https://cdn.iiit.ac.in/cdn/irel.iiit.ac.in/uploads/CloudComputingForEGovernance.pdf. Retrieved on March 8 2018.

Karimunnisa, S., Vijaya, d. & Kompalli, S. (2019). Cloud computing: review on recent research progress and issues international, *Journal of Advanced Trends in Computer Science and Engineering, 8*(2), 216–223.

Kirschnick, J., Alcaraz, C., Edwards, N. (2010). Toward an architecture for the automated provisioning of cloud services. *IEEE Communications Magazine, 48*(12), 124–131. https://doi.org/10.1109/MCOM.2010.5673082.

Lanfear, T. (2019). Shared responsibility in the cloud. *Microsoft Azure.* https://docs.microsoft.com/en-us/azure/security/fundamentals/shared-responsibility.

Liu, F., Tong, J., Mao, J., Bohn, R., Messina, J., Badger, M., & Leaf, D. (2011). NIST Cloud Computing Reference Architecture, *National Institute of Standards and Technology,* Gaithersburg. https://doi.org/10.6028/NIST.SP.500-292.

Microsoft Azure (2020). *Azure Documentation.* http://azure.microsoft.com/en-gb/azure/?product=featured. Retrieved on December 5 2020.

Ministry of Electronics & Information Technology (MeitY), Government of India (2014). *Digital India Programme.* http://www.digitalindia.gov.in. Retrieved on February 24 2017.

Peterson, G. (2010). Don't trust and verify: A security architecture stack for the cloud. *IEEE Transaction on Security & Privacy, 8*(5), 83–86.

Samimi, F., McKinley, P., Sadjadi, S. et al. (2007). Service clouds: Distributed infrastructure for adaptive communication services. *IEEE Transactions on Network and Service Management, 4*(2), 84–95.

Takabi, H., Joshi, J. (2010). Security and privacy challenges in cloud computing environments, *IEEE Computer and Reliability Societies*, 1540-7993/10.

U.S. Department of Commerce. (2010). *NIST Cloud Computing Program–NCCP.* https://www.nist.gov/programs-projects/nist-cloud-computing-program-nccp. Retrieved on March 2018.

4 Performance Analysis of a Hypervisor and the Container-based Migration Technique for Cloud Virtualization

Aditya Bhardwaj

Bennett University (Times of India Group), Greater Noida, Uttar Pradesh, India

CONTENTS

4.1 Introduction 57
4.2 Background and Motivation 58
  4.2.1 Need for Virtualization in Cloud Data Centers 59
  4.2.2 Evolution of Cloud Virtualization from Hypervisor to Containerization 59
4.3 Related Work 60
4.4 LXD/CR: A Container Migration Technique 61
4.5 Experimental Setup 62
4.6 Performance Evaluation of LXD/CR 66
  4.6.1 Downtime and Migration Time 66
  4.6.2 Number of Pages Transferred 66
4.7 Open Research Issues and Challenges 68
4.8 Conclusion and Future Scope 68
Bibliography 69

4.1 INTRODUCTION

The advantages of cloud computing, such as on-demand resource provisioning, multi-tenancy, elasticity, pay-per-use, and resource pooling, have encouraged IT industries to deploy business projects. In this era of technology, with the popularity of cloud computing, industrial data is moving to cloud computing platforms (Al-Ruithe et al. 2019). To achieve cloud services users need not worry about the maintenance overheads of the physical server (Younas et al. 2019). Virtualization is the backbone technology that works for cloud computing in a data center. This is because virtualization technology enables the deployment of multiple instances of virtual

DOI: 10.1201/9781003146711-4

servers, thus increasing the efficiency and flexibility of computer hardware. However, virtualization deployment using a hypervisor-based technique creates performance overheads. This is because an iso image of a virtual machine (VM) comprises a complete set of OS files. This disadvantage hinders its applicability to today's industrial applications. Recently, container-based virtualization gained significant popularity to address these issues. In containerization, a running application can share the kernel files of the underlying OS and contain only the required packages. A container can boot in just 50 ms while a VM might take 30 to 40 s to start, though its restart operation does not need rebooting. Therefore, applications were deployed using containerization to increase the utilization of resources (Casalicchio 2019). Thus, containers start faster compared to VMs with a performance that closely resembles a native computer hardware system (Cerny et al. 2018).

To explore containerization, recent publications (Chae et al. 2017) have explored the performance evaluation between the virtualization techniques of VMs and containers. From this work, it was found that, compared to a VM, the container provided a simple and resource-utilized platform to deploy cloud applications. Further, it is interesting to explore how to provide support for a container migration mechanism.

The main contributions of this chapter are:

1. An experimental testbed was developed to facilitate migration support in a container-based virtualization technique.
2. Experiments were conducted by considering the wide variety of industry oriented workload benchmarks.
3. Performance evaluation was conducted of the proposed container migration technique with the existing hypervisor-based VM migration technique.
4. Consolidated existing state-of-the-art research challenges for the container-based virtualization technique.

The remaining sections of this chapter are organized as follows. The background and motivation for hypervisor to container-based virtualization technology is presented in Section 4.2. Section 4.3 presents the literature on the container and hypervisor-based virtualization system. In Section 4.4 the architecture framework for the proposed container migration technique is discussed. In Section 4.5, a testbed setup with a workload specification is presented. Section 4.6 presents the performance evaluation with respect to the existing hypervisor-based VM migration technique. Finally, research issues, challenges, and the conclusion of this study are presented in Sections 4.7 and 4.8.

4.2 BACKGROUND AND MOTIVATION

A hypervisor-based virtualization framework deploys users' services using monolithic architecture. Therefore, such systems are not suitable for the current industrial environment. This has arisen due to the requirement for a new virtualization technology called containerization. This section first highlights the necessity of virtualization technology to deploy cloud applications and then discusses the evolution and popularity of containerization as the future of virtualization is presented.

4.2.1 Need for Virtualization in Cloud Data Centers

The conventional framework of computing is limited to deploying single instances only. This results in the under-utilization of system resources. In 1990, IBM (2019) introduced the concept of abstracting computer resources into various virtualized resources (Bugnion et al. 2012). Virtualization is the backbone technology for the working of cloud computing servers. This is because, using a virtualization technique, a large number of virtual instances can be deployed over a single physical server. This property of virtualization technology has led to hardware utilization.

4.2.2 Evolution of Cloud Virtualization from Hypervisor to Containerization

In a traditional cloud data center framework, applications have been deployed through hypervisor-based virtualization. Such a virtualization is suitable when cloud applications require services from heterogeneous types of OSs. For example, a recently automated system requires both a real-time and a general application OS. In hypervisor technology, KVM and XEN are the popular solutions (Bhardwaj and Krishna 2018a). However, virtualization deployment using a hypervisor-based technique creates performance overheads. This is because an iso image of a VM comprises a complete set of OS files (Mavridis and Karatza 2019). This has become a major issue for cloud administration because a hypervisor-based virtualization system results in performance overheads (Felter et al. 2015). To address this issue, container-based virtualization technology has gained in popularity. In containerization technology, the host OS is shared between the container's instances, thus making them lightweight in nature as depicted in Figure 4.1.

In containerization technology, LXC, Docker, and rkt are the popular categories. The LXC container is an OS-level virtualization mechanism that allows deploying multiple services using a single computer hardware OS (LXC 2020). The Docker container is an open source project with an executable package of software and was

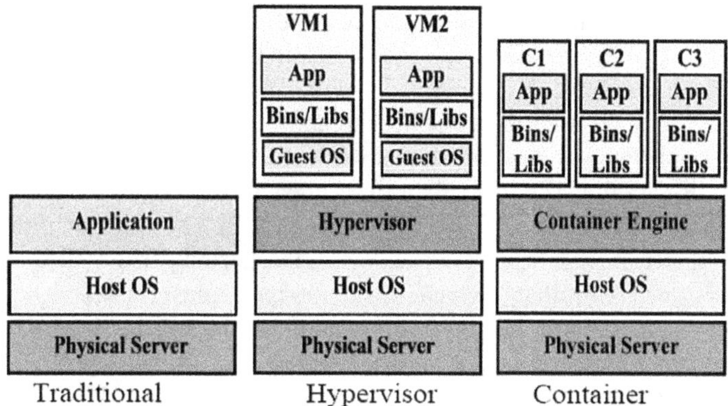

FIGURE 4.1 Architecture comparison between traditional, hypervisor, and container technology.

introduced in 2013 by dotCloud. The Docker container is particularly aimed at single app platforms which lack the ability to run multiple environment applications (Docker 2020). The major issue with the Docker container is the lack of a security feature (Xie et al. 2017). To resolve this issue a new container category called the rkt container was launched by CoreOS in 2014. In an rkt container, security is provided by using the feature of image signature verification (CoreOS 2019).

4.3 RELATED WORK

A hypervisor-based virtualization system causes significant performance overheads. Therefore, the existing literature has sought to enhance the VM migration scheme. In our previous work, we proposed a technique to improve the bandwidth allocation for VM migration (Bhardwaj and Krishna 2018b). We then explored a technique to facilitate a migration mechanism in container-based virtualization. However, in this study we provided a performance evaluation of the proposed container-based migration technique by executing different sets of benchmarks. This is because these days the cloud computing platform is gaining in popularity for deploying real-time industrial applications. Recent relevant literature studies are summarized in the following.

Morabito et al. (2015) explored a performance evaluation study for hypervisor and a containerization platform. They explored Docker and LXC from the container category. In their testing environment, they used Y-cruncher (Y-cruncher 2019), NBENCH as CPU intensive benchmarks, Bonnie++ (Bonnie++ 2018), Netperf, and disk write (dd test) for the network-intensive workload. The testbed results show that, compared to hypervisor technology, container virtualization introduced fewer overheads. Therefore, in a cloud data center, container virtualization is preferable.

Herrera-Izquierdo and Grob (2017) and Kozhirbayev and Sinnott (2017) also explored a performance evaluation study for VM and Docker virtualization platforms. But, as compared to Morabito et al. (2015), their study considered VirtualBox software for launching VM and used different types of workload benchmarks. The authors considered Linpack, Sysbench, Iperf, and Bonnie++ as benchmarking tools. In the first type, the Linpack benchmark (Linpack 2019) was used to test CPU performance by generating a linear equation with an input variable matrix size $N = 24{,}000$. Secondly, the efficiency of RAM storage was measured using Sysbench benchmark tools with block size = 8 and 12 GB. Thirdly, to measure the system network bandwidth they used the Iperf benchmarking tool with a client-server model and finally the Bonnie++ benchmark tool was used to test the performance of a disk with an 8 GB data file size. From their experimental results, it was found that for a CPU Linpack test case, Docker performed better than VM, with a difference of 1 Gflop. For the Sysbench benchmark test case, Docker outperformed by 137 MB/s and was 4.5 times faster than VM in terms of the Iperf transfer rate.

Pickartz et al. (2018) explored the migration framework for high performance computing (HPC) and discussed the challenging issues. They also proposed a protocol to enable migration support for message passing interface (MPI) processes in a distributed application environment. The authors stated that in an HPC environment, the migration of VMs give rise to challenges of load imbalance and residual dependencies. In residual dependency, processes are dependent on the resource available at the source node. To copy up with these challenges, the authors suggested that a migration mechanism needs to provide a solution for residual dependencies by providing an isolation mechanism at the system level virtualization. This can be achieved using a namespace mechanism, which provides isolation for the processes running in the system. From this study it was found that, compared to container technology, a hypervisor-based system creates performance overheads. Therefore, further studies are required to explore the container-based virtualization solution for future generation cloud data centers.

Kristiani et al. (2018) utilized container technology but from the Internet of Things (IoT) platform. The authors in this study highlighted that IoT application deployment in cloud computing has gained significant attention from various enterprises. Recent IoT applications such as robotics, ubiquitous computing, and small healthcare devices require only low latency and immediate processing capabilities. Traditional cloud data center architecture failed to meet this requirement. This is because of the large network delay between the cloud and the IoT edge device. To solve these issues, the authors studied edge computing and implemented this system between the cloud and the IoT device. Here, the processing for the sensor data is done near to the source of the data. The authors used Raspberry Pi 3++ for the edge computing architecture and a Docker container to deploy the IoT application using a temperature sensor device. Finally, Grafana plugin was integrated with InfluxDB for analysis and monitoring of sensor data using a graphical output. Their experimental results indicate that, compared to a VM based cloud data center, IoT application deployment using edge computing provides fast processing and low latency.

From these literature studies, it can be inferred that, compared to VM, container technology applications can be deployed in an easy way. Therefore, in recent cloud data centers, a container-based virtualization system can play a significant role. Further, with the recent popularity of container virtualization, it would be interesting to explore a study based on migration support in container-based virtualization. Compared to existing studies in this field I proposed an experimental testbed setup to facilitate migration support in a container-based virtualization system.

4.4 LXD/CR: A CONTAINER MIGRATION TECHNIQUE

This section illustrates the implementation details of the LXD/CR container migration technique. The major limitation when using hypervisor virtualization is that it creates significant overheads due to the large VM image size. To mitigate this, I explored virtualization deployment using the container-based technique. There exist different container virtualization technologies, which include Docker (Docker 2020), rkt (CoreOS 2019), and LXD (an extension of LXC). All these three container-based virtualization technologies follow the same working principle that applications share system kernel files and contain only the required packages, thus making it a lighter framework.

In our study, we adopted LXD as a lightweight virtualization technology. Both Docker and rkt containerization technologies use an application delivery framework that provides limited control over configuration files, while LXD uses a virtualization framework where administration related activities can be easily managed. The checkpoint/restore userspace (CRIU) technique (CRIU 2018) has been explored for

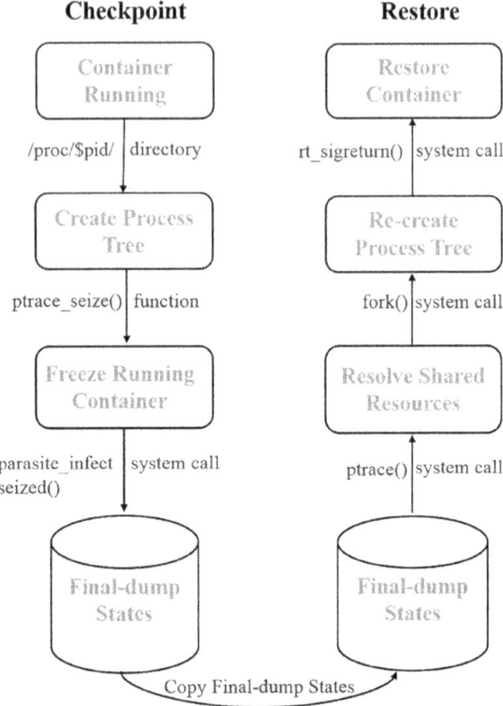

FIGURE 4.2 Workflow of the checkpoint/restore module.

implementing migration support in container-based virtualization. This works on a checkpoint/restore mechanism from the source to the destination server. Further, synchronization overheads during the container migration are eliminated by implementing the Z File System (ZFS) (LXD 2020).

LXD is an extension of Linux Containers (LXC) which provides add on features and functionality for deployment. Container-based virtualization improves the resource utilization for application deployment in cloud data centers. Therefore, to stabilize, create, and manage containers, liblxc and Go language functionalities have been used in LXD containers. Containerization, since the host OS is shared by running applications, may be vulnerable to attack. Thus to address the issue of security concerns, we used an AppArmor Linux security feature and a namespace, cgroups, for management and isolation of resources among containers. The namespace property is used to provide isolation for containers and the control group (cgroup) is used to configure resource management. Figure 4.2 shows the workflow of a configured checkpoint/restore module of a CRIU technique.

4.5 EXPERIMENTAL SETUP

The testbed environment includes two HP physical systems with a modified Ubuntu Linux kernel whose configurations are highlighted in Figure 4.3.

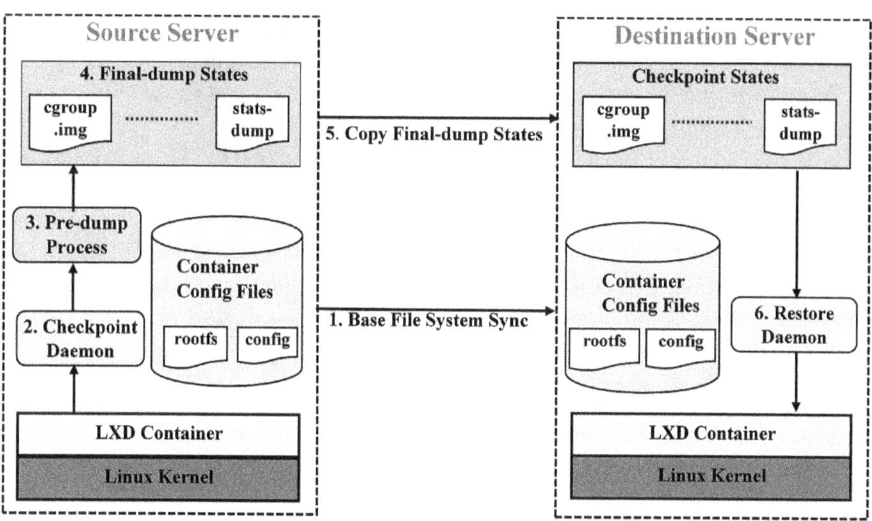

FIGURE 4.3 System architecture of the proposed LXD/checkpoint restore (CR) technique.

FIGURE 4.4 Experimental testbed developed to implement the LXD/CR container migration technique.

This experimental testbed setup, shown in Figure 4.4, was developed in the research and innovation lab at the Department of CSE, NITTTR, Chandigarh.

To demonstrate the solution for a cloud scenario, we performed a migration process using 15 iterations when the container was running Idle, UnixBench (Unixbench 2020), kernel-compile, or memtester benchmark (Memtester 2020). Idle flags up the test case when no external application is running. In the second category, UnixBench, which is a Linux benchmark, was executed with multiple tasks. In the third experiment, kernel-compile was used to test a system call by compiling the kernel source code. Finally, a memtester benchmark was used to test memory work loads.

To perform the migration of running containers, source and destination servers should have a Linux kernel version higher than 4.4, a CRIU with at least 2.0, Golang 1.10 or later, and an LXD container-hypervisor with liblxc API. Thus, in our work host's system the Linux kernel is modified to a 4.4.12-generic version and a CRIU is built from its source repository. In our experimental setup, the Go programming language environment is used to build LXD from its binary code. The Go programming environment path is set in the */.bashrc* file and its configuration is checked using the command:

$ go env.

By considering the above system requirements, the testbed environment was deployed using source and destination Ubuntu 16.04 servers with a modified Linux kernel. On both source and destination server, we first installed an LXD container engine to be deployed in a container-based virtualization environment. The following command was used to check the system supported features of the installed Linux kernel:

$ sudo lxc-checkconfig.

If the output of the above command shows kernel configuration as enabled, this means the successful configuration of the LXD hypervisor. Otherwise, there is a need to retest the configuration file of the computer system. A sample screenshot for the command is depicted in Figure 4.5. The file location */etc/lxc/default.conf* is used to store the configurations for all the running containers. Further, configuration files specific to the container can be located in the */var/lib/lxd/containers/c1* directory. The following command is used for the initial configuration of the LXD container:

$ sudo lxd init.

The above command is used to configure backend storage (ZFS) and the networking setting of the running container. The IPv4 and IPv6 networking configurations for the running container are set up using the *lxdbr0* bridge feature. Finally, configured Ubuntu containers are run using the *$ lxc launch* command.

Table 4.1 shows the state of the running container in terms of IP address allocation; its full configuration can be tested using the following command:

$ sudo lxc info container name.

For migration to work and allow communication between source and destination servers, TCP port 8443 was used.

The connection between these systems is maintained using the remote add server cmd. Here, server3 and server4 host names have been used for the source and destination server:

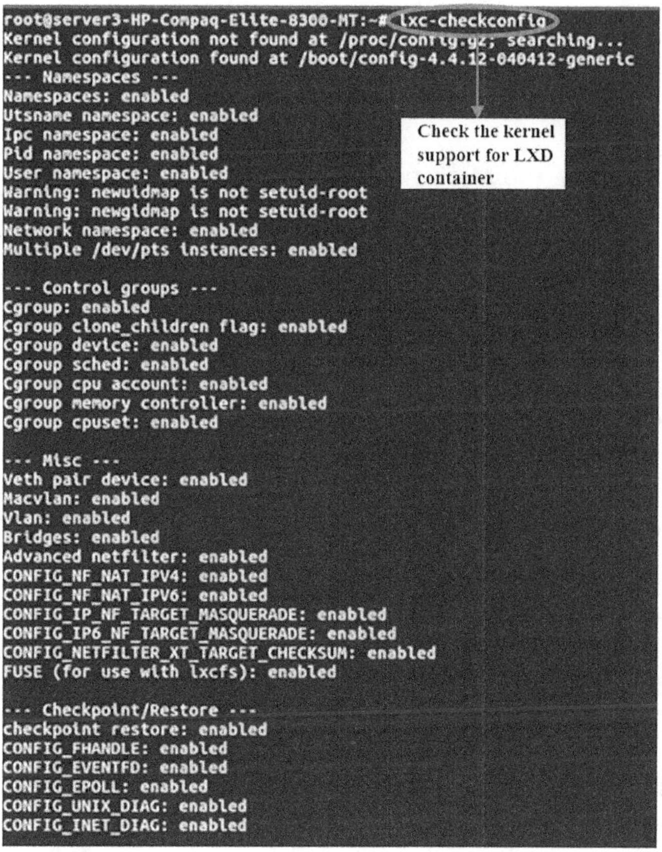

```
root@server3-HP-Compaq-Elite-8300-MT:~# lxc-checkconfig
Kernel configuration not found at /proc/config.gz; searching...
Kernel configuration found at /boot/config-4.4.12-040412-generic
--- Namespaces ---
Namespaces: enabled
Utsname namespace: enabled
Ipc namespace: enabled
Pid namespace: enabled
User namespace: enabled
Warning: newuidmap is not setuid-root
Warning: newgidmap is not setuid-root
Network namespace: enabled
Multiple /dev/pts instances: enabled

--- Control groups ---
Cgroup: enabled
Cgroup clone_children flag: enabled
Cgroup device: enabled
Cgroup sched: enabled
Cgroup cpu account: enabled
Cgroup memory controller: enabled
Cgroup cpuset: enabled

--- Misc ---
Veth pair device: enabled
Macvlan: enabled
Vlan: enabled
Bridges: enabled
Advanced netfilter: enabled
CONFIG_NF_NAT_IPV4: enabled
CONFIG_NF_NAT_IPV6: enabled
CONFIG_IP_NF_TARGET_MASQUERADE: enabled
CONFIG_IP6_NF_TARGET_MASQUERADE: enabled
CONFIG_NETFILTER_XT_TARGET_CHECKSUM: enabled
FUSE (for use with lxcfs): enabled

--- Checkpoint/Restore ---
checkpoint restore: enabled
CONFIG_FHANDLE: enabled
CONFIG_EVENTFD: enabled
CONFIG_EPOLL: enabled
CONFIG_UNIX_DIAG: enabled
CONFIG_INET_DIAG: enabled
```

Check the kernel support for LXD container

FIGURE 4.5 Configuration testing of the LXD container.

TABLE 4.1
State of Running Container "c1"

ID	Status	IPv4	IPv6	Type
c1	Running	10.13.198.64	fd42..43ff	Persistent

$ lxc remote add server3 IP address:8443
$ lxc remote add server4 IP address:8443

Once the server's connectivity is made the CRIU technique is used to facilitate migration support for the running container "c1" from source to destination. Here, the CRIU technique is used to freeze the running application into a collection of checkpoint image files. These collected images are further resumed at the restore procedure in the destination server.

4.6 PERFORMANCE EVALUATION OF LXD/CR

This section shows the performance evaluation of the container and hypervisor-based migration technique.

4.6.1 Downtime and Migration Time

Figures 4.6 and 4.7 show the downtime and migration times for the container migration approach LXD/CR compared to the pre-copy VM migration technique. For the idle test case and workload execution, our approach reduces the downtime by 59.48, 73.07, 77.56, and 78.24%, an average of 72.08%. Similarly, the amount of migration time is decrease by 33.35, 44.18, 66.43, and 75.81%, an average of 54.94%. This is because data transfer using VM employs full binary libraries, codes, and OS files, but LXD/CR only migrates memory checkpoint dump states which require less migration duration compared to a VM.

4.6.2 Number of Pages Transferred

Figure 4.8 presents a performance comparison for the number of pages transferred between LXD/CR and the hypervisor-based VM migration technique. The figure reveals that, compared to the pre-copy technique, container migration (LXD/CR) reduces the number of pages transferred significantly by 95.08, 97.91, 98.26, and

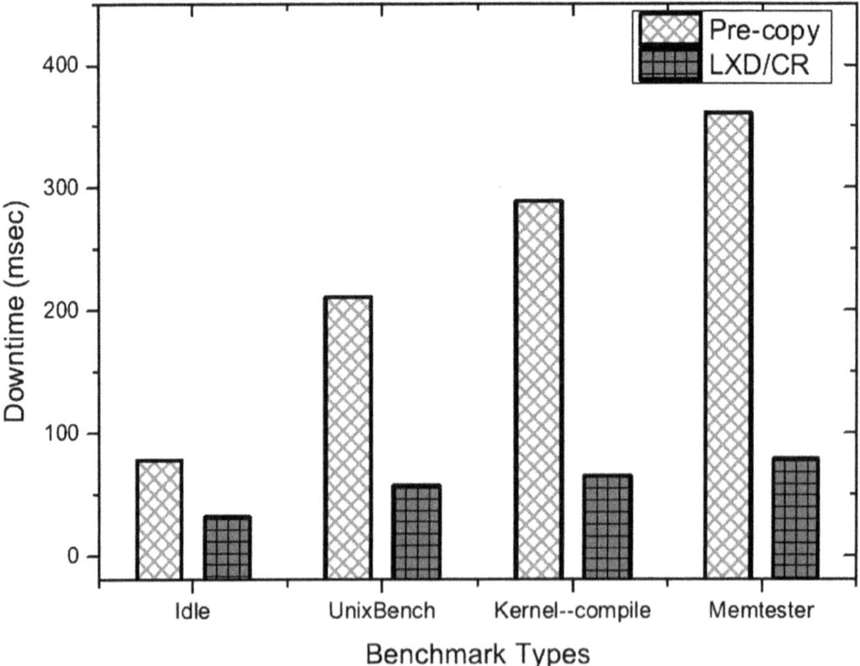

FIGURE 4.6 Performance evaluation for downtime (T_d).

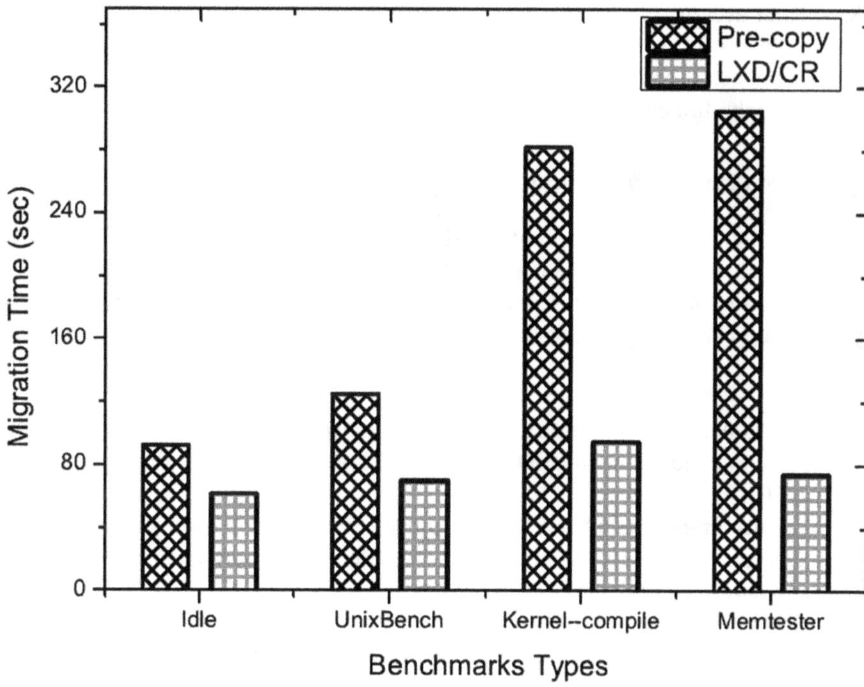

FIGURE 4.7 Performance evaluation for migration time (T_m).

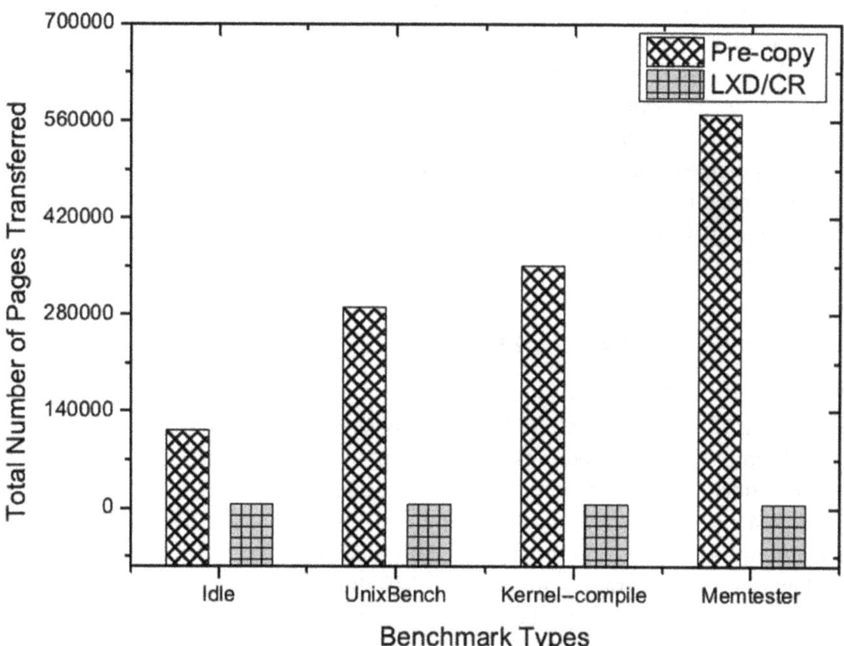

FIGURE 4.8 Performance evaluation for the number of pages transferred (T_{pages}).

98.94%, an average of 97.54%. There is a significant difference for this parameter because, in LXD/CR, the number of pages transferred equal the container's checkpoint dump states, while pre-copy transfers are all dirtied memory pages along with a heavyweight, full edge, OS VM image size.

4.7 OPEN RESEARCH ISSUES AND CHALLENGES

In the era of cloud computing, container-based virtualization plays a significant role. A traditional hypervisor virtualization system suffers from performance overheads. This section highlights the research challenges in a container-based virtualization system. The challenges that need to be addressed in such a virtualization are the following.

1. **Container auto-scaling:** In container virtualization there is a need to explore the automatic scaling of container instances (Karn et al. 2018; Rovnyagin et al. 2020).
2. **Failure management:** From a container failure management perspective it is necessary to address log monitoring and failure prediction schemes (Rodriguez and Buyya 2019).
3. **Efficient resource allocation:** Container resource allocation is a key mechanism for system performance. Further, with the popularity of cutting edge technologies (Kaur et al. 2017; Morabito 2017) there is under-utilization of resources in static allocation schemes. Therefore, there is a need to address dynamic resource allocation (Haji et al. 2020; Pitanga et al. 2020).
4. **Need for a fault tolerance mechanism:** In container virtualization platforms, a malicious container may cause the failure of a host's OS. So, there is a need to explore a fault tolerance isolation mechanism (Rodriguez and Buyya 2019).
5. **Energy consumption:** In containerization it is also interesting to explore things from an energy consumption perspective (Chen et al. 2020).
6. **Challenges in affinity-based placements:** In container affinity placement there is a need to address the issues in terms of (i) concurrent container scheduling (Yang et al. 2018), (ii) the configuration of network conditions (Hu et al. 2018), and (iii) identifying a target node for container placement (Ludwig et al. 2019).

4.8 CONCLUSION AND FUTURE SCOPE

A container-based solution to implement a migration technique for data centers would be a good basis for cloud service providers. In this chapter, an experimental testbed setup was developed to facilitate a migration mechanism for container-based virtualization. The experimental test cases reveal that, compared to a hypervisor-based VM migration technique, the proposed LXD/CR container-based migration scheme shows significant improvement, with a reduction range from 72.08 to 54.94% for downtime and migration time, and 97.54% for the number of pages transmitted. Thus, container migration can be used to migrate the running application workload, given the challenges of server overloaded, fault management, and when system maintenance activities are carried out.

Further, container-based virtualization is gaining significant popularity and there is a need to address the research issues of (i) auto-scaling, (ii) failure management, (iii) dynamic resource allocation, (iv) fault tolerance, (v) energy consumption, (vi) efficient scheduling of concurrent containers, and (vii) interference of containers.

BIBLIOGRAPHY

Al-Ruithe, M., Benkhelifa, E., & Hameed, K. (2019). A systematic literature review of data governance and cloud data governance. *Personal and Ubiquitous Computing*, 23(5), 839–859. https://doi.org/10.1007/s00779-017-1104-3.

Alves, M. P., Delicato, F. C., Santos, I. L., & Pires, P. F. (2020). LW-CoEdge: a lightweight virtualization model and collaboration process for edge computing. *World Wide Web,* 23(2), 1127–1175. https://doi.org/10.1007/s11280-019-00722-9.

Bhardwaj, A., & Krishna, C. R. (2018a). Performance evaluation of bandwidth for virtual machine migration in cloud computing. *International Journal of Knowledge Engineering and Data Mining*, *5*(3), 139–152. http://www.inderscience.com/offer.php?id=94743.

Bhardwaj, A., & Krishna, C. R. (2018b). Efficient multistage bandwidth allocation technique for virtual machine migration in cloud computing. *Journal of Intelligent & Fuzzy Systems*, *35*(5), 5365–5378. https://doi.org/10.3233/JIFS-169819.

Bhardwaj, A., & Krishna, C. R. (2019). A container-based technique to improve virtual machine migration in cloud computing. *IETE Journal of Research*, 1–16. https://doi.org/10.1080/03772063.2019.1605848.

Bonnie++. (2018). Online from http://www.brendangregg.com/ActiveBenchmarking/bonnie++.html. Retrieved on June 15 2018.

Bugnion, E., Devine, S., Rosenblum, M., Sugerman, J., & Wang, E. Y. (2012). Bringing virtualization to the x86 architecture with the original vmware workstation. *ACM Transactions on Computer Systems (TOCS)*, *30*(4), 1–51. https://doi.org/10.1145/2382553.2382554.

Casalicchio, E. (2019). Container orchestration: a survey. *Systems Modeling: Methodologies and Tools*, 221–235. https://doi.org/10.1007/978-3-319-92378-914

Cerny, T., Donahoo, M. J., & Trnka, M. (2018). Contextual understanding of microservice architecture: current and future directions. *ACM SIGAPP Applied Computing Review*, *17*(4), 29–45. https://doi.org/10.1145/3183628.3183631.

Chae, M., Lee, H., & Lee, K. (2019). A performance comparison of linux containers and virtual machines using Docker and KVM. *Cluster Computing, 22*(1), 1765–1775. https://doi.org/10.1007/s10586-017-1511-2.

Checkpoint/Restore In Userspace. (2018). Online from https://criu.org/Main_Page. Retrieved on August 12 2018.

Chen, W. Y., Ye, K. J., Lu, C. Z., Zhou, D. D., & Xu, C. Z. (2020). Interference analysis of co-located container workloads: A perspective from hardware performance counters. *Journal of Computer Science and Technology*, 35, 412–417. http://jcst.ict.ac.cn/EN/Y2020/V35/I2/412

Containers on google cloud platform. (2018). Online from https://cloud.google.com/containers/. Retrieved on July 24 2018.

CoreOS running containers with rkt virtualization technology. (2019). Online from https://coreos.com/rkt/docs/latest/. Retrieved on Sept 1 2019.

Docker. (2020). Online from https://www.docker.com/resources/what/container. Retrieved on August 3 2020.

Felter, W., Ferreira, A., Rajamony, R., & Rubio, J. (2015, March). An updated performance comparison of virtual machines and linux containers. In *2015 IEEE international symposium on performance analysis of systems and software (ISPASS)* (pp. 171–172). IEEE. https://ieeexplore.ieee.org/document/7095802.

Haji, L. M., Zeebaree, S. R., Ahmed, O. M., Sallow, A. B., Jacksi, K., & Zeabri, R. R. (2020). Dynamic resource allocation for distributed systems and cloud computing. *TEST Engineering & Management*, *83*, 22417–22426. https://www.ijert.org/dynamic-resource-allocation-in-cloud-computing

Herrera-Izquierdo, L., & Grob, M. (2017). A performance evaluation between Docker container and Virtual Machines in cloud computing architectures. *Maskana, 8*, 127–133. https://d1wqtxts1xzle7.cloudfront.net/

Hu, Y., Zhou, H., de Laat, C., & Zhao, Z. (2020). Concurrent container scheduling on heterogeneous clusters with multi-resource constraints. *Future Generation Computer Systems*, *102*, 562–573. https://doi.org/10.1016/j.future.2019.08.025

Hu, Y., Zhou, H., de Laat, C., & Zhao, Z. (2018, August). Ecsched: Efficient container scheduling on heterogeneous clusters. In *European Conference on Parallel Processing* (pp. 365–377). Springer, Cham. https://doi.org/10.1007/978-3-319-96983-1_26.

IBM desktop virtualization system. (2019). Online from https://www.ibm.com/cloud/learn/virtualization. Retrieved on Nov 7 2019.

Karn, R. R., Kudva, P., & Elfadel, I. A. M. (2018). Dynamic autoselection and autotuning of machine learning models for cloud network analytics. *IEEE Transactions on Parallel and Distributed Systems*, *30*(5), 1052–1064. https://ieeexplore.ieee.org/document/8500348.

Kaur, K., Dhand, T., Kumar, N., & Zeadally, S. (2017). Container-as-a-service at the edge: Trade-off between energy efficiency and service availability at fog nano data centers. *IEEE Wireless Communications*, *24*(3), 48–56. https://ieeexplore.ieee.org/document/7955911.

Kozhirbayev, Z., & Sinnott, R. O. (2017). A performance comparison of container-based technologies for the cloud. *Future Generation Computer Systems*, *68*, 175–182. https://doi.org/10.1016/j.future.2016.08.025.

Kristiani, E., Yang, C. T., Wang, Y. T., Huang, C. Y., & Ko, P. C. (2018, October). Container-based virtualization for real-time data streaming processing on the edge computing architecture. In *International Wireless Internet Conference*. Springer. 203–211. https://doi.org/10.1007/978-3-030-06158-6_2.1.

Linpack benchmark for distributed-memory computers. (2018). Online from https://www.top500.org/project/linpack/ Retrieved on June 5 2018.

Ludwig, U. L., Xavier, M. G., Kirchoff, D. F., Cezar, I. B., & De Rose, C. A. (2019). Optimizing multi-tier application performance with interference and affinity-aware placement algorithms. *Concurrency and Computation: Practice and Experience*, *31*(18), 5098. https://doi.org/10.1002/cpe.5098.

LXC. (2020). Online from https://linuxcontainers.org/lxc/introduction/. Retrieved on August 3 2020.

LXD container configuration.(2020). Online from https://lxd.readthedocs.io/en/latest/storage/. Retrieved on Oct 28 2020.

Mavridis, I., & Karatza, H. (2019). Combining containers and virtual machines to enhance isolation and extend functionality on cloud computing, *Future Generation Computer Systems*, 94, 674– 696. https://doi.org/10.1016/j.future.2018.12.035.

Memtester memory test benchmark. (2020). Online from https://linux.die.net/man/8/memtester. Retrieved on September 10 2020.

Morabito, R. (2017). Virtualization on internet of things edge devices with container technologies: A performance evaluation. *IEEE Access, 5*, 8835–8850. https://ieeexplore.ieee.org/document/7930383.

Morabito, R., Kjällman, J., & Komu, M. (2015, March). Hypervisors vs. lightweight virtualization: a performance comparison. In *2015 IEEE International Conference on Cloud Engineering,* 386–393). IEEE. https://ieeexplore.ieee.org/document/7092949/.

Pickartz, S., Clauss, C., Breitbart, J., Lankes, S., & Monti, A. (2018). Prospects and challenges of virtual machine migration in HPC. *Concurrency and Computation: Practice and Experience*, *30*(9), 4412–4419. https://doi.org/10.1002/cpe.4412.

Pitanga, Marcelo Alves, Delicato, F. C., Santos, I. L., & Pires, P. F. (2020). LW-CoEdge: a lightweight virtualization model and collaboration process for edge computing. *World Wide Web*, *23*(2), 1127–1175. https://doi.org/10.1007/s11280-019-00722-9

Rodriguez, M. A., & Buyya, R. (2019). Container-based cluster orchestration systems: A taxonomy and future directions. *Software: Practice and Experience*, *49*(5), 698–719. https://arxiv.org/abs/1807.06193.

Rovnyagin, M. M., Hrapov, A. S., Guminskaia, A. V., & Orlov, A. P. (2020, January). ML-based heterogeneous container orchestration architecture. In *2020 IEEE Conference of Russian Young Researchers in Electrical and Electronic Engineering (EIConRus)*, 477–481. IEEE. https://ieeexplore.ieee.org/document/9039033.

Running containers with rkt virtualization technology. (2019). Online from https://coreos.com/rkt/docs/latest/. Retrieved on Sept 1 2019.

UnixBench. (2020). Online from https://www.ostechnix.com/unixbenchbenchmarksuiteunix-like-systems/. Retrieved on September 8 2020.

Xie, X. L., Wang, P., & Wang, Q. (2017, July). The performance analysis of Docker and rkt based on Kubernetes. In *2017 13th International Conference on Natural Computation, Fuzzy Systems and Knowledge Discovery (ICNC-FSKD)*, 2137–2141. IEEE. DOI: https://ieeexplore.ieee.org/document/8393101/.

Yang, Hu, Zhou, H., de Laat, C., & Zhao, Z. (2020). Concurrent container scheduling on heterogeneous clusters with multi-resource constraints. *Future Generation Computer Systems*, 102, 562–573. https://doi.org/10.1016/j.future.2019.08.025

Y-cruncher: a multi-threaded pi program. (2019). Online from http://www.numberworld.org/y-cruncher/. Retrieved on May 8 2019.

Younas, M., Jawawi, D. N. A., Ghani, I., Shah, M. A., Khurshid, M. M., & Madni, S. H. H. (2019). Framework for agile development using cloud computing: A survey. *Arabian Journal for Science and Engineering*, *44*(11), 8989–9005. https://doi.org/10.1007/s13369-019-03923-6.

5 Segmentation of Fine-Grained Iron Ore Using Deep Learning and the Internet of Things

Ada Cristina França da Silva
Universidade Estadual do Maranhão (UEMA), Brazil

Omar Andres Carmona Cortes
Instituto Federal do Maranhão (IFMA), Brazil

CONTENTS

5.1 Introduction 73
5.2 Related Work 76
5.3 Problems and Challenges in Image Segmentation 76
5.4 Deep Neural Networks 77
5.5 Experiments and Results 79
5.5.1 The Dataset 79
5.5.2 Hardware and Software Configuration 80
5.5.3 Model Evaluation: U-Net 82
5.5.4 Edge Device Solution Evaluation 83
5.6 Conclusions and Future Work 83
Note 84
References 84

5.1 INTRODUCTION

Industry 4.0 introduces several changes to the original approach of industrial automation (Dalzochio et al., 2020). It has been associated with a new industrial model that might lead companies to increase performance results by adopting innovative data and technologies. We may note that the Internet of Things (IoT) and machine learning algorithms are among those new technologies.

The IoT was an evolving process. It started using sensors in a can vendor machine (Gupta, 2019) and has evolved to single-board computers, such as Raspberry, Jetson, Google Coral, and Khadas VIM, which are powerful devices that can connect various sensors and storage accessories. Thus, it is reasonable to move applications, such as

DOI: 10.1201/9781003146711-5

machine learning algorithms, from servers to IoT devices, allowing servers to focus on other relevant tasks.

Machine learning algorithms, associated with a sub-field called deep learning, provide computers with the ability to make predictions or recognize patterns in complex data, especially those associated with image prediction or image segmentation, which are essential applications in Industry 4.0. The usage of deep learning has been encouraged following the development of AlexNet (Krizhevsky, Sutskever, & Hinton, 2017), which is a convolutional neural network (CNN) that has been broadly adopted for different applications in industries for object detection (D'Avella, Tripicchio, & Avizzano, 2020) and optical quality inspection (Sassi, Tripicchio, & Avizzano, 2019). After AlexNet many CNNs have been successfully proposed and applied, including that called U-Net. As we can see, image segmentation became an essential tool in industry in real-time applications. For example, Vision Rock (Guyot et al., 2004) is a computer vision industrial mining system that can be used in various processing stages to calculate granulometry. Also, studies regarding the moister, which is designed to improve the iron ore's quality, exist. For instance, Lage et al. (2018) present a sensor design for real-time moisture analysis. Other works deal with solutions involving the iron ore's supply chain's main activities, such as plants, mines, railways, and ports. Regardless of the purpose, most solutions to plants are related to decisions made in real time.

Thus, there is no turning back after Industry 4.0. In this context, this work proposes using U-Net to perform a real-time image segmentation in iron ore production, a primary extracted raw material for producing steel that can be used to make, for instance, vehicles, construction, railways, and many engineering products. According to the World Steel Association, crude steel production has increased significantly since 2010, as presented in Figure 5.1, mainly due to China. Consequently, the iron

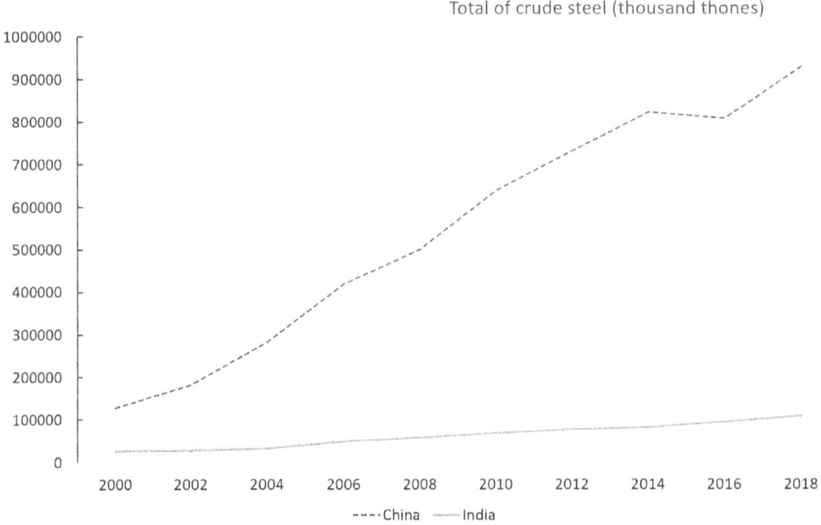

FIGURE 5.1 Total production of crude steel since 2000.

market is investing in many strategies to increase production and improve product quality.

Most iron ore types pass by a beneficiation process consist of two stages: crushing and screening. Crushing uses large machines to decrease the grain size. The screening process separates the ore between fine-grained particles (less than 19 mm) and coarse-grained particles (greater than 19 mm). Then, during the crushing process, it is critical to obtain information about the granulometry. This information allows the operators to stop the machinery from damaging any equipment in the next stages in real-time operation. After the screening process, the operator can analyze the separation and use a supervisory system that changes the amount of iron ore going into the system, based on the screening efficiency. As obtaining the granulometry is a critical part of producing iron, several types of research and products have been implemented to compute it on the conveyor belts in real time.

Analyzing fine-grained particles in each sieve provides a way of identifying which sieve lacks the correct efficiency and then of acting in a specific way. Developing a computer vision solution to perform this granulometry calculation provides gains when acting after quick decisions are made because it optimizes the circuit in real time when checking the material, allowing the flow to increase. Also, the computer vision solution allows removing operators from nearby equipment, thus improving process safety.

Even though computer solutions are attractive for dealing with image segmentation in the industry, some challenges still exist. There are problems such as dust, no network communication, reduced space, and temperature. Therefore, IoT devices play an essential role in providing a more comfortable instalation and better opportunities for portable solutions involving machine learning algorithms. Hence, as previously stated, our proposal analyzes the iron ore material in real time and automates the decision of flow in a sieve, communicating with an open platform communication (OPC) server interface, as presented in Figure 5.2.

In this context, this chapter is divided as follows. In Section 5.2 we present some research that deals with computer vision, IoT devices, and deep learning; in Section 5.3 we introduce some problems and challenges in computer vision-based applications; in Section 5.4 we describe the neural network subject and the U-net network that was used in the experimental section; in Section 5.5, we present the experiments and describe how we obtained the dataset, including the hardware and software setup, and results; finally, in Section 5.6 we conclude and suggest future work.

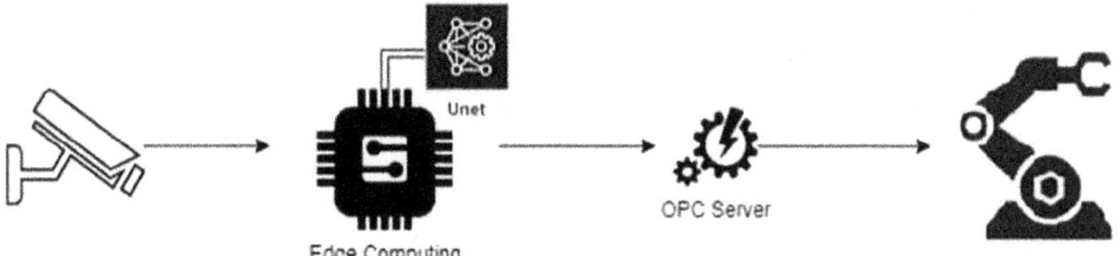

FIGURE 5.2 Purpose of video analytics for analysis of material in real time.

5.2 RELATED WORK

As previously mentioned, the idea of using IoT devices is to move the processing requirements to the edge, freeing servers to perform other relevant tasks. Moreover, there is an intrinsic philosophy in Industry 4.0 of joining together IoT systems and deep learning, especially those associated with computer vision tasks. In this context, Table 5.1 presents recent work involving the IoT and deep learning models in different applications. The table also shows which single-board computer has been used in the application.

In Ukaegbu et al. (2020), the authors use a CNN called ResNet-50 embedded in a Raspberry PI 3 to detect deficiency in a red grapevine. Stankov et al. (2019) used MobileNet to identify humans in operational areas using a Raspberry PI to improve the safety and collaboration between humans and robots. Chen et al. (2021) used a Tiny YOLO V3 embedded in a Jetson TX2 placed in a drone to identify an insect named *Tessaratoma Papillosa* to provide accurate pesticide spraying in crops. Chou et al. (2019) implemented a robot system using a Raspberry PI, a Movidius Stick,[1] and a GAN to control a robot that identifies and removes from the production line defective coffee beans. Li et al. (2020) used a Jetson Xavier running a CNN called ICNet in a harvester machine for avoidance of collisions.

5.3 PROBLEMS AND CHALLENGES IN IMAGE SEGMENTATION

Several computing applications need image segmentation to understand an image with the classification, position, and size of the segmentation. This technique is useful for calculating regions of interest and separating objects from a scenario. It has an essential role for various applications, such as analyzing medical images, calculating granulometry, and driving autonomous vehicles.

There are a lot of traditional image segmentation algorithms, such as thresholds, watershed (Kornilov & Safonov, 2018), k-means clustering, and region-growing techniques. There has been a growth in using deep learning techniques for image detection, classification, and segmentation in recent years. Generally, deep learning approaches, especially CNNs, present better results than more well-known techniques for specific problems. In other words, compared to traditional computer vision techniques, deep learning enables computer vision engineers to achieve greater accuracy in tasks such as image classification, semantic segmentation, and object detection (Walsh et al., 2019).

TABLE 5.1
IoT Devices with Deep Learning Models

Use Cases	IoT Device	DNN Model
Ukaegbu et al. (2020)	Raspberry PI 3 (ARM V7)	ResNet-50
Stankov et al. (2019)	Raspberry PI	MobileNet
Chen et al. (2021)	Jetson TX2	Tiny-YOLOv3
Chou et al. (2019)	Raspberry PI Movidius Stick	GAN network
Li et al. (2020)	Jetson Xavier	ICNet

Segmentation can be considered a classification problem, in which each pixel has a class that represents an object (e.g., dogs, cats, cars). With the principles of similarity and proximity, segmentation techniques are used to perform pixel grouping. The grouping of a pixel base by similarity uses certain attributes, such as color, texture, movement, and lighting. In real-world images, it is difficult to find examples with edges well defined or objects with constant colors (i.e., there is much noise in the image, therefore many solutions of segmentation are combined techniques of image processing).

For the granulometry problem, it is common to find solutions based on the watershed algorithm, which uses a morphology technique and identifies an interior mark on each object to separate them. In Thurley (2013), a watershed is used to identify the size of particles in mine piles, from large rocks to areas of fines.

In this context, the main objective of this work is to identify regions of fine-grained iron ore. These regions differ from those with larger stones because they have a completely different texture and do not have well-defined edges, representing a challenge to any segmentation algorithm, as previously mentioned.

5.4 DEEP NEURAL NETWORKS

A neural network can be viewed as a computational graph of elementary units in which great power is gained by connecting them in particular ways (Aggarwal, 2018). Each basic unit resembles a logistic regression algorithm; it calculates the probability of a class or event fail or pass. Neurons are trained and there is a dependancy between them throughout the entire neural network architecture. This combination improves the model's ability to learn complex functions to map the problem. Deeper neural network architectures can improve knowledge representation and learn any mathematical function with sufficient data. These models advance because current hardware has the computational power for quick experimentation.

Deep neural networks solve problems that have a large amount of data to be analyzed. Non-textual data, such as images, sounds, and other types of signals, require more complex analysis. Several neural network architectures are proposed for signal processing. CNNs are an example in the area of computer vision, in which these networks are designed to work with grid-structured inputs, which have strong spatial dependencies in local regions of the grid, as an image. In other words, the characteristics of more distant connections are weaker; this makes the network able to learn proximate information; however, essential relationships that are not close are lost.

CNNs use the layer structure of neural networks to identify primitive shapes, such as straight and curved lines, and in deeper layers, more complex shapes can be identified. This type of network's input structure is 3D, representing the height, width, and depth of the data being inserted. In the case of images, the depth represents the color channels. Each layer of the network needs to analyze this 3D dimension. The structure for a CNN is devised by convolution layers, pooling, and the rectified linear unit (ReLU) activation layers:

- Convolution: in a CNN, the parameters to be learned are called a filter or kernel. Each kernel acts in the full depth of the characteristics map. Generally, this

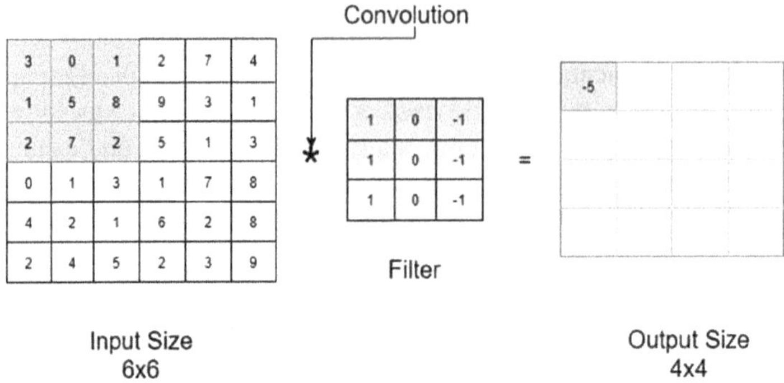

FIGURE 5.3 Convolution operation.

kernel has values parameterized according to the size of the input. An example is values of 3 × 3 or 16 × 16. When passing through convolution layers, as shown in Figure 5.3, the product between the input values and each kernel generates an overlap that allows connectivity between the layers.

- Pooling: this layer is responsible for downsampling, in which the vector of characteristics is transformed into a vector of condensed characteristics. For example, a map with a dimension of 24 × 24 can be summarized, after grouping, as 12 × 12.
- The ReLU activation layer is similar to the activation functions of traditional neural networks. This activation layer continues to guarantee the non-linearity of the model. Compared to other functions such as sigmoid or tanh, ReLU has advantages over speed because it has a simple function. Nevertheless, it has an efficiency equivalent to more complex functions.

The U-Net was proposed by Shelhamer, Long, and Darrell (2017) to produce segmented medical images. Even though some research using this segmentation network are associated with the medical area, some articles show that this neural network's capacity is vast, ranging from segmentation of urban areas in satellite images (McGlinchy et al., 2019) to segmentation of rice lodgings (Zhao et al., 2019). Therefore, it is suitable for any type of segmentation task in which the desired location and the predicted classes need to be associated with each image's pixels.

The U-Net was built on the architecture of the fully convolutional network. The network comprises two parts: a hiding part (contracting path) and a decoder (expansive path). The main idea is that instead of pooling operators, there are upsampling operators in expansive paths, which are essential to learning where the classification is located. There are many techniques to upsampling the image, such as unpooling, cubic interpolation, and transposed convolution (Dumoulin & Visin, 2018). In U-Net segmentation, the transposed convolution is used in expansive paths, also called decoders.

Figure 5.4 presents the original paper's image that represents a "U" shape. In this architecture, it can be seen that each process consists of two convolution layers,

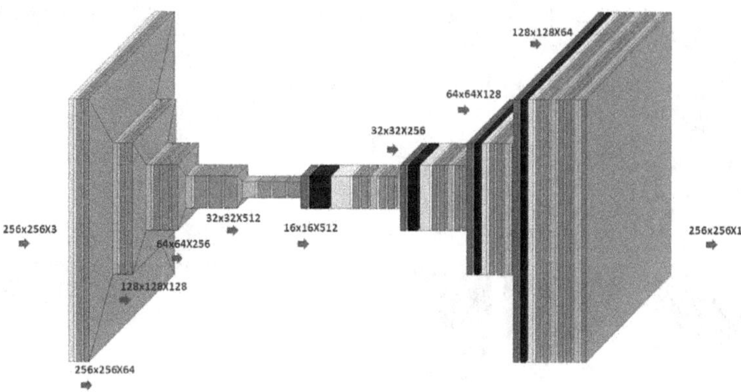

FIGURE 5.4 Basic architecture of the U-Net.

which increase the depth of the characteristics map. These convolution operations are distributed along with pooling layers, which decrease the image size. On the way to upsampling, there are successive processes of convolutions interchanged with transposed convolution techniques.

5.5 EXPERIMENTS AND RESULTS

In this section, we show how we obtained the dataset, the setup of the training computer and the IoT device, and the model's evaluation based on the Intersect of the Union (IoU).

5.5.1 The Dataset

The data for this experiment was acquired from a mining company, in which the main product is iron ore. The images were taken with a camera installed inside the plant. Further, the images were obtained in all periods of a day to characterize all material types and different luminosity degrees. Figure 5.5 shows some typical images of the captured material with the installed camera. There are apparent differences in particle size distribution and material texture. These characteristics are essential features for differentiating the region of interest for the problem. However, the colors of the images are similar, making it challenging to use segmentation color-threshold-based techniques. The region of interest for the problem is characterized by similarity to the texture of sand and to very thin granulated ore.

The collected images were taken at a distance of at least 5 meters from the material due to equipment vibrations that made it difficult to install the camera closer to the ore. Thus, the distance between the camera and the iron ore is also a challenge in the segmentation task.

Two hundred and fifty images collected from the camera were used to train the model. We used data augmentation techniques to increase the database size and guarantee enough images to train and test the neural network. We performed operations such as rotations, color variation, contrast, and brightness, using a library for data

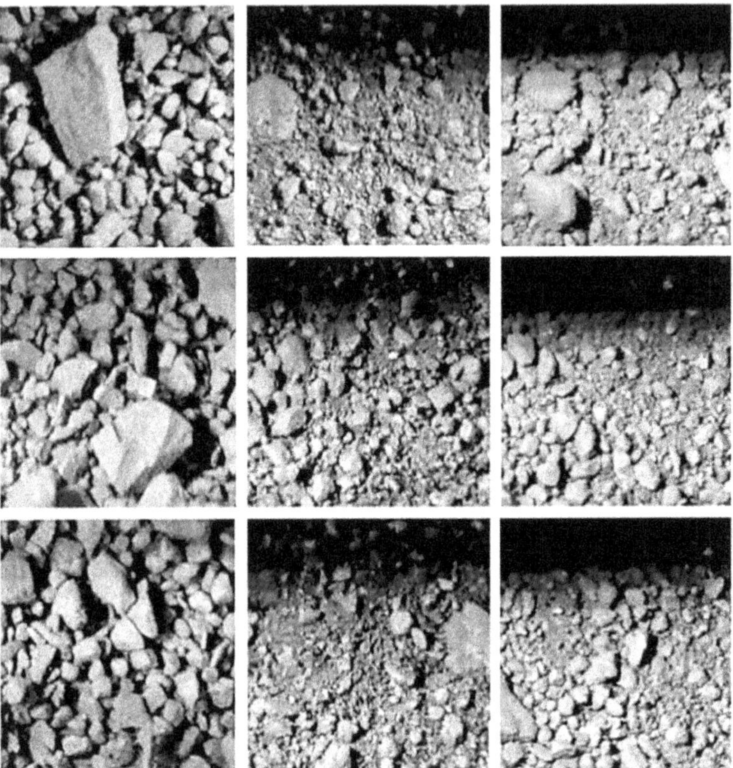

FIGURE 5.5 Example of iron ore image frames.

augmentation called imgaug (https://imgaug.readthedocs.io/en/latest/). This technique ended up doubling the size of the database.

Hence, 500 images were used for training, validating, and testing the model. The validation process was used to separate the training data and the test and validation data. Each sample (256 × 256) was manually segmented using labeling software. We divided the 500 images into three parts: 70% for the training set, 20% for the testing set, and the remaining 10% for the validation set. A specialist in the area helped identify fine-grained ore regions as presented in Figure 5.6. This process helped us in creating masks to train the model and compare the results against those images segmented by U-Net.

5.5.2 Hardware and Software Configuration

Several devices support computer vision solutions that require real-time responses. One solution that has received considerable attention is the usage of energy-efficient multi-core platforms, which are equipped with graphics processing units (GPUs) that can speed mathematical computations inherent to signal processing, image processing, motion planning, and so on (Otterness et al., 2017). Usually, these equipment

FIGURE 5.6 Example of labeled iron ore image frames.

types are small in size and powerful at processing data, comprising what we call IoT equipment.

One of the latest IoT equipment is the Jetson TX1 architecture, an NVIDIA platform used for applications that require portability. Yan Han evaluated the platform's performance with convolution networks to detect traffic signs and managed to execute an average of 1.6 frames per second (Han & Oruklu, 2017). The article "Deep Convolutional Neural Networks for Pedestrian Detection" (Tomè et al., 2015) validates a light-weight version of a solution to detect pedestrians using an AlexNet deep learning model, whose code was optimized for a version of the algorithm in C and tested in a Jetson Tx1 which requires 405 ms per frame.

The Jetson TX1, shown in Figure 5.7, is a board computer with a quad-core 1.91 GHz 64-bit ARM CPU, with 4 GB of DRAM memory. The architecture comes with NVIDIA and microarchitecture Maxwell, an integrated GPU comprising 256 CUDA cores and 256 KB of L2 cache. The integrated GPU shares DRAM memory with the CPU, which is typical for IoT devices, while enabling SWaP (strict Size, Weight, and Power) since it uses less power, less heat, and results in longer battery life. The dedicated GPU is separate from the processor chip. On the one hand, it provides better performance, but on the other hand, it requires a more efficient cooling system (Figure 5.7).

FIGURE 5.7 Jetson TX1 development kit.

The development kit includes many hardware interfaces such as Ethernet, camera expansion, USB, and HDMI, which facilitates testing different embedded solutions. This configuration is ideal for the developers of deep learning models. Furthermore, the development kit provides a Linux environment named L4T, a version of Linux based on the Tegra project, compatible with an NVIDIA SDK jetpack. This includes many libraries for deep learning and computer vision solutions, such as the CUDA toolkit that comes up with an environment to optimize GPU applications.

5.5.3 Model Evaluation: U-Net

The results presented in this section demonstrate the effectiveness of the model for the identification of fines. The metric was used to evaluate the test results for the U-Net model, which is called the IoU and consists of an area of overlap or intersection between what was predicted and the ground truth, divided by the union of the segmentation prediction and the ground truth. The IoU compares the area segmented by the neural network with the expert's mask.

The results of the IoU ranges between 0 and 1, with 0 meaning that there was no correct segmentation, and 1 meaning that there was total correct segmentation. The threshold value above 50% in this metric was adopted to classify whether the segmentation was correct for each tested frame.

In the U-Net model's training process, the IoU metric reached a value of 0.85 after 150 iterations for training the model, and for the dataset validation it reached above 0.9. Figure 5.8 shows the training value results.

As previously mentioned, it used 20% of the data to test and compare the results. All in all, the image segmentation achieved adequate results when targeting regions of fine-grained iron ore. The segmentation was not entirely accurate, but it is noticeable that the difference between the material presented high granularity and fine-grained texture in the frames.

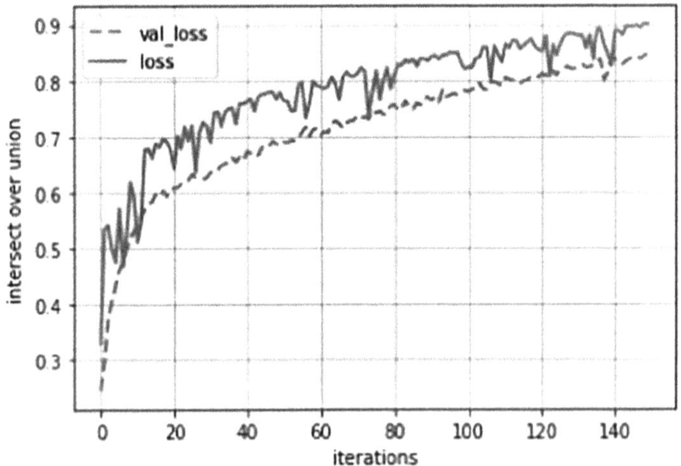

FIGURE 5.8 Training results for each iteration.

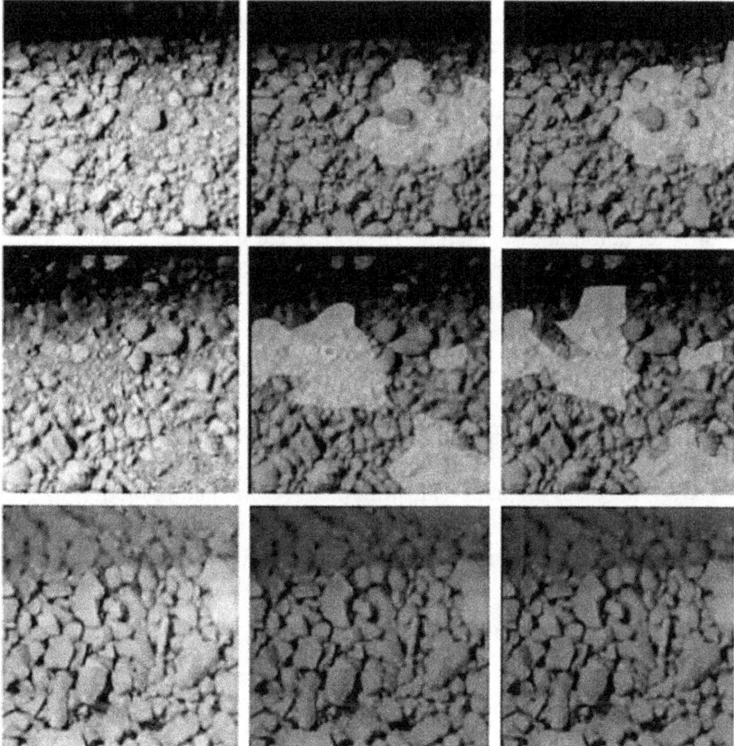

FIGURE 5.9 (a) Original Image, (b) predicted segmentation mask (c), ground truth.

Figure 5.9 shows some examples of how the solution acts with a region with fine-grained ore and when it presents only large rocks. In other words, the figure shows the image, the predicted segmentation, and the masks.

5.5.4 Edge Device Solution Evaluation

The Python program running collected frames in real time from the camera executes the image segmentation with the U-Net model and communicates with the controllers in 6 seconds, which is enough time for a minimum response of 1 minute required for the problem. The processing speed of frames in a resolution of 256 × 256 achieved a mean of 3.46 fps for the segmentation process, which is adequate time to stop the machinery if necessary.

5.6 CONCLUSIONS AND FUTURE WORK

We have proposed a solution capable of acting in automation and promptly in 1 minute, capturing images in real time from a camera, and performing image processing using an edge IoT device. We have demonstrated the steps needed to achieve and implement the solution that reaches a segmentation performance of 3.45 fps, which is enough time to make decisions during the production period.

The solution has some limitations related to the hardware jetson, which the IoT device enables to decentralize the heavy processing, though this brings difficulties related to the maintenance of hardware in places with varying weather conditions. The IoT device needs to be encapsulated so as not to be in contact with the dirt and water of the environment. Another limitation of the solution is that one device can only support one instance of image processing because of the jetson memory limitation which is only 4 GB.

Providing an automatic segmentation process led us a step closer towards safer, standardized, and optimized production, even though industrial environments make it challenging because of their particular communication infrastructures. Moreover, the use of IoT equipment at the edge facilitates and moves the processing requirements from servers and data centers to the IoT device, decentralizing the data processing. Such devices, such as the Jetson TX1, provide super-computing power to the edge, making it possible to embed artificial intelligence solutions that were impossible previously.

Future work needs to include: (i) testing other segmentation models to evaluate if the response time could be faster and present better results at segmenting the iron ore; and (ii) ensemble other machine learning techniques with neural networks for more accurate segmentation.

NOTE

1 A Movidius Stick or Neural Compute Stick is an Intel IoT device comprised of a visual processing unit (VPU) whose architecture implements a neural computing engine and 16 programmable cores.

REFERENCES

Aggarwal, C. C. (2018). *Neural networks and deep learning*. Cham: Springer. doi:10.1007/978-3-319-94463-0

Chen, C. J., Huang, Y. Y., Li, Y. S., Chen, Y. C., Chang, C. Y., & Huang, Y. M. (2021). Identification of fruit tree pests with deep learning on embedded drone to achieve accurate pesticide spraying. *IEEE Access, 9,* 21986–21997. doi:10.1109/ACCESS.2021.3056082

Chou, Y.-C., Kuo, C.-J., Chen, T.-T., Horng, G.-J., Pai, M.-Y., Wu, M.-E., … Chen, C.-C. (2019). Deep-learning-based defective bean inspection with gan-structured automated labeled data augmentation in coffee industry. *Applied Sciences, 9,* 4166. doi:10.3390/app9194166

D'Avella, S., Tripicchio, P., & Avizzano, C. A. (2020). A study on picking objects in cluttered environments: Exploiting depth features for a custom low-cost universal jamming gripper. *Robotics and Computer-Integrated Manufacturing, 63,* 101888.

Dalzochio, J., Kunst, R., Pignaton, E., Binotto, A., Sanyal, S., Favilla, J., & Barbosa, J. (2020). Machine learning and reasoning for predictive maintenance in industry 4.0: Current status and challenges. *Computers in Industry, 123,* 103–298.

Dumoulin, V., & Visin, F. (2018). *A guide to convolution arithmetic for deep learning*. arXiv preprint arXiv:1603.07285v2

Gupta, A. (2019). *The iot hacker's handbook*. Apress.

Guyot, O., Monredon, T., LaRosa, D., & Broussaud, A. (2004). Visiorock, an integrated vision technology for advanced control of comminution circuits. *Minerals Engineering, 17,* 1227–1235. doi:10.1016/j.mineng.2004.05.017

Han, Y., & Oruklu, E. (2017). Traffic sign recognition based on the nvidia jetson tx1 embedded system using convolutional neural Networks. *2017 (MWSCAS). IEEE 60th International Midwest Symposium on Circuits and Systems (MWSCAS),* 184–187.

Kornilov, A., & Safonov, I. (2018). An overview of watershed algorithm implementations in open source libraries. *Journal of Imaging, 4,* 123. doi:10.3390/jimaging4100123

Krizhevsky, A., Sutskever, I., & Hinton, G. E. (2017, May). Imagenet classification with deep convolutional neural networks. *Commun. ACM, 60* (6), 84–90.

Lage, V., Rêgo Segundo, A. K., Pinto, T. V. B., Silva, C., Pinto, E., Silva, M., … Monteiro, P. (2018). Bench system for iron ore moisture measurement. In (139–142). doi:10.1109/ICSensT.2018.8603578

Li, Y., Iida, M., Suyama, T., Suguri, M., & Masuda, R. (2020). Implementation of deep-learning algorithm for obstacle detection and collision avoidance for robotic harvester. *Computers and Electronics in Agriculture, 174,* 105499. doi:10.1016/j.compag.2020.105499

McGlinchy, J., Johnson, B., Muller, B., Joseph, M., & Diaz, J. (2019). Application of UNnet fully convolutional neural network to impervious surface segmentation in urban environment from high resolution satellite imagery. In (3915–3918). doi:10.1109/IGARSS.2019.8900453

Otterness, N., Yang, M., Rust, S., Park, E., Anderson, J., Smith, F. D., … Wang, S. (2017). An evaluation of the nvidia tx 1 for supporting real-time computervision workloads.

Sassi, P., Tripicchio, P., & Avizzano, C. A. (2019). A smart monitoring system for automatic welding defect detection. *IEEE Transactions on Industrial Electronics, 66* (12), 9641–9650.

Shelhamer, E., Long, J., & Darrell, T. (2017). Fully convolutional networks for semantic segmentation. *IEEE Transactions on Pattern Analysis and Machine Intelligence, 39* (4), 640–651.

Stankov, S., Ivanov, S., & Todorov, T. (2019). An application of deep neural networks in industrial robotics for detection of humans. In *2019 IEEE xxviii international scientific conference electronics (et)* (1–3). doi:10.1109/ET.2019.8878583

Thurley, M. (2013). Automated image segmentation and analysis of rock piles in an open-pit mine. In (1–8). doi:10.1109/DICTA.2013.6691484

Tomè, D., Monti, F., Baroffio, L., Bondi, L., Tagliasacchi, M., & Tubaro, S. (2015). Deep convolutional neural networks for pedestrian detection. *CoRR, abs/1510.03608.* Retrieved from http://arxiv.org/abs/1510.03608

Ukaegbu, U., Tartibu, L., Laseinde, T., Okwu, M., & Olayode, I. (2020). A deep learning algorithm for detection of potassium deficiency in a red grapevine and spraying actuation using a raspberry pi3. In *2020 international conference on artificial intelligence, big data, computing and data communication systems (icabcd)* (1–6). doi:10.1109/icABCD49160.2020.9183810

Walsh, J. O., O'Mahony, N., Campbell, S., Carvalho, A., Krpalkova, L., Velasco-Hernandez, G., … Riordan, D. (2019). *Deep learning vs. traditional computer vision.* CVC: Advances in Computer Vision. Springer. doi: 10.1007/978-3-030-17795-9_10

Zhao, X., Yuan, Y., Song, M., Ding, Y., Lin, F., Liang, D., & Zhang, D. (2019). Use of unmanned aerial vehicle imagery and deep learning UNet to extract rice lodging. *Sensors, 19,* 3859. doi:10.3390/s19183859

6 Amalgamation of Blockchain Technology and Cloud Computing for a Secure and More Adaptable Cloud

Himanshu V. Taiwade
Priyadarshini Institute of Engineering & Technology, Nagpur

Premchand B. Ambhore
Government College of Engineering, Amravati

CONTENTS

6.1 Basics of Cloud Computing 88
 6.1.1 Some of the Essential Characteristics of Cloud Computing 89
 6.1.2 Cloud Service Model 89
 6.1.3 Types of Cloud Deployment Model 91
 6.1.4 Advantages and Disadvantages of Cloud Computing 91
6.2 Existing Cloud Computing Security and Related Issues 92
 6.2.1 Confidentiality 92
 6.2.2 Integrity 93
  6.2.2.1 Data Integrity 93
  6.2.2.2 Virtualization Integrity 94
 6.2.3 Availability 94
  6.2.3.1 Data/Service Availability 94
  6.2.3.2 Virtualization Availability 95
6.3 Limitations in Existing Cloud-based Security 95
6.4 Basics of Blockchain Technology 96
6.5 Advantages of Blockchain 97
6.6 The Working of Blockchain 98
6.7 Amalgamation of Blockchain Technology and Cloud Computing 99
6.8 Conclusion, Future Scope, and Limitations 100
 6.8.1 Conclusion & Future Scope 100
 6.8.2 Limitations 100
References 101

DOI: 10.1201/9781003146711-6

6.1 BASICS OF CLOUD COMPUTING

Cloud computing can be explained as a mechanism that allows requirement-based access of network and related services for its users. Cloud computing also provides users with numerous benefits (i.e., high performance, cost benefits, enhanced productivity, and efficiency) (Mell & Grance, 2011). It also facilitates on-demand usage of resources, and this access can be in the form of an entire infrastructure known as Infrastructure as a Service (IaaS) or as part of a service which may be either an operating system, a server, or a database; and this service is better known as Platform as a Service (PaaS) or Software as a Service (SaaS), where a part (i.e., an instance) of software is provided as a service where all these services are on the basis of pay for what you use. It has been claimed that this technology has not only transformed the world of information but has also provided ease of use (Kaur et al., 2018). Figure 6.1 shows an illustration of a basic cloud computing framework consisting of servers, databases, applications, application software, and so on.

In simple terms, cloud computing can be explained as a process where the distribution of various computing services is provided over the Internet. These services largely consist of, but are not limited to, databases, networks, servers, software, and storage. The best part of cloud computing is that it is a pay-for-what-you-use model, and hence the user has to typically pay only for those cloud-based services which he or she wishes to use. The cloud imparts accessible, elastic, and shared computing

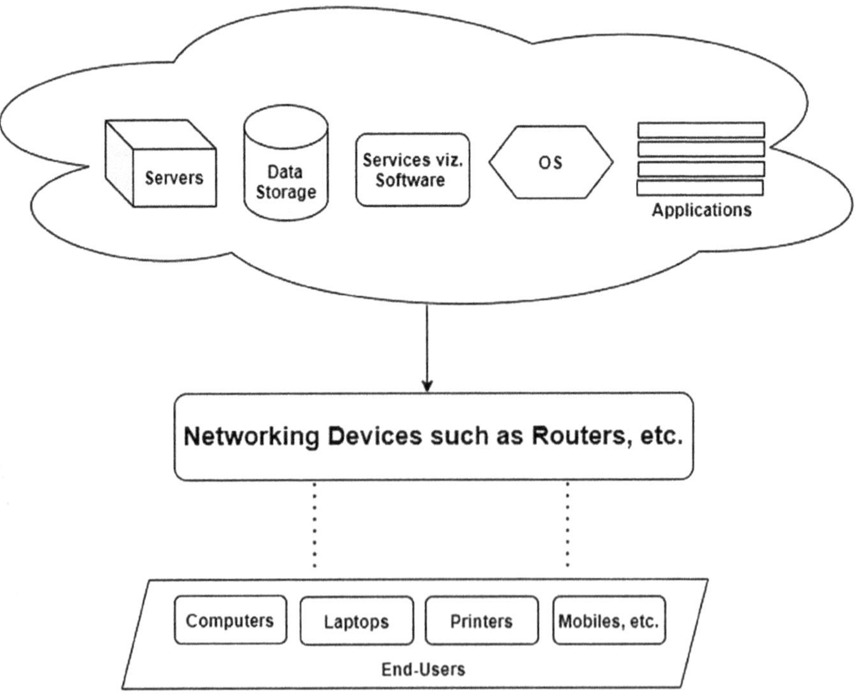

FIGURE 6.1 Framework of cloud computing.

services. The cloud's precedence is mainly due to its features, which include low operating costs and efficiency.

The cloud is appealing to business owners because it has several convincing features, such as scalability, reliability, easy access, and low cost. Some of the best-known examples of cloud services providers (CSPs) are Amazon AWS and Google Cloud (Nirmal, 2018).

It is safe to say that, even though all clouds are not the same, one or the other type of cloud computing might be right for everyone. There are several different types, models, and services that have evolved for offering the right solution to every need.

6.1.1 Some of the Essential Characteristics of Cloud Computing

Resource pooling: In the case of resource pooling, all or the required computing resources of the CSP can be combined together for serving a number of consumers together who can use a multi-tenant model. It is important to know that resources, which in this case can be either in physical or virtual form, are distributed to customers as per their demand.

On-demand service: The user here is provided with computational capabilities in accordance with requirements and without the need for interaction with every individual service provider.

Swift elasticity: Cloud computing power can be allotted and taken back (once the use is over) swiftly. This is done for the rapid scaling of services, enabling the use of computing resources to meet a varying workload (Mell & Grance, 2011).

Measured service: Cloud based systems are known for automatically optimizing and controlling resources, and because of their metering capability. Monitoring and controlling of resource usage are reported in a timely way and then provided to both the service provider and the consumer to maintain transparency (Zhou et al., 2017).

6.1.2 Cloud Service Model

As per the National Institute of Standards and Technology (NIST), cloud models can be classified by the three types of services they provide. The detailed service model architecture of cloud computing technology is shown in Figure 6.2.

- **Infrastructure as a Service (IaaS):** Here delivery of resources to customers, which can be in the form of either storage services, hardware, is done virtually. The best part of IaaS is that even though the infrastructure is managed by the CSP, customers can deploy different services, selected parts of the network, or operating systems as per their requirement (Tabrizchi & Rafsanjani, 2020). An advantage of IaaS is that without requiring the installation of any type of hardware on their premises customers, including businesses, can choose from the variety of computing resources as per their requirement. A few of the well-known examples of CSPs include: AWS (Amazon Web Services) from

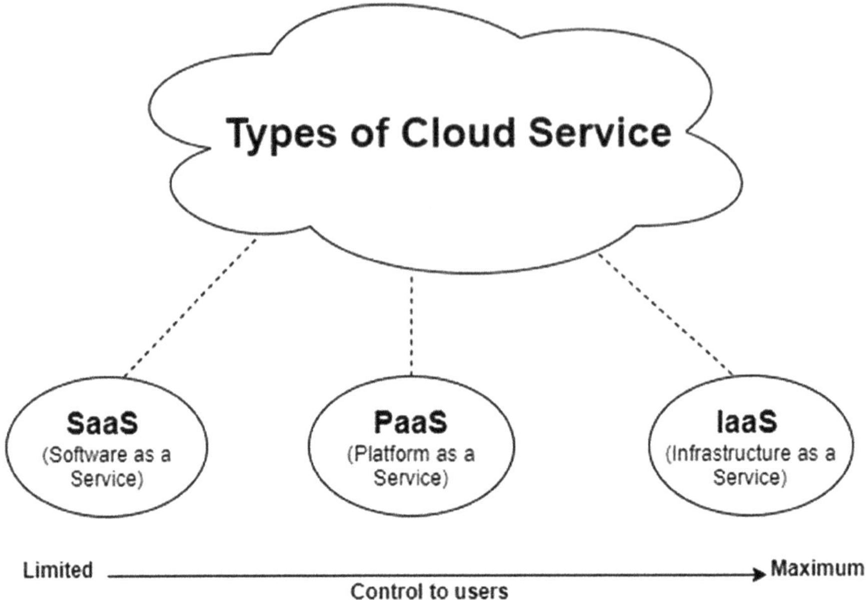

FIGURE 6.2 Different service models of cloud computing.

Amazon, MetaCloud from Cisco, Rackspace, and MS-Azure from Microsoft. Better security and governance is the key here.

- **Platform as a Service (PaaS):** PaaS offers complete assets which can be either in the form of servers or storage or it can be the entire network which is managed either by the organization or a service provider. Features including development, maintenance, and management of applications are utilized in PaaS. The control of this entire cloud infrastructure, which includes operating systems, a network, and servers rests with the CSP. On the other hand, customers have little computing power themselves. A suitable space for developing and testing applications is provided by the virtual runtime environment which is enabled by the PaaS. Typical examples of PaaS include the well-known Windows Azure and Google App Engine.
- **Software as a Service (SaaS):** SaaS is a class of service where customers can have quick access to cloud-based Web applications running over the Internet or in a cloud infrastructure. These applications running on the cloud can be used for free with limited access or as a paid service with licensed subscription. Since the maintenance part and overall support is taken care of by the CSP, there's no need for installing applications on customers' computers and thus there is only minimal control by the customer, especially in the case of security (Mell & Grance, 2011). Cisco WebEx, Google G Suite, Dropbox, and Microsoft Office 365 are some examples of SaaS (Tabrizchi & Rafsanjani, 2020).

6.1.3 Types of Cloud Deployment Model

As per the service they offer and the requirements which they fulfil, the cloud is divided into four models: public, private, hybrid, and community cloud.

- **Public cloud:** The best feature of a public cloud is scalability, as it has a vast amount of space available. Cloud services are provided to common users as well as enterprises with large businesses. Here the cloud storage is with the service provider. A public cloud not only guarantees scalability but it also provides reliability. The biggest advantage of a public cloud lies in its versatility along with the pay-as-you-go facility, allowing its customers to have capacity on demand. On the other hand, the disadvantages of a public cloud are that customers generally are unaware of the variety of storage being used by the service providers. Hence a few security aspects persist while using this type of cloud service.
- **Private cloud:** In the case of a private cloud, services are generally used by an individual. A private cloud can be useful for resolving dependability and security related issues that are generally present in the case of a public cloud. Another benefit here is that the user is given more control of and more accessibility to data. On the other hand, a private cloud instigates overheads related to capacity watching, storage management, provisioning, and so on.
- **Community cloud:** In the case of community clouds, assets are shared amongst the users which can be different organizations, including various communities, (e.g., financial organizations or trading firms). The setup is shared and managed between a number of organizations which can belong to a specific group having the same or related computing concerns, which may include the same type of security requirements, policies, or compliance considerations. The main motivation in a community cloud is to achieve related business goals. It can either be hosted internally or externally as per the requirements and provisions.
- **Hybrid cloud:** The unification of two or more types of cloud which can include public, private, or community clouds is known as a hybrid cloud. It may be noted that clouds which are integrated here remain as unique entities which are still bound together with the help of a proprietary technology that enables application and data exchange. Hybrid clouds cannot be easily distinguished as private, public, or hybrid clouds; in fact all the resources here are provided by external or internal providers (Tabrizchi & Rafsanjani, 2020). This type of cloud is pay on demand. One of the threats in the case of a hybrid cloud is related to integrity and data privacy, especially in the case of data transmission from a public to a private environment or vice versa (Goyal, 2014).

6.1.4 Advantages and Disadvantages of Cloud Computing

Even though cloud computing offers a lot of benefits over traditional computing services, a major concern which still exists while shifting towards its usage is related to security. It may be noted that CSPs do provide a high level of inbuilt security features and a large amount has been spent on its enhancement; however, security still remains a challenge for both CSPs and users.

6.2 EXISTING CLOUD COMPUTING SECURITY AND RELATED ISSUES

It is safe to assume that the cloud is an emblem of the Internet. By effective use of hardware virtualization and channelized delivery of services, cloud computing has arisen as a principal option for storing large chunks of data along with features such as fault tolerance, availability, and scalability. Not only business houses but also moderate users and individuals have started shifting to using the cloud. With the data of financial services, educational materials, social networking websites, and so on being stored on the cloud, CSPs have evidentially targeted security as one of the primary features to be provided for its users.

It is noteworthy that with recent data breaches in focus, doubt arises in the minds of users, and they are forced to have second thoughts and wonder if they can really trust the cloud. Whenever users invoke the Internet or any phone-based application, data is accessed from cloud infrastructure. The consumer here completely relies on his CSP for issues related with the privacy of his information. This further emphasizes the fact as to why CSPs are accountable when it comes to data protection and privacy related issues (Tabrizchi & Rafsanjani, 2020).

The prima facie of cloud security is some way or the other concerned with Data **C**onfidentiality, Data **I**ntegrity and Data **A**vailability, also known as CIA. Hence CIA is the foremost concern for any person dealing with the cloud. From the users' point of view it is important to trust a CSP, as the reason behind choosing any service provider is driven by what sort of security features are provided and what level of support is provided to its users. The relation between a CSP and its users is based on trust, and that trust is dependent on whether the CSP has covered all the security related risks or not.

6.2.1 Confidentiality

The protection of data from unauthorized usage and access can be referred to as confidentiality. Confidentiality is roughly equivalent to privacy. It is related to measures which are undertaken to ensure confidentiality and can lead to the prevention of sensitive information reaching unauthorized persons, plus also ensuring that data reaches the authorized people. It should not be the case that any insider or person working within the CSP's office can access or tamper with the private data of the user. A virtual machine image, other than the client's or a virtual machine network, also has confidentiality requirements which should be taken care of.

There are various confidentiality related issues in cloud systems which can be categorized as:

1. **Data confidentiality:** Personal and sensitive users' data is frequently stored in remote machines which are generally operated by third-party service

providers. This may lead to the *unauthorized disclosure of data*. To avoid this, many techniques are applied, but at present there is no effective way of protecting this sensitive data.

2. Another issue related to data confidentiality is **client data segregation** when multiple users are on the same platform and services, and it becomes important to segregate the data of one user from that of other users, which must be handled by CSP explicitly (Goyal, 2014).
3. **Virtualization confidentiality:** Issues of data confidentiality also arise during *virtual machine migration*. This feature is supported by cloud computing systems for the purpose of hardware maintenance, load balancing, and fault tolerance. It is very important for the CSP to adopt all necessary means for preserving the confidentiality of all the VM instances and VM metadata at the time of live migration because the multiple workloads are in parallel, sharing the single hardware environment that may create an issue where the workload has to be isolated, as it is a primary and foremost requirement to keep data secure and separate from each other. Hence the necessary means and methods should be applied for governing the process of resource sharing between the workloads of a data center.
4. **Data encryption:** Another challenge is at times with those CSPs which don't permit its users (data owners) to encrypt data. Even though users own the data, they are not allowed to protect their own information. This further raises serious challenges related to confidentiality of data. CSPs are always trusted by their users, but in a few cases they have been curious. When data, such as health records, medical records, defense related data, and government data, is priceless, and with recent incidents where in some cases users' data has been compromised, it is important to know and understand whether or not the facility of encryption is available to users (Park & Park, 2017).

6.2.2 Integrity

Integrity can be defined as a policy that guarantees the security of an asset and confirms the fact by ensuring that no modifications or alterations have been made by anyone who is not authorized to do so. Thus this property ensures correctness and accuracy for an asset which is advantageous for its owner. The operations and triggers which can hamper the integrity of data can be editing, deleting, and appending.

With respect to integrity of data, different requirements are mandatory in the case of cloud systems. These can be discussed as:

6.2.2.1 Data Integrity

A huge number of data-centric operations is performed in the cloud. These operations include large data transfers which can be in the order of terabytes (TB) or even petabytes (PB). This directly leads to challenges related to data integrity (Nirmal, 2018). These challenges need to be addressed very carefully. A few data integrity issues related to the cloud are as follows.

Data outsourcing is the most serious threat at the CSP end, as a CSP has the authority to modify or delete valid entries from the data (Huckle et al., 2016). Even incomplete data can be sometimes sent to a client, knowingly or unknowingly, and this can go entirely unnoticed by the client.

Cross-scripting attacks is a type of attack where malware is inserted. Here the hacker inserts some kind of malicious script, which can be JavaScript or any other format, into any of the Web pages which are most of the time dynamic in nature. This inserted code then executes itself on the browser of the user, which in turn gives unauthorized control to the account of the authenticated user, ultimately leading to a situation where data integrity is jeopardized (Tabrizchi & Rafsanjani, 2020).

An attack that can modify the contents in the database of a customer is ***SQL injection***. This is known as one of the most remarkable attacks from the category of Web-based attacks. It exploits the vulnerabilities of the concerned Web servers and then injects the malicious code into the system (Sarmah, 2019).

6.2.2.2 Virtualization Integrity

By now it is already clear that a virtualization layer induces quite a few security related issues by itself. These issues are not at all confined inside the boundaries of confidentiality alone which means that not only confidentiality but integrity is also an important issue in the case of virtual machines (VMs) and virtual machine introspection (VMI).

CSP administrators are those who have all the access and control but work in the background on the assigned VMs, and it is because of this that stringent security policies are a must for securing VMs especially from attackers who are within the system itself.

VM replication is caused by unnecessary data leakage due to improper handling – which is another issue, and therefore there is always motivation for users to handle data properly and pause/temporarily suspend VMs when not in use, hence ensuring data integrity. Policies are enforced for limiting the duplication of important VMs as well as for controlling their flow inside and outside of the system (Bin & Xiaoyi, 2019).

VM rollback introduces a phenomenon where quite a few integrity related problems, especially in the case of VMs, occur. The process of rollback in VMs can again reintroduce security vulnerabilities which are generally fixed in the case of previously disabled accounts and passwords. In this case, it becomes important to take and maintain VM snapshots for this purpose (Goyal, 2014).

6.2.3 Availability

One of the most critical features with respect to security for any person related to cloud computing is *availability*. For any type of user, be it an individual user or an enterprise, data availability is the most important aspect when considering a CSP. The slightest downtime or any minute lag can lead to large monetary loss. Hence, at the time of contracting for service, data availability and its response time needs to be mentioned. The following are issues related to data availability in the cloud.

6.2.3.1 Data/Service Availability

One of the biggest threats in the case of cloud computing which can cause data unavailability is *denial of service*. Here, a huge number of indistinct requests are sent for a specific service. After noticing the huge workload on the servers, the operating system provides extra computational power in order to handle this additional workload. In this situation the CSP is struggling with vague requests and trying to manage them and at the same time it is encouraging the attacker by allowing him or her to exploit its own resources. This leads to a situation where legitimate users are deprived of their intended services (Taiwade et al., 2019).

Another type of threat is *indirect denial of service*, when the resources of the server are exhausted due to the continuous processing of requests caused by a flooding attack. In this type of attack the server is with flooded with services which are placed on the same server which is already under attack, hence this new instance may also become unavailable. This attack gets even more dangerous when the cloud system notices the inability to process requests and then attempts to shift those instances of service which are affected to other servers. The outcome here is an unnecessary load being assigned to these servers which indirectly affects the entire cloud system.

Natural disasters, that may include cases of flood, in the area where a data center is placed, can likely affect both copies of data stored in the centers. Hence appropriate measures should be guaranteed for handling these situations (Feng et al., 2019).

6.2.3.2 Virtualization Availability

Making sure that high availability is always provided requires that lots of areas be monitored by a CSP. These may include storage failure and network related issues. IP failover is also a very important aspect which should be taken care of and for which VM instances must be carefully managed, especially against the failure of one instance, and related issues should be overcome immediately by some alternative instance (Sarmah, 2019).

From the study so far, it can be concluded that cloud computing undeniably offers major enhancements over traditional networks and also offers major benefits to its users; but it can't be denied that security related issues are still a major concern for both CSPs and their users. One of the main reasons for security concerns over cloud computing is its dependency on the Internet connection and the fact that it is itself endangered with multiple threats and security issues like DOS attacks, injection of malware, loss of data, and data breaches (Ambhore & Meshram, 2014). Even though researchers have been providing regular solutions and updates related to the security risks of cloud computing, a systematic approach is still far from developed.

6.3 LIMITATIONS IN EXISTING CLOUD-BASED SECURITY

After going through the traditional cloud infrastructures it can be concluded that there are loop holes in the system mainly because traditional clouds and even IoT-based services are extremely dependent on the models based on centralized communication and the presence of a third party. There is a big chance that models supporting such systems are generally incapable of handling ever increasing traffic and congestion because a small failure occurring in the architecture may lead to disruptions in the entire network (Atlam et al., 2017). One major cloud computing model concern is related to SaaS, due to the number of applications offered to users and Web servers, email being the best example. Hence it is very important to provide services to authorized clients only. For this, maintaining proper login details and logs with the application server process is a must, though it is difficult in the case of systems dependent on third parties (Huang et al., 2018).

Dependency on those systems that are administered centrally and which are dependent on a third party more or less raises lots of concerns, especially those related to data privacy. Due to the fact that all cloud servers can easily obtain confidential information without the permission of the authentic users, further leads to issues which are related to system security and confidential information. One more limitation in the existing system is that related to users, where the users find it extremely difficult to keep records of data exchange and also have only limited control over their personal data (Subramanian & Jeyaraj, 2018). Due to this leakage of sensitive information, user data can be easily disclosed, causing monetary damage and physiological concerns. Hence user anonymity and confidentiality of user data are far from achieved in cloud based technologies, thwarting the way to a far more secure, adaptable, and efficient technology (Goyal, 2014).

6.4 BASICS OF BLOCKCHAIN TECHNOLOGY

Blockchain technology in simple words is a chain containing multiple blocks which are used for storing details along with some digital signatures. It should be noted that these blocks are connected in a decentralized manner and stored as a distributed network. The attributes of blockchain are transparency, decentralization, auditability, and immutability. Blockchain uses a peer-to-peer (P2P) transmission of data, which is done without any centralized dependency (Zheng et al., 2017). This technology was originally created for supporting the famous cryptocurrency called "bitcoin" (Park & Park, 2017). Bitcoin was initially introduced in around 2008 and was proposed by Nakamoto (Bonneau et al., 2015).

Because of its many unique features, blockchain is used in lots of applications. The advantage of this technology revolves around the concept of decentralization where the transactions are done on the basis of a P2P model. This process is undertaken without the requirement of any type of centralized monitoring and hence trust relies on individual nodes. Another benefit is its distributed consensus, based on supporting features like data verification, time-stamping, and encryption of data, providing security and privacy (Bonneau et al., 2015).

Blockchain as discussed previously is a sequence of blocks which holds data, such as transaction records and public ledgers (Zheng et al., 2017). Figure 6.3 shows the architecture of a blockchain. It is clearly visible that a preceding block hash is carried

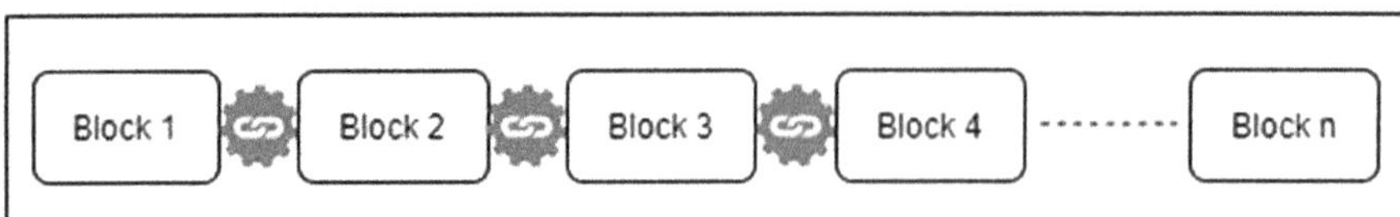

FIGURE 6.3 Blockchain consisting of a sequence of blocks.

in the header block. It should be noted that there is only one parent block (Wood, 2017). An initial block of a blockchain is known as a "genesis block," which does not have a parent block.

A blockchain is a *P2P network* supported e-currency, enabling online transactions between two parties without requiring the need to go through a financial institution. The use of *digital signatures* and a P2P network are adopted for compensating the requirement related to trusted third parties, though one of the main benefits is to prevent the otherwise required double-spending. Further enhancing security is the use of time-stamped transactions which are hashed into an ongoing chain which forms a kind of record that is impossible to rollback unless proof of work (PoW) is verified again (Huckle et al., 2016).

Here the largest sequence of chain is used to verify the progress of events which are observed. Until that time, CPU power will be controlled by the nodes which themselves do not compete and hence generate the largest chain and thus outnumber any attackers (Subramanian & Jeyaraj, 2018). The network here has the smallest requirement because communication is now broadcast on the basis of its best possible efforts, hence the nodes are allowed to exit and subsequently rejoin the network as per their requirements and thus verifying the chain with the largest PoW as proof in case anything goes wrong (Zhang & Wen, 2017).

6.5 ADVANTAGES OF BLOCKCHAIN

1. **Decentralization:** One of the main issues which may result in increased cost and performance related to bottle-necks at the central server was the requirement for all transactions to be validated by some trusted party. This not only increased the cost but also led to dependency on the central server. Decentralization is the method used by blockchain to maintain data consistency and also reduce the bottle-neck-related issues. Decentralization also gives users the power of storing their assets on a network which can be easily accessed by the use of the Internet. These assets can be a document or a contract.
2. **Increased capacity:** This is one of the most important features of blockchain, where the most remarkable thing is the increase of the capacity of the entire network. The reason for this is that lots of computers are working together, and they offer great power, especially when the devices are not centralized.
3. **Persistency:** In the case of blockchain, all the transactions are validated very quickly and those which are invalid are disallowed. To perform any kind of rollback or delete operation is not possible. Also the discovery of invalid transactions and the blocks containing them is initiated quickly (Zheng et al., 2017).
4. **Anonymity:** The feature of anonymity provides users with the ability to interconnect with the blockchain itself, and this is done with the help of an initiated address (Bonneau et al., 2015). Genuine identity isn't disclosed here, thus guaranteeing user anonymity. But one important thing here is to note that because of intrinsic constraints, blockchain alone is unable to guarantee absolute privacy (Wood, 2017).

6.6 THE WORKING OF BLOCKCHAIN

The key benefit of blockchain is based on the concept of decentralization where transactions are done through the P2P mode and cooperation without centralized monitoring and where mutual trust is established between individual nodes (Huang et al., 2018). Some other benefits supported are an algorithm-based distributed consensus which, in return, supports data verification, time-stamping, and encryption of data, providing security and privacy (Huckle et al., 2016).

The basic functioning of a blockchain is shown in Figure 6.4 where, once a transaction is requested, it is then broadcast to a P2P network consisting of nodes. These

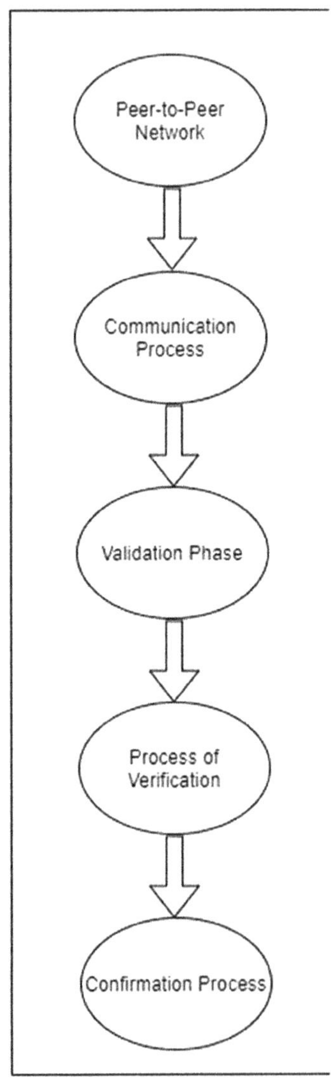

FIGURE 6.4 Working of a blockchain.

nodes are used for validation. This verified transaction can include a cryptocurrency or any record or data. Once verified this transaction can now be amalgamated with remaining transactions, hence creating the next block of data which can be used for a ledger; at the end, this block can be combined with an existing blockchain which would be unalterable (Zheng et al., 2017).

It should be noted that the new transactions are collected by the miners present in the P2P network until and unless a solution is found for a specific piece of work. After this a block which contains the solution and related transactions is then broadcast to all the participating parties and the block chain, which is also known as a distributed ledger, is finally updated. For determining ownership, traversing the block chain is done and this process continues until the latest transaction is found (Sharma et al., 2018). Consensus can now be identified by resolving the malicious alterations and delays caused by propagations and then by considering the longest fork as the consensus (Zhang et al., 2017).

Whenever a blockchain saves any kind of data, it is very hard to modify or alter the data. Some of the major reasons for this type of security measure are characteristic of a blockchain, which are P2P in character and require the consent of a minimum of 50% or more peers before accepting any change. Owing to these, applications of a blockchain already vary greatly and include their use in digital identities, registries, smart contracts, healthcare, supply chain management, and so on (Dabbagh et al., 2019).

6.7 AMALGAMATION OF BLOCKCHAIN TECHNOLOGY AND CLOUD COMPUTING

As per the above research, the concept of a blockchain is based on a decentralized mechanism, which is not the case with cloud computing. Hence to amalgamate both the technologies or rather for extracting the best out of both it must somehow incorporate decentralized cloud enabling frameworks operating individually (Liang et al., 2017). Quite a lot can be achieved by integrating the two mentioned technologies, especially in terms of cloud computing as the cloud itself offers a sustainable model; however, security is always an issue with it. Below are a few out of many possible areas where integration rather than amalgamation of blockchain technology with cloud computing can potentially offer a robust solution to the already mentioned security issues in Sections 6.2 and 6.3, that is denial of service and indirect denial of service types of threats, which can be largely reduced just because of the blockchain's properties, such as (i) *persistency*, where transactions can be validated very quickly and invalid ones are disallowed; and (ii) *decentralization*, where the requirement of trusted third parties can be reduced drastically, enabling smoother and faster transactions.

Attacks caused by *cross-scripting*, which otherwise can be an issue in the cloud, can be restricted rather completely removed through a blockchain, as any new entry/block has to be verified through other existing blocks and once they are approved only and only then a new block or in this case a new entry into the existing system can be validated. Similarly, issues which are related to *data integration* and *data confidentiality* can be overcome easily.

One very important thing to consider regarding the amalgamation of the above mentioned technologies is to understand all the types of tools used by cloud providers which require a clear infrastructure and a variety of development platforms. Hence the integration requires consideration of every detailed aspect, such as type model and environment. Once this can be properly achieved, a blockchain in the cloud can open up much scope, which will not be limited to only security issues.

Some of the major players like Amazon and Google have already been exploring this amalgamation in the form of *AWS Blockchain*, which initially works with templates that provide ways to build as well as deploy blockchain dependent networks that are more secure. *Azure Blockchain* is a service that uses Microsoft's own cloud resources but with better and enhanced security (Coutinho et al., 2020).

6.8 CONCLUSION, FUTURE SCOPE, AND LIMITATIONS

6.8.1 Conclusion & Future Scope

In this chapter, we have seen that cloud computing is a collection of large networks providing virtualized services (i.e., software and hardware resources). There are quite a few ways for possibly amalgamating blockchain with cloud systems. It is clear that if blockchain technology and the related security can be integrated into a cloud based system then the resulting security would be enhanced, viable, and more reliable. Thus the resulting model would surely guarantee much more confidence to the current as well as potential cloud user, further promising enhanced growth of the revenues of businesses.

Large numbers of businesses are already using cloud storage and extracting benefits from cloud computing technology. To further enhance cloud capability and security, blockchain capability can be added to the mix. A blockchain, if properly integrated in the cloud computing scenario, enables the existing security of a cloud based system to be enhanced to a great extent. Furthermore, the elimination of third parties can be used to increase confidence and trust. Overall, it can be concluded that, if properly used, a blockchain can easily upgrade an existing cloud based security mechanism. For this to happen, a systematic and practical approach needs to be undertaken with the proper aim of achieving the desired objectives.

6.8.2 Limitations

From the study covered in this chapter, it is easy to say that amalgamation of blockchain and cloud based systems can enhance the existing structure of cloud computing security, though it is far more difficult to do than say. Lots of limitations and considerations need to be taken care of. The ever growing pace of technologies such as IoT and 5G extend pressure on cloud based systems, and with such divergence in technologies it becomes even more complex to amalgamate these models with blockchain technology. Processing and storing of data from such diversified technologies will bring more vulnerabilities and issues related to security. Blockchain itself is not an easy technology to implement: the creation of a blockchain and management of nodes and then finally implementing the same in a cloud environment may require extensive study and an extended budget. Hence these points are a must to be addressed so as to move ahead with the amalgamation of blockchain technology and cloud computing systems.

REFERENCES

Ambhore, P. B., & Meshram, B. B. (2014). Intrusion detection system for intranet security. *International Journal of Advanced Research in Computer Science & Software Engineering*. http://ijarcsse.com/Before_August_2017/docs/papers/Volume_4/7_July2014/V4I7-0287.pdf

Atlam, H. F., Alenezi, A., Alharthi, A., Walters, R. J., & Wills, G. B. (2017). Integration of cloud computing with internet of things: Challenges and open issues. *IEEE International Conference on Internet of Things (iThings) and IEEE Green Computing and Communications (GreenCom) and IEEE Cyber, Physical and Social Computing (CPSCom) and IEEE Smart Data (SmartData)*. 670–675. https://doi.org/10.1109/iThings-GreenCom-CPSCom-SmartData.2017.105

Bin, C., & Xiaoyi, Y. (2019). Research on the application and security of cloud computing in smart power grids. *4th International Conference on Mechanical, Control and Computer Engineering (ICMCCE)*.784–786. https://doi.org/10.1109/ICMCCE48743.2019.00180

Bonneau, J., Miller, A., Clark, J., Narayanan, A., Kroll, J. A., & Felten, E. W. (2015). SoK: Research perspectives and challenges for bitcoin and cryptocurrencies. *IEEE Symposium on Security and Privacy.* 104–121. https://doi.org/10.1109/SP.2015.14

Coutinho, E. F., Paulo, D. E., Abreu, A. W., & Carla, I. M. B. (2020). Towards cloud computing and blockchain integrated applications. *IEEE International Conference on Software Architecture Companion (ICSA-C)*, 139–142. https://doi.org/10.1109/ICSA-C50368.2020.00033

Dabbagh, M., Sookhak, M., & Safa, N. S. (2019). The evolution of blockchain: A bibliometric study. *IEEE Access*, *7*, 19212–19221. https://doi.org/10.1109/ACCESS.2019.2895646

Feng, Q., He, D., Zeadally, S., Khan, M. K., & Kumar, N. (2019). A survey on privacy protection in blockchain system. *Journal of Network and Computer Applications, 126*, 45–58. https://doi.org/10.1016/j.jnca.2018.10.020

Goyal, S. (2014). Public vs private vs hybrid vs community - cloud computing: a critical review. *I.J. Computer Network and Information Security 2014*, *3*, 20–29. Published Online in MECS. https://doi.org/10.5815/ijcnis.2014.03.03

Huang, H., Chena, X., Wu, Q., Huang, X., & Shen, J. (2018). Bitcoin-based fair payments for outsourcing computations of fog devices. *Journal Future Generation Computer Systems, 78*(2), 850–858. https://doi.org/10.1016/j.future.2016.12.016

Huckle, S., Bhattacharya, R., White, M., & Beloff, N. (2016). Internet of Things blockchain and shared economy applications. *Procedia Computer Science, 98*, 461–466. https://doi.org/10.1016/j.procs.2016.09.074

Kaur, A., Singh, V. P., & Gill, S. S. (2018). The future of cloud computing: Opportunities, challenges and research trends. *IEEE Xplorer Proceedings of the 2nd International conference on I-SMAC (IoT in Social, Mobile, Analytics and Cloud)*. 213–219, https://doi.org/10.1109/I-SMAC.2018.8653731

Liang, X., Shetty, S., Tosh, D., Kamhoua, C., Kwiat, K., & Njilla, L. (2017). Provchain: A blockchain-based data provenance architecture in a cloud environment with enhanced privacy and availability. *17th IEEE/ACM International Symposium on Cluster, Cloud and Grid Computing (CCGRID)*. 468–477. https://doi.org/10.1109/CCGRID.2017.8

Mell, P., & Grance, T. (2011). The NIST definition of cloud computing. *NIST Special Publication* 800–145. https://nvlpubs.nist.gov/nistpubs/Legacy/SP/nistspecialpublication800-145.pdf

Nirmal, S. K. (2018). Security in cloud computing: A review. *International Journal of Electronics Engineering, 10*(2), 927–936. http://www.csjournals.com/IJEE/PDF10-2/110.Sanju.pdf

Park, J. Ho., & Park, J. Hyuk. (2017). Blockchain security in cloud computing: Use cases, challenges, and solutions. *Symmetry, 9*, 164. https://doi.org/10.3390/sym9080164

Sarmah, S. S. (2019). Application of blockchain in cloud computing. *International Journal of Innovative Technology and Exploring Engineering (IJITEE), 8*(12), 4698–4704. https://doi.org/10.35940/ijitee.L3585.1081219

Sharma, S. G., Ahuja, L., & Goyal, D. P. (2018). Building secure infrastructure for cloud computing using blockchain. *IEEE Xplore, Proceedings of the Second International Conference on Intelligent Computing and Control Systems (ICICCS 2018). 7*, 1985–1988. https://doi.org/10.1109/ICCONS.2018.8663145

Subramanian, N., & Jeyaraj, A. (2018). Recent security challenges in cloud computing. *Computers & Electrical Engineering, 71*, 28–42. https://doi.org/10.1016/j.compeleceng.2018.06.006

Tabrizchi, H., & Rafsanjani, M. K. (2020). A survey on security challenges in cloud computing: issues, threats, and solutions. *The Journal of Supercomputing.* https://doi.org/10.1007/s11227-020-03213-1

Taiwade, H. V., Kaley, B. T., & Raoot, A. R. (2019). Three level advanced cloud computing security mechanism. *International Journal of Research in Electronics and Computer Engineering, 7*(1), 1590–1592. http://nebula.wsimg.com/18c5c2dad900ac74808ac615477e8c17?AccessKeyId=DFB1BA3CED7E7997D5B1&disposition=0&alloworigin=1

Wood, G. (2017). Ethereum: A secure decentralised generalised transaction ledger. *Psychology of Popular Media Culture*, *8*(3), 207–217. https://gavwood.com/paper.pdf

Zhang, Y., & Wen, J. (2017). The IoT electric business model: Using blockchain technology for the internet of things, *Peer-to-Peer Netw. Appl, 10*, 983–994. https://doi.org/10.1007/s12083-016-0456-1

Zheng, Z., Xie, S., Dai, H., Chen, X., & Wang, H. (2017). An overview of blockchain technology: Architecture, consensus, and future trends. *IEEE 6th International Congress on Big Data, 2017.* 557–564. https://doi.org/10.1109/BigDataCongress.2017.85

Zhou, Y., Zhang, D., & Xiong, N. (2017). Post-cloud computing paradigms: A survey and comparison. *Tsinghua Science and Technology. 22*(6), 714–732. https://doi.org/10.23919/TST.2017.8195353

7 Cloud, Edge, and Fog Computing

Trends

Siddhant Thapliyal and Lisa Gopal
Graphic Era University, Dehradun

Piyush Bagla and Kuldeep Kumar
Dr. B.R. Ambedkar National Institute of Technology, Jalandhar

CONTENTS

7.1 Introduction 103
7.2 Background 106
7.3 Cloud Computing 106
  7.3.1 Advantages of Cloud Computing 108
  7.3.2 Limitations of Cloud Computing 108
7.4 Edge Computing 108
  7.4.1 Advantages of Edge Computing 109
  7.4.2 Limitations of Edge Computing 110
7.5 Fog Computing 110
  7.5.1 Advantages of Fog Computing 110
  7.5.2 Limitations of Fog Computing 111
7.6 Edge Computing vs Fog Computing 111
7.7 Smart Devices 112
7.8 Trends of Cloud, Edge, and Fog Computing 113
7.9 Conclusion 114
Notes 114
References 114

7.1 INTRODUCTION

In the last few years, the emergence of artificial intelligence and the Internet of Things (IoT) has enhanced the working capability and digital powers of sectors such as healthcare, banking, and education. The benefits of these technologies are faster transmission, security, and integrity of data. These features can be achieved through cloud computing, which is a virtual storage concept for processing and storing data.

DOI: 10.1201/9781003146711-7

But every second the amount of data is increasing, which affects the analyzing and processing speed of data and which will become more complicated over the next few years. As a solution to these problems, there are two enhancements that can be made to cloud computing: edge computing and fog computing. These computing technologies try to reduce the size of the data to be transmitted to the cloud by performing processing closer to the entry point of the data. The strategy behind these technologies is to provide such enhancement that only necessary and important data, which are used frequently or which will affect real-time conditions in the future, are stored in the cloud, and all temporary and intermediate data should be analyzed and processed at the data feeding end.

This chapter presents the different technological aspects of data storage and fast retrieval as is the case at present. Each day, industries are adopting more and more technologies that can perform enhanced and real-time user interaction, and can produce more error-free results in a minimum amount of time. These activities also generate a large volume of data, so the issue of traditional data storage is the biggest problem here. For overcoming these there is the concept of virtual space in the system. As shown in Figure 7.1, in earlier times, devices had less storage, and data entering the system increased the load on it. As a result, the device was not able to access files or software which had high storage capacities, as it also generates intermediate data which results in computation and data storage problems.

Each minute, society is consuming a gigantic amount of data because almost everything that we use is smarter than the earlier versions or models of itself; to enhance things, there is a need for data. In the current scenario, data has become one of the most important utilities because everything is based upon data and its uses. Let us take the example of the stock market, where every day many transactions take place. Goods, stocks, and shares change their prices each day. No one can predict manually what will be the prices of a particular share on a given day, but it is possible to predict them using data with high accuracy and efficiency. This can be achieved by

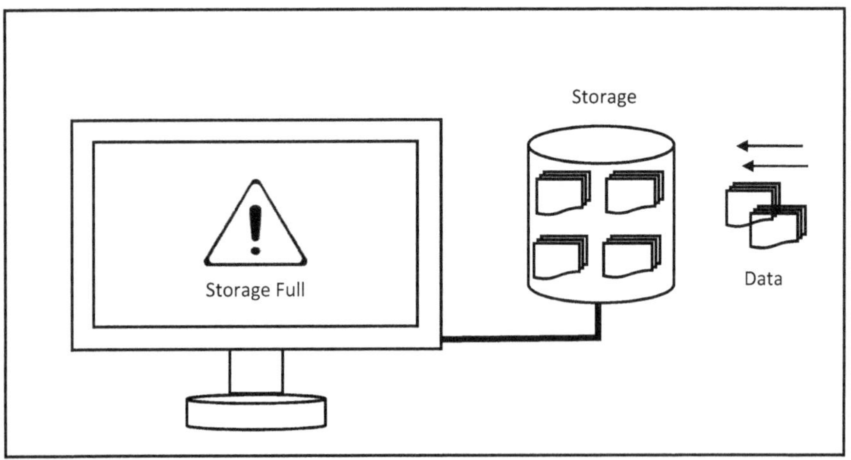

FIGURE 7.1 Storage issue with devices having limited memory: in traditional data storage, due to limited storage, the user is unable to utilize data having large space requirements.

collecting past data, extracting useful data, and performing some mathematical and logical calculations. This requires machines with high efficiency and a large amount of storage at each point in time of the calculations.

Note that current systems can perform considerable calculations with high efficiency because of their enhanced processing powers, though sometimes lack of storage does not allow them to perform such actions. Input can be of any size. Further, during such calculations, some intermediate data that can affect the final results also need to be stored. Thus, there may not be enough space to store input or intermediate results during the computation. To overcome these situations where there may occur some obtrusion, due to storage capacity, the concept of virtual storage comes into the picture. This provides virtual space for the user where he or she can perform all the tasks and calculations without any disturbance. This concept allows the user to take and use some virtual space for performing calculations, then returning it to the service provider. The reason behind the blooming of this technology is that there is no need for maintenance at the user's end, only the service provider has to take care of it. All these concepts are adopted by the technology called cloud computing, which is enhanced further by edge computing and fog computing.

As shown in Figure 7.2, cloud computing provides virtual storage to users which is not physically present in the user's device but is present at the provider's end. This produces the illusion that the virtual storage is present in the user's device. It also reduces costs as well as maintenance-related problems for the user. Cloud services aim to provide data everywhere, always with high efficiency, and with less processing time.

Both fog computing and edge computing work almost the same but there are some differences between them. In fog computing, the data is processed in the node or

FIGURE 7.2 Cloud computing extends the storage capability of a system by providing the virtual storage.

gateway which is present in the same network, whereas in edge computing the processing is done in the device or sensor. Only the necessary data is delivered to the cloud in these two technologies.

This chapter covers various aspects of cloud, edge, and fog computing, their requirements, and their usages. The chapter is organized as follows. Section 7.2 provides the background. Various aspects of cloud, edge, and fog computing along with their advantages and disadvantages are covered in Sections 7.3–7.5 respectively. The differences between fog and edge computing are highlighted in Section 7.6. Section 7.7 presents an overview of smart devices that can send or receive data in almost real time. Recent trends in cloud, edge, and fog computing are explored in Section 7.8. Finally, Section 7.9 concludes the chapter.

7.2 BACKGROUND

It is expected that a variety of future creative computer services will use fog computing systems (FCSs) connected to IoT resources. Based on the fusion of many distinct innovations, these modern services need to execute time-sensitive tasks, have variable degrees of environmental integration, and combine data collection, computing, communications, sensing, and power. However, before these services are found to be fit to function, there are many problems that need to be solved. Such a scenario of fog computing is presented by Moura and Hutchison (2020). The emergence and need for fog and edge computing, and how the industries are adopting them in the real world along with virtualization in the network is described by Vaquero and Rodero-Merino (2014). Fog computing is a technology that is growing much faster in industry to deliver faster computation at the user's end. The future of growth in fog computing is discussed by Mouradian et al. (2018) and Firdhous et al. (2014). Bonomi et al. (2012), Atlam et al. (2018), and Bellavista et al. (2019) mainly focus on how fog computing is changing the efficiency of IoT devices and how it only sends the amount of data to the server that has to be stored; all other computational and intermediate calculations are done at the fog nodes themselves.

The potential and the working capability of edge technologies are presented by Bilal et al. (2018). They showed that edge computing is enhancing working capabilities as well as providing high user-interactive abilities. The foundation of cloud computing and the trends emerging with the evolution of cloud computing applications are presented by Hosseinian-Far et al. (2018). Tordera et al. (2016, 2017) mainly focus on the core functionalities of fog computing nodes, along with the risks and other opportunities that arise in practical scenarios.

Fog computing is much more flexible as compared to cloud computing as it provides a flexible infrastructure and efficient data processing by consuming the minimum bandwidth instead of placing all the data in the cloud (Kumar et al., 2019). The prospective risks, trends, and applications of edge computing are presented by Naha et al. (2018).

7.3 CLOUD COMPUTING

From mainframe computing to the hybrid cloud, we have come a long way in 70 years. The cloud has become ubiquitous, with almost all businesses today using some form of cloud computing technologies. It started with the idea of one individual operating a machine in terms of the concept of virtualization and multiple individuals operating a machine. IBM never thought that there might be such a huge demand for such systems, but they were quick to respond. They started building this type of mainframe. The virtual machine (VM) operating system took the shared access mainframe to another level by making more than one distinct computing environment work on a single machine.

The term "cloud computing" was coined in 1996, and has now taken its share of the market. Everything is related to the cloud in one way or another. Companies such as Amazon[1] specialize in IaaS, which provides an entire Infrastructure as a Service. Google[2] focuses on PaaS, which is a Platform as a Service. There is also SaaS, Software as a Service, which is provided by Salesforce.[3]

ARPANET (Advanced Research Projects Agency Network), which was introduced in the late 1960s, marked the growth of today's Internet.[4] Using ARPANET in 1969, four computer network nodes located at four universities communicated among themselves. That communication was much faster than the time of mode of communication. Virtual machines and virtual desktop infrastructures (VDIs) were introduced in the late 1970s, but in the beginning, they were just on a single machine. With the introduction of UNIX-based systems, everything changed. In 1999, Salesforce was the first company to provide enterprise applications from a website.[5]

In the 2000s, the cloud took a major leap forward as we saw the rebirth of Netflix. The arrival of AWS, Dropbox, and Google Apps changed the cloud as we know it. The private cloud entered as a concept which is more secure than the public cloud. The need for these changes arose due to the high amount of data to be stored. The cloud was initially costly but public cloud providers such as Google reduced the pricing drastically, which made the cloud accessible to almost every individual. Talking about statistics, the data we produce every day is beyond our imagination. We produce approximately 2.5 quintillion bytes of data every day. Breaking this down into simpler terms, 1.7 MB/second of data is produced per individual.[6] Almost all of that is stored in the cloud. For all that, another change has been necessary, which is brought about by edge computing. This is a distributed computing paradigm that brings computation and data storage closer to the end user or wherever the services of the cloud are needed. Not only does it reduce the traffic but it also increases the response times and bandwidth optimization.

The cloud enables the system to store data of high storage value without changing any hardware components. This is a much more efficient mechanism. For most people, this transition is such a smooth one that highlighting a breakpoint in it is very difficult. For businesses and network architects, this transition is a major boon, a game changer. It has made everything into a service concept. Things that can be run on local machines are running on the cloud just because it is secure and readily available from just about anywhere. The pace of change in cloud technology is exponential. If you ask a student what is the cloud, the majority will say data storage or streaming. But there is more to the cloud than first appears. It has been growing so rapidly that a new concept comes into the picture every year. The cloud provides such flexibility and reliability to the client that he or she needs only to be focused on the project or process rather than worrying about storage-related issues.

7.3.1 Advantages of Cloud Computing

- There is no need to worry about storage when we are using heavy data-related projects because on the cloud we can use and occupy as much storage as we need in the project.
- In the cloud, users can access resources without installing them into the physical system: it is all available at the single cloud server.
- Resource pooling techniques allow users to access resources more frequently and increases the throughput and use of resources that are available at the server.
- The client only needs a system with a good Internet connection for accessing cloud storage, which the client can do anywhere, anytime, and with anything available in the cloud.
- Another advantage of cloud services is that you have to pay only for your usage; this means only the storage you access and the time duration for which you use the services.
- At the client's end, there is no need for maintenance, as the cloud service provider will take care of it completely.

7.3.2 Limitations of Cloud Computing

- Cloud computing is completely dependent on the Internet: accessing or any action involving data is not possible without the Internet.
- Privacy and security are the main drawbacks of cloud computing. As cloud services are provided by various service providers and they have access to it, so there might be a possibility that they can manipulate, change, or delete data which is very essential to the user.
- It is time consuming: since the cloud stores its data on its server located outside the local area, it needs some time to effect transmission.
- Any failure at the server side can be a reason for data loss.

7.4 EDGE COMPUTING

Edge computing is an emergence of cloud computing, where computing power is provided to the cloud so that it can compute data at the edge of the device. In this technology, the main concern is to reduce the distance between the data and their generation point so that we can make the transmission more quickly.

Edge computing is growing much faster in today's technologies because of its feature of performing calculations at the local server. Then the question arises as to why we need calculations at the local server when it works perfectly at the cloud server. The answer is that in today's world, where at each point in time a large amount of data is created or used, all data is stored by the server cloud. But sometimes all the data being stored by the cloud is not useful to the user. This might be data that can be used only once or it might be intermediate data involved in the process or calculation.

All data which is unused for a longer time and not accessed by the user is erased by the service provider, though the duration of storing that data might be more than

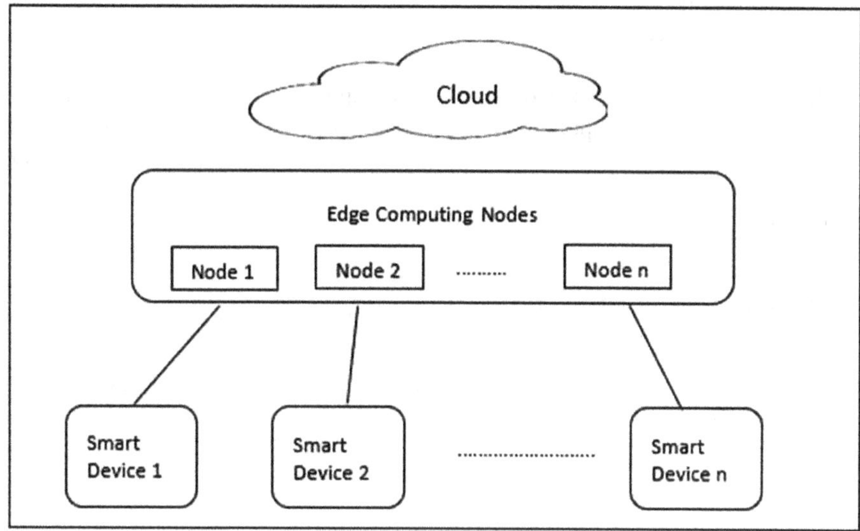

FIGURE 7.3 Edge computing architecture.

a decade, which would also be a total wastage of storage. To overcome these problems with earlier technologies, in edge computing we send only that data to the cloud which may be used in the future by the user or which may affect the actual result of a process. All the intermediary data which is not useful and only being used up until the completion of a process is stored locally where the process is taking place; only then will it be erased by the local server.

As shown in Figure 7.3, there are three layers in any edge-computing-enabled network. At the first layer, a cloud service provider provides the services to the nodes or systems of edge computing. Then, smart devices are connected through the edge nodes.

Edge computing is a boon for IoT or smart devices, which are now capable of doing calculations locally that increase their speed and reduce the response time. This is because previously there might have occurred a scenario where sensors that were being used in various areas were connected to a global server which is situated at a great distance from the sensors, so whenever a transmission of data occurs, it takes too much time to send data from one end to the other. This affects the speed and proficiency of the smart device.

7.4.1 Advantages of Edge Computing

- Edge technology enables the system to compute data locally as well as store that data on the local server.
- Local storage and servers enable the system to compute a process even when it is not connected to the Internet which increases its work time and enhances the throughput of the machine.

- It reduces process failures in certain programs due to connection loss or an unreachable server.

The latest 5G technology also works on edge computing. Because of local computation and storage, edge computing makes the process much faster. It is also beneficial when there is so much traffic in the network, because it only needs the essential and the final data to be stored in the global server, and all other computation and intermediate data is stored locally (Hu et al., 2015; Yu, 2016).

7.4.2 Limitations of Edge Computing

- In edge computing, one disadvantage is security. This is because data are computed locally, and it is a possibility that there might be some bugs or viruses present in the local server.
- It requires high bandwidth.
- It needs high controlling management.

7.5 FOG COMPUTING

Fog computing is a concept in which the computation and the data lie somewhere in between the data entry point and the cloud storage. It is like edge computing in infrastructure; it also tries to locate the computation closer to the data entry point. Further, it is a decentralized and distributed computing technology. Fog computing tries to supply a channel that needs storage of data only for the final or essential data, and all other mathematical calculations or any other calculations are done at the edge of the data entry point (Wang et al., 2015).

In fog computing, the power of computation is available at the local server, and data that is available at the endpoints need to be carried at the fog gateways. Compared to that in edge computing, the power of computation can be available at the data endpoint or the gateway. This is a benefit of edge computing because it decreases the possibility of transmission failure (Shirazi et al., 2017).

The basic concept of fog computing is to make those devices capable of processing locally that are not connected to any physical sensor nearby. Figure 7.4 shows the three-layer architecture of fog computing. The first and the last layers are similar to edge computing; here the second layer concerns fog which is used as intermediate storage.

7.5.1 Advantages of Fog Computing

- Fog computing sends only the necessary or essential data to the server, so it reduces the total amount of data sent.
- It increases the throughput of the network.
- Because of its low response time, it makes the network more and more reliable for the user.
- Due to local data storage, it reduces the chance of transmission failure due to Internet connection loss.

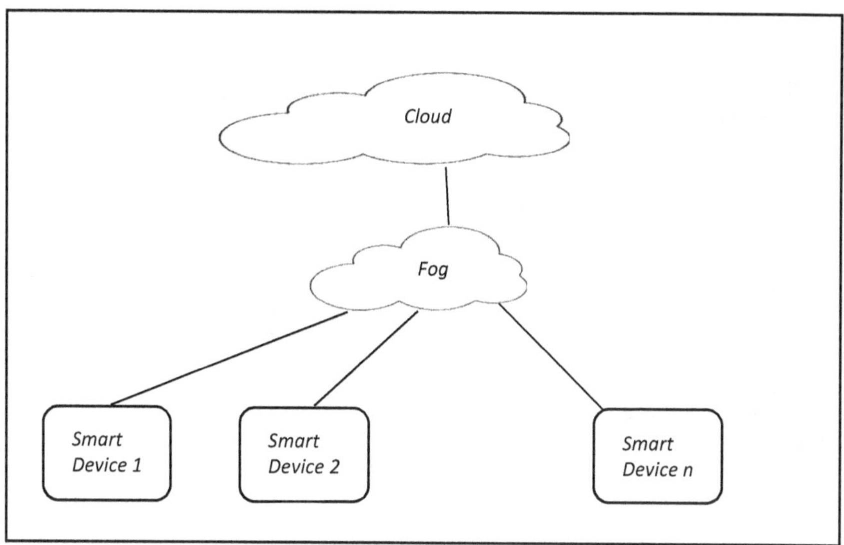

FIGURE 7.4 Fog computing architecture.

Like edge computing, fog computing is also beneficial for the IoT or smart devices because of the reduced distance between the computation and the data endpoint. Let us take the example of a showroom that is designed in a fully smart manner in terms of technology. There are many sensors, CCTV cameras, and some other appliances which can respond according to the environment, condition, time, or any emergency. All these devices are connected to the one global server which is present somewhere in the world at the headquarters of the showroom. Further, each of the decisions for or against action is taken according to the orders of the global server, since the response time for the node of the network might be slow due to some network congestion. In that case, the node has to wait to perform an action which can cause hazardous results in an emergency situation. For overcoming such situations, we have fog computing which brings the decision-making capabilities closer to the data endpoint or node.

Along with edge computing, fog computing is also a part of the new 5G technology. One can use edge and fog computing simultaneously according to the purpose and conditions of the network.

7.5.2 Limitations of Fog Computing

- Its structure is complex to implement.
- It has more connections which increases the chances of reduced privacy.

7.6 EDGE COMPUTING VS FOG COMPUTING

Both technologies, edge computing and fog computing, look very similar in the working but there are some key differences between the two.

TABLE 7.1
Key Differences between Fog Computing and Edge Computing in Terms of Implementation

Fog Computing	Edge Computing
Deployed at the user end or the node acting as a data source.	Just another version of a cloud data center that has much more capability than the traditional approach.
Absent of a central server.	Presence of control edge components.
Uses network devices like routers, switches, and gateways as node devices. These devices transmit data to the local server or node for processing.	Does not require any additional network device because it is used where data sources have their own sensors or availability nearby.
The location of nodes is mobile because of variations in devices and their processing requirements.	The location of nodes is at the radio network controller.
Communication between internodes is possible.	Communication between internodes is partially possible.

Edge computing works when the device that is generating the data or acting as a data source has its own sensor or is physically available nearby, whereas fog computing provides the feature of processing or analyzing data at its fog nodes or any gateway related to smart devices that are present at the local server. In terms of their implementation, Table 7.1 highlights the key differences between fog computing and edge computing.

7.7 SMART DEVICES

Smart devices are those that can generate data and signals in real time through sensors and various other peripheral devices that can work even when there is no physical structure present locally. A smart device can send or receive data anytime, anywhere through an Internet connection to the server; it acts as an endpoint of data and performs various actions according to the orders and directions of the server.

The concept behind the use of smart devices is to reduce human interaction with devices, which increases the throughput efficiency and decreases transmission time. In this modern era, we all desire the use of quick and real-time services. For example, if we are driving a car and suddenly, at a crossroads, we see a red light on our side, we have to stop and wait, even if there is no vehicle on the other side. But, if we use a smart traffic light device that decides the red and green lights according to the traffic on a particular road, traffic flow would be enhanced and waiting times reduced.

Figure 7.5 shows this scenario of a crossroads. In this case, the traditional traffic signal will allow vehicles to pass on each road equally, but the smart traffic signal will allow more time for the vehicles of route 4 than other routes because route 4 has more vehicles currently on the road.

Let us take another example, of a smart fire extinguisher that can begin showering water whenever it senses fire in the room or building; it does this without the need for human intervention. It is beneficial because sometimes there are cases where people only get to know about the fire when it is too late and the fire has almost destroyed everything.

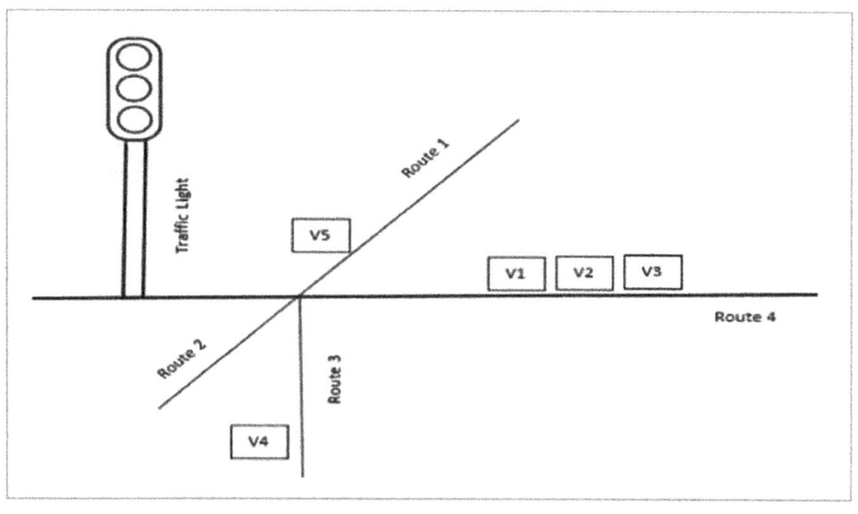

FIGURE 7.5 Smart traffic light system.

7.8 TRENDS OF CLOUD, EDGE, AND FOG COMPUTING

Cloud computing is changing the industry drastically. This is because it is the most important thing in today's scenario: the diamond of today is "data." Data is one of the key essentials of today's industry. Each business or firm is based on data. Everyone is dealing with data at any point in time. Hence, we need to store that data because even a small part of a particular data file can change the whole outcome of a particular process. Here the cloud comes into the picture. The cloud stores and maintains that data, and the user can access it at any point in time anywhere.

Today, when there are a lot of new industries, cloud service providers play a very crucial role in the emergence of these industries. Let us take the example of a new growing industry that does not have any base or foundation for the establishment of the business. It needs a lot of resources along with data storage capacity. But, because there is no customer base and a reduced workload, the industry needs variable storage at every period. If the industry borrows a large amount of space, it will suffer from a wastage of storage and unnecessary costs. On the other hand, if the industry borrows less space, it might suffer from inadequate storage at some other time. For such a scenario, cloud service providers allow a facility for cloud space that is needed for a particular duration of time at a much reduced rate.

Cloud services also provide fast and efficient data transmission services so that the response looks like a local server response. Small scale organizations and startups obtain many advantages from these services because of the reduced revenue charges. Further, new future technologies will implement cloud services as a foundation because of their reliable features and the worry-free maintenance.

Cloud technologies are also playing a very crucial role in health sectors. Researchers and scientists can store the past data of many years and analyze it and utilize it in developing a new medicine, treatment, or research. 5G technology which has a higher data transmission speed than the previous is completely involved in cloud technologies (using the enhancement of cloud, edge, and fog computing).

Edge computing is the basic component for future technologies like 5G mobile services, real-time response systems, and autonomous cars. Real-time response systems need quick data transmission as well as some intermediate data that needs to be stored in the cloud. So, for that, we can use edge computing. This is because we can compute some intermediate calculations at the edge nodes which will decrease the burden on the actual service provider. It can also be used for autonomous cars where we only need quick responses to signals produced by sensors.

Fog computing plays a very crucial role when we are using IoT devices, which can check their data at each point of time without any delay in between, which increases the throughput of devices. Fog computing is also useful for the implementation of 5G technology which needs more user interaction with a minimum of access time.

7.9 CONCLUSION

The demand for cloud, fog, and edge computing has been increased with the rise of the IoT. The cloud is useful at quickly sending data between local nodes and data centers, and also improves the communication between applications and devices. The cloud will always be a priority for storing data. Much reduced cost is offered by fog and edge computing that helps to segregate critical from generic information. Performance optimization is the hardcore that can effectively be achieved through edge and fog computing. Telecoms and middleware companies widely prefer edge computing. Edge computing and fog computing will be the base for upcoming technologies like 5G which will enhance the working capability at the node and will reduce the overheads at the server end. Also, network traffic will be reduced by implementing them.

NOTES

1 Amazon Web Service, https://aws.amazon.com/
2 Google Cloud, https://cloud.google.com/
3 Salesforce, https://www.salesforce.com/
4 On the birth of the Internet, see https://www.cs.utah.edu/birth-of-the-internet/
5 On the History of Salesforce, see https://www.salesforce.com/news/stories/the-history-of-salesforce/
6 https://www.domo.com/solution/data-never-sleeps-6

REFERENCES

Atlam, H., Walters, R., & Wills, G. (2018). Fog computing and the internet of things: A review. *Big Data and Cognitive Computing, 2*(2), 10. https://doi.org/10.3390/bdcc2020010

Bellavista, P., Berrocal, J., Corradi, A., Das, S. K., Foschini, L., & Zanni, A. (2019). A survey on fog computing for the internet of things. *Pervasive and Mobile Computing, 52*, 71–99. https://doi.org/10.1016/j.pmcj.2018.12.007

Bilal, K., Khalid, O., Erbad, A., & Khan, S. U. (2018). Potentials, trends, and prospects in edge technologies: Fog, cloudlet, mobile edge, and micro data centers. *Computer Networks, 130*, 94–120. https://doi.org/10.1016/j.comnet.2017.10.002

Bonomi, F., Milito, R., Zhu, J., & Addepalli, S. (2012). Fog computing and its role in the internet of things. *Proceedings of the First Edition of the MCC Workshop on Mobile Cloud Computing - MCC '12*, 13. https://doi.org/10.1145/2342509.2342513

Firdhous, M., Ghazali, O., & Hassan, S. (2014). Fog computing: Will it be the future of cloud computing?. *The Third International Conference on Informatics & Applications (ICIA2014)*, 8–15.

Hosseinian-Far, A., Ramachandran, M., & Slack, C. L. (2018). Emerging trends in cloud computing, big data, fog computing, IoT and smart living. In *Technology for Smart Futures* (pp. 29–40). Springer. https://doi.org/10.1007/978-3-319-60137-3_2

Hu, Y. C., Patel, M., Sabella, D., Sprecher, N., & Young, V. (2015). Mobile edge computing—A key technology towards 5G. *ETSI white paper, 11*(11), 1–16.

Kumar, V., Laghari, A. A., Karim, S., Shakir, M., & Anwar Brohi, A. (2019). Comparison of fog computing & cloud computing. *International Journal of Mathematical Sciences and Computing, 5*(1), 31–41. https://doi.org/10.5815/ijmsc.2019.01.03

Moura, J., & Hutchison, D. (2020). Fog computing systems: State of the art, research issues and future trends, with a focus on resilience. *Journal of Network and Computer Applications, 169*, 102784. https://doi.org/10.1016/j.jnca.2020.102784

Mouradian, C., Naboulsi, D., Yangui, S., Glitho, R. H., Morrow, M. J., & Polakos, P. A. (2018). A comprehensive survey on fog computing: state-of-the-art and research challenges. *IEEE Communications Surveys & Tutorials, 20*(1), 416–464. https://doi.org/10.1109/COMST.2017.2771153

Naha, R. K., Garg, S., Georgakopoulos, D., Jayaraman, P. P., Gao, L., Xiang, Y., & Ranjan, R. (2018). Fog computing: Survey of trends, architectures, requirements, and research directions. *IEEE Access, 6*, 47980–48009. https://doi.org/10.1109/ACCESS.2018.2866491

Shirazi, S. N., Gouglidis, A., Farshad, A., & Hutchison, D. (2017). The extended cloud: review and analysis of mobile edge computing and fog from a security and resilience perspective. *IEEE Journal on Selected Areas in Communications, 35*(11), 2586–2595. https://doi.org/10.1109/JSAC.2017.2760478

Tordera, E. M., Masip-Bruin, X., García-Almiñana, J., Jukan, A., Ren, G.-J., & Zhu, J. (2017). Do we all really know what a fog node is? Current trends towards an open definition. *Computer Communications, 109*, 117–130. https://doi.org/10.1016/j.comcom.2017.05.013

Tordera, E. M., Masip-Bruin, X., Garcia-Alminana, J., Jukan, A., Ren, G. J., Zhu, J., & Farré, J. (2016). What is a fog node a tutorial on current concepts towards a common definition. ArXiv:1611.09193 [Cs]. http://arxiv.org/abs/1611.09193

Vaquero, L. M., & Rodero-Merino, L. (2014). Finding your way in the fog. *ACM SIGCOMM Computer Communication Review, 44*(5), 27–32. https://doi.org/10.1145/2677046.2677052

Wang, Y., Uehara, T., & Sasaki, R. (2015). Fog computing: Issues and challenges in security and forensics. *2015 IEEE 39th Annual Computer Software and Applications Conference*, 53–59. https://doi.org/10.1109/COMPSAC.2015.173

Yu, Y. (2016). Mobile edge computing towards 5G: Vision, recent progress, and open challenges. *China Communications*. https://doi.org/10.1109/cc.2016.7405725

8 Progression in Cyber Security Concerns for Learning Management Systems

Analyzing the Role of Participants

Tejaswini Apte and Sarika Sharma

Symbiosis Institute of Computer Studies and Research,
Symbiosis International (Deemed University), Pune, India

CONTENTS

8.1 Introduction ... 117
8.2 Literature Review ... 121
  8.2.1 Cyber-Physical Systems for Education ... 121
  8.2.2 Cyber Security in Learning Management Systems ... 121
8.3 Adoption of Cyber Security in LMSs by Instructors: A Case Study ... 122
8.4 Methodology ... 126
  8.4.1 Measurement Model ... 126
  8.4.2 Data Analysis Using Structural Equation Modeling (SEM) ... 129
8.5 Results and Discussion ... 130
  8.5.1 Theoretical Implications ... 130
  8.5.2 Practical Implications ... 130
8.6 Limitations and Future Scope ... 130
References ... 131

8.1 INTRODUCTION

To facilitate education in larger communities the tradition of face-to-face connection is often overruled. This was where the instructor and the learners were meeting synchronously for discussion and knowledge dissemination. However, the growing need for skill based education and the inability to attend classroom sessions with work hours is restricting. The delivery and dissemination of Cyber Physical Systems (CPSs) by way of learning management systems (LMSs) as a platform necessitates

DOI: 10.1201/9781003146711-8

attention (Torngren et al., 2015). Leia et al. (2013) suggested the development of smart learning environments by using CPSs. e-learning systems have also witnessed a major boost via the Internet of Things, which can then be utilized to implement the concept of virtual educational space (Glushkova et al., 2018). Online LMSs are developed by exploring the technology and Internet for virtual training or training by electronic means. The major objective of the LMS platform is to share, disperse, and group the study or reference material with a larger group of participants who are heterogenetically located across the globe (Alkhalaf et al., 2012). To support the load of several participants and hence the transactions every LMS is distributed over the network, and thus is complex in nature. This complexity is built according to the requirement for scalability and availability, which in turn spawns multiple entry points in LMSs. These entry points develop numerous interactions among the diverse components and respective service endpoints in the underlying infrastructure of LMSs. This complexity results in several vulnerabilities that can be exploited by non-legitimate users or applications.

Thus, it is a responsibility of developers and designers to curtail the number of entry points by bringing forth only privileged access to the service endpoints. The service endpoint access can be restricted by the security of the network, communication protocol, and identity management services. The amount of security implementation is driven by the risk associated with each unauthorized access and potential data loss (Nickolova & Nickolov, 2007). The following briefly describes three pillars for understanding security and privacy: confidentiality, integrity, and availability (CIA). In addition, we also outline some of the attacks which affect the CIA triangle.

Cyber security monitors data transmission visibility for electronic media. It implements the CIA triangle to maintain security and privacy. Confidentiality preserves the authorization of access to information. A breach in confidentiality is the sharing of information with non-legitimate users. Integrity protects the authenticity of information while in transmission. A breach in integrity enables the manipulation of information whilst being communicated. The role of availability is to cater for user demands in the required time. A breach in availability means the respective service is not accessible to legitimate users (Gupta & Sharman, 2008). Inadequately designed LMS architecture and coding style compromises the CIA triangle by allowing attacks, for example cross-site scripting (XSS) attacks, SQL injection attacks, and session hijacking attacks. These attacks also significantly affect confidentiality and integrity during collaborative learning, assessment, evaluation, and authentication of a legitimate user on LMSs.

XSS is application layer attack. XSS exploits the vulnerability of the client-side scripting language for execution by malicious users. Every page load proceeds with execution of the script. Such attacks are performed so as to impersonate authenticated users and further terminate the required service. Every LMS platform designer practices and extends the development process for reducing vulnerabilities. Every designer and additionally every developer is responsible for investigating or examining the input received before it can reveal itself as output to the end user. Poor coding practices and the respective buffer attacks fabricate the space for the attacker to overwrite the stack space by passing control to the malicious code. Furthermore, application-level security always honors the permission required for database access. Secondly, deficient database security design leads to SQL injection attacks in which non-legitimate users insert unauthorized data to gain database access. Additionally, the session expiration time and weak encryption for user credentials also allow sufficient time for attackers to use brute force credentials or hijack the session (Luminita & Magdalena, 2012). These attacks are also triggered by interactions among various users and technologies by design and social engineering (Elmaghraby & Losavlo, 2014). The after-effects of each attack is determined by the LMS design and the number of entry points an LMS has. The next section concisely describes the architecture of an LMS.

The design of the LMS follows a component-based architecture to construct the service. These components are customizable and are stored in a code repository. Each component is designed and based on the interaction among the instructors and the learners; the major interactions are the consumption of service (assignment and classwork submission, downloading the notes and study material) by the learner, and the consumption of service (creating classes, assessments, attendance, etc.) by the instructors. These interactions are required to be monitored regularly for any suspicious activity. Furthermore, the instructor and the learner engage in many-to-many interactions with each consumption over the network and thus the rate of data transmission is increased. The frequency of data transmission constitutes a ground for much vulnerability, which is common to other Internet applications as well. However, the impact differs as the associated risk varies for each interaction according to the functionality of LMS modules. It is necessary to identify the risks and threats so as to segment the interaction and mitigate the data loss (Rabai et al., 2012). The possible reasons for data loss are eavesdropping, stealing of identity information, and monitoring behavior by social engineering. Social engineering skillfully manipulates individual weakness and launches the attack on the targeted participants of the LMS, who are within the circumference of each individual (Ariu et al., 2016).

Course management services with their respective learning objectives and collaborative services to map the educational delivery and learning outcome is an essential part of every LMS. To design collaborative services in every LMS, the threat of elevation of privileges and thus understanding a legitimate user is required to be addressed carefully before implementation. Every LMS has a repository for the course objectives, learning objectives, assessment questions, and marking scheme according to an instructor. The access to each repository is by a service which is executed on a designated endpoint. Erroneous design of LMSs may introduce a breach of identity, thereby impersonation, falsification of course assessments, or alteration of the timestamp, which by nature reduces the importance of the assessment and respective certificate (Bandara et al., 2014). Figure 8.1 presents the general architecture of interaction among participants and LMS components.

With the architecture portrayed in Figure 8.1, the participant requests the authentication token from the authentication server, which performs the proxy and authentication layer validation. Based on the token received the user is allowed to access the authorized service endpoints. The service endpoints then are responsible for

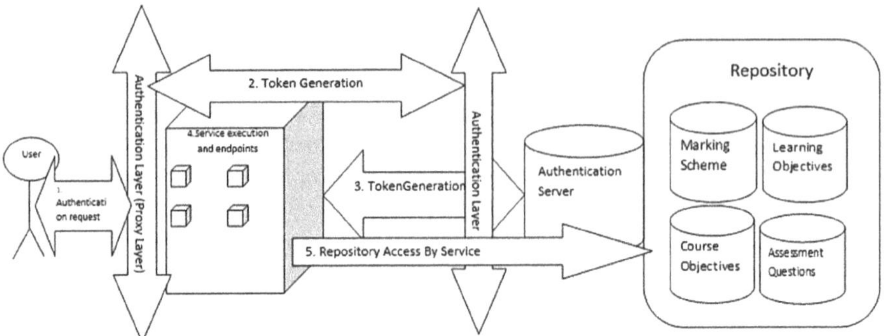

FIGURE 8.1 Generalized architecture of learning management system for accessing services and the repository.

accessing the requested database repository. The authentication server again authenticates the said request by the service endpoint and the required token is generated.

Hence, the possible entry points to access these services and the respective attacks associated with them are:

- During token request and response; attacks: session hacking, impersonation.
- Accessing the endpoints for authorized service; attacks: denial of service.
- Accessing the authentication server for database access; attacks: SQL injection.

To maintain the interconnectivity among the components with scalability and performance, the distributed nature of an underlying infrastructure is increased, which by delivery results in increased threats to security and privacy (Elmaghraby and Losavlo, 2014).

Many LMSs today use secure socket layer or transport layer security for the secure interaction among various components in order to protect the infrastructure and sensitive data from unauthorized access. The identity and access management service for each communication is a major focus for any attacker to impersonate as a legitimate user. The article published by security intelligence shows the elevating of permissions by participants and thereby the accessing of data owned by other participants in the analysis of various LMSs (i.e., WordPress, LMS, plugins, LearnPress, LearnDash, and LifterLMS) (Bisson, 2020). These attacks were executed due to misconfigured DNS records and the way the authentication tokens interacted. The impact of the attacks are multifold for collaborative learning and assessment evaluation. Based on the current need for security and privacy in collaborative services and assessment, LMSs are redesigned to meet the demands of confidentiality and integrity for each participant. The interaction, lack of awareness, and casual approach of the instructor or the learner with the LMSs are the major causes of session hijacking or impersonation. There exists a need to explore LMSs and identify the vulnerabilities and risks associated with them.

8.2 LITERATURE REVIEW

8.2.1 Cyber-Physical Systems for Education

Although researchers have attempted to progress through the proposed models in CPSs for educational domains (Santos et al., 2017), there is still scope for further research. Bachir and Abenia (2019) illustrated various components of the Internet of Everything (IoE) with respect to the delivery and dissemination processes. Mourtzis et al. (2018) implemented CPSs in education through the concept of "teaching factory 4.0", whereas Stankovic et al. (2017) explored the possibilities of CPSs in 21st-century education. There is a scarcity of studies in cyber-physical systems for education and specifically for the security related issues in LMSs. Table 8.1 summarizes some of the relevant studies mentioned in this and the next section.

8.2.2 Cyber Security in Learning Management Systems

A security issue for an LMS has been acknowledged as an emerging area (Salimovna et al., 2019). Burov et al. (2021) identified the risks of cyber security in the present digital system of education. Khan et al. (2019) produced a multilayered security model. Rahima et al. (2020) carried out a survey of existing literature in this emerging area to obtain better insight into cyber security for education.

There exist many LMSs, for example Moodle, Dokeos, and Sakai. Moodle is an LMS widely used among the instructors and the participants of various universities and colleges. It has developed various functionalities to build a common platform for interaction among various users in terms of assignment creation, multiple-choice questions, and availability of discussion platforms (Luminita & Magdalena, 2012). Many of the participants also leverage Moodle functionalities for collaborative services and assessment submission. Dokeos, on the other hand, is used by many organizations for collaborative learning and content creation (Cuervo et al., 2016). Sakai, very similar to Moodle, has also embedded the functionality to attach many learning tools, streaming, and conference services to provide better services to participants (Trustradious, 2021).

We have examined these platforms for collaborative services, assessments, authentication, and reported vulnerabilities. Moodle is designed with a collaborative

TABLE 8.1
Literature on CPSs in Education and LMSs

Serial No.	Author(s) (Year)	Research Findings
1	Santos (2017)	Proposed model for CPS in education
2	Bachir and Abenia (2019)	Components of Internet of Everything in education
3	Mourtzis et al. (2018)	Suggested teaching factory 4.0
4	Stankovic et al. (2017)	CPS in 21st-century education
5	Salimovna et al. (2019)	Issues in LMSs
6	Burov et al. (2021)	Identified risks in cyber security in digital mode of education
7	Khan et al. (2019)	Proposed multilayered physical model
8	Rahima et al. (2020)	Systematic review on cyber security in education
9	Cuervo et al. (2016)	Collaborative learning methods in education

service to create or upload a document as collaborative work. During the shared assignment or presentation, any changes are immediately reflected to the team. Next, the assessment evaluation is acquitted with a threshold as a weight for being acceptable and satisfactory. The remaining assignments are evaluated against the threshold. The authentication plugin of Moodle is compatible with LightWeight Directory Access Protocol (LDAP) and Security Assertion Markup Language (SAML2) for a Single Sign On (SSO) authentication service. The mentioned discussion of this paragraph is referred from the links mentioned in Table 8.2 for Moodle.

Secondly, Dokeos incorporates collaborative services via video conferencing and a platform to gather and share knowledge through discussion forums. For supporting assessment and evaluation, functionality is developed to assist the instructor to establish the evaluation criteria and certification for learners, considering the required skill achieved. Dokeos provides a central authentication service and single sign on through SAML. The mentioned discussion of this paragraph is referred from the links mentioned in Table 8.2 for Dokeos.

Sakai, the next LMS, has the functionality to inject Dropbox and assignment tools to support collaborative services. These tools help the participants to upload the files and prepare the assignments in a team. Sakai also facilitates discussion about course content asynchronously. The assessment and the evaluation are supported by game elements, focused quizzes, and the portfolios of participants. For the authentication, Sakai provides a feature to create a record of users by embedding external authentication services. Sakai can potentially use LDAP-enabled directories for authentication and authorization. In addition, Sakai has a feature to implement a centralized authentication service via an apache module. The Sakai portal login web service is used to allow a user to have a single sign on. Table 8.2 briefly summarizes this discussion of Moodle, Dokeos, and Sakai for collaborative, assessment, and authentication services with the required link references. The mentioned discussion of this paragraph is referred from the links mentioned in Table 8.2 for Sakai.

Table 8.3 presents some of the vulnerabilities reported in Moodle, Dokeos, and Sakai. The table can also be used as an outline for every participant to handle the vulnerabilities, mitigate the challenges, and increase the integrity of an LMS.

Many of the instructors executing deliveries on LMS are not aware of cyber threats and associated risks. However, the measures to prevent threats can be implemented effectively by the active involvement of the instructor and their intention to adopt them. Therefore, the role of instructors is vital while delivering the sessions and assessing learner performance with LMSs. In the next section on adoption, the intention of the instructor is presented and their role is explored in the implementation of cyber security measures.

8.3 ADOPTION OF CYBER SECURITY IN LMSs BY INSTRUCTORS: A CASE STUDY

The role of the instructor in cyber security measures and approval is crucial to understand and realize. The adoption of cyber security through instructors is affected by various factors: computer/IT knowledge, teaching experience, casual behavior, familiarity with LMSs, and so on. Such factors lead to the formation of a positive or negative attitude, thereby affecting the intention to adopt it further in respective

TABLE 8.2
Description of Moodle, Dokeos, and Sakai

Serial No.	Parameter	Moodle	Dokeos	Sakai
1.	Collaborative work	Create or upload document as a team https://moodle.org/plugins/mod_collabora	Video conferencing and discussion forum https://www.dokeos.com/elearning-collaborative-learning/	Inject Dropbox or assignment tool for collaborative learning, supports discussion about course content asynchronously https://confluence.sakaiproject.org/display/PED/Sakai+Learning+Capabilities+v+1.0 https://sakai.claremont.edu/portal/site/5760451c-50fa-4762-ad6f-856e8896a679/tool/408e7644-9913-4b13-afe4-e74c7a758bc5
2.	Assessment evaluation	Marking a threshold for assessment evaluation https://moodle.org/plugins/workshopeval_weightiest	Instructor establishes the evaluation criteria and a certificate is generated to the participant based on required skills achieved https://www.dokeos.com/dokeos-evaluation/	Uses focused quizzes and participant portfolios to evaluate holistically https://www.sakailms.org/post/alternative-assessment-in-sakai-part-3-portfolios-and-peer-review
3.	Authentication plugin	LDAP or SAML2 https://moodle.org/plugins/auth_ldap_syncplus,https://moodle.org/plugins/auth_userkey	Central authentication service or single sign on with SAML2 https://intercom.help/dokeos-lms/en/articles/2964674-the-authentication-options-of-your-dokeos-portal	Provides support for external authentication, LDAP, and single sign on. https://confluence.sakaiproject.org/display/DOC/Sakai+Admin+Guide+-+Advanced+Configuration+Topics

TABLE 8.3
Vulnerabilities in Moodle, Dokeos, and Sakai

Serial No.	LMS	Vulnerability	Description
1	Moodle	CVE-2019-3849	Participants were able to escalate roles within courses or content accessed. This has a partial impact on the CIA https://www.cvedetails.com/cve/CVE-2019-3849/
2		CVE-2019-3847	Participants with the other user capability were able to access the dashboard of other users, and were able to see the content which has a partial impact on CIA https://www.cvedetails.com/cve/CVE-2019-3847/
3		CVE-2019-10186	Session key token was not utilized by XML for admin tool. The authentication was not required to exploit the vulnerability https://www.cvedetails.com/cve/CVE-2019-10186/
4	Dokeos	CVE-2009-2005	Cross site forgery attacks: where remote attackers were able to hijack the authentication of legitimate users and modify the agenda for execution. There was partial CIA impact reported https://www.cvedetails.com/cve/CVE-2009-2005/
5		CVE-2009-2004	The SQL injection attack vulnerability is reported: CVE-2009-2004 to execute the SQL commands allow attackers to execute arbitrary SQL commands https://www.cvedetails.com/cve/CVE-2009-2004/
6		CVE-2005-2598	Allows attackers to remove or move the location of the files for functionalities https://www.cvedetails.com/cve/CVE-2005-2598/ The authentication is not required to exploit these functionalities.
7	Sakai	CVE-2019-16148	XSS vulnerability via chat. There was a partial impact on integrity. Authentication is not required to exploit the vulnerability https://www.cvedetails.com/cve/CVE-2019-16148/

behavior. A technology acceptance model (TAM) by Davis (1989) is widely used in technology adoption studies, and in the present study it is adopted as a basis of development of conceptual models for LMSs. A TAM was developed in the context of information systems theory and user acceptance of a particular technology. To assess the user acceptance for a given technology the model extensively focuses on perceived usefulness (PU) and perceived ease of use (PEOU).

PU is characterized as "the degree to which a person believes that using a particular system would enhance his or her job performance." PEOU is interpreted as "the degree to which a person believes that using a particular system would be free from effort." Jones et al. (2010) noted the impact of PU and PEOU on the intention to use security measures in information systems. Seuwou et al. (2016) suggested in a study that PU and PEOU affect the decision to invest in information security. According to Davis (1989), PEOU has a significant impact on PU. This discussion leads to positing the following hypotheses:

H1: Perceived usefulness has a significant impact on attitude towards using cyber security in LMSs.

H2: Perceived ease of use has a significant impact on attitude towards using cyber security in LMSs.
H3: Perceived ease of use has a significant impact on perceived usefulness.

Instructors' awareness about cyber security can play a major role in cyber security acquisition and thereby maintenance while managing the teaching learning process. Awareness about cyber attacks by non-legitimate users stimulates an environment for a robust and secure LMS (Al-Janabi & Al-Shourbaji, 2016). Wang (2010) reveals that knowledge of information security can lead to the intention to adopt for the given technology, that is for LMSs as well. Every application measures the security in terms of CIA. Bada et al. (2015) put forward findings about the cyber security awareness campaign and its effect on attitude and behavior. With this literature support we propose the following hypothesis:

H4: Cyber security awareness affects the attitude towards using cyber security in LMSs.

For collaborative work, assessment evaluation, and authentication, we consider cyber security awareness and the attitude to it in the adoption of LMSs. Knowledge of and attitude to cyber security influences its adoption by participants (Kok et al., 2020). Hadlington (2017) investigated human factors in cyber security and identified the role of attitude to it in its adoption. Therefore, we propose the following hypothesis:

H5: The attitude towards using cyber security in LMSs has a significant and positive impact on the intention to adopt it.

With these hypotheses we propose a conceptualized theoretical model (Figure 8.2).

There are three independent variables namely, perceived usefulness, perceived ease of use, and cyber security awareness. The two dependent variables are attitude towards using cyber security awareness in LMSs and the intention to adopt cyber security awareness in LMSs. Perceived usefulness is both an independent and dependent variable.

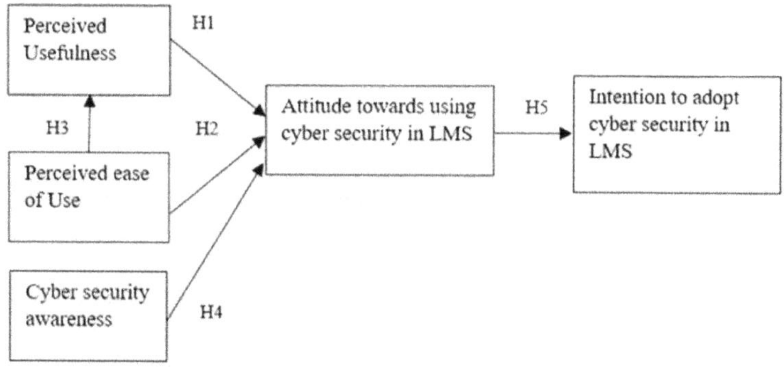

FIGURE 8.2 Conceptualized theoretical model.

TABLE 8.4
Demographics

Item	Number	Percentage
Computer/IT knowledge	Yes: 151	77.8
	No: 43	22.2
Teaching experience	0–5 years: 23	11.8
	5–10 years: 98	50.5
	>10 years: 75	38.6
Experience in using LMSs	0–5 years: 102	52.5
	5–10 years: 89	45.8
	>10 years: 3	1.5
Level of teaching	Undergraduate: 85	43.8
	Postgraduate: 109	56.1
Total	**194**	

8.4 METHODOLOGY

The proposed conceptual model (Figure 8.2) was empirically tested using primary data collected through a self-designed and structured questionnaire. The scope of the questionnaire was the adaptability and thereby efficiency to endorse cyber security measures by instructors in LMSs. Previous similar studies were considered for the formulation of items/questions (Burnett et al., 2007; Davis, 1989). The questionnaire was conducted quantitatively first on 20 varied profile instructors for undergraduate and postgraduate programs. Likert's five-point agreement scale was used to collect responses (5: strongly agree; 1: strongly disagree). A total of 194 instructors participated in the survey. The profile of the respondents is presented in Table 8.4.

8.4.1 Measurement Model

To test the proposed model, structural equation modeling (SEM) was applied with IBM AMOS 20.0 and SPSS 22.0. Initially, first-order confirmatory factor analysis (CFA) was carried out to test measure the indices of the proposed model. Secondly, path analysis using SEM was organized. The goodness of fit was calculated using chi-square (CMIN/DF), the Tucker-Lewis index (TLI), and the comparative fit index (CFI). For badness of fit, the root mean square error of approximation (RMSEA) as suggested by Hair et al. (2010) was used (Table 8.5). It can be seen that all obtained values are within the suggested range. Figure 8.3 exhibits the confirmatory factor analysis for the model.

The validity of the proposed model was tested using various validity parameters and comparing them either with the standard acceptable values or through comparisons:

- Content validity of the model was achieved by taking the views of two experts and using a thorough review of the relevant literature for model development.
- Internal reliability is measured using Cronbach's alpha (>0.8), which is achieved for all constructs.

TABLE 8.5
Model Fit Indices

Measure	Range	Values obtained in model
CMIN/DF	<5	1.052
TLI	>0.95	0.988
CFI	>0.95	0.990
RMSEA	<0.1	0.027

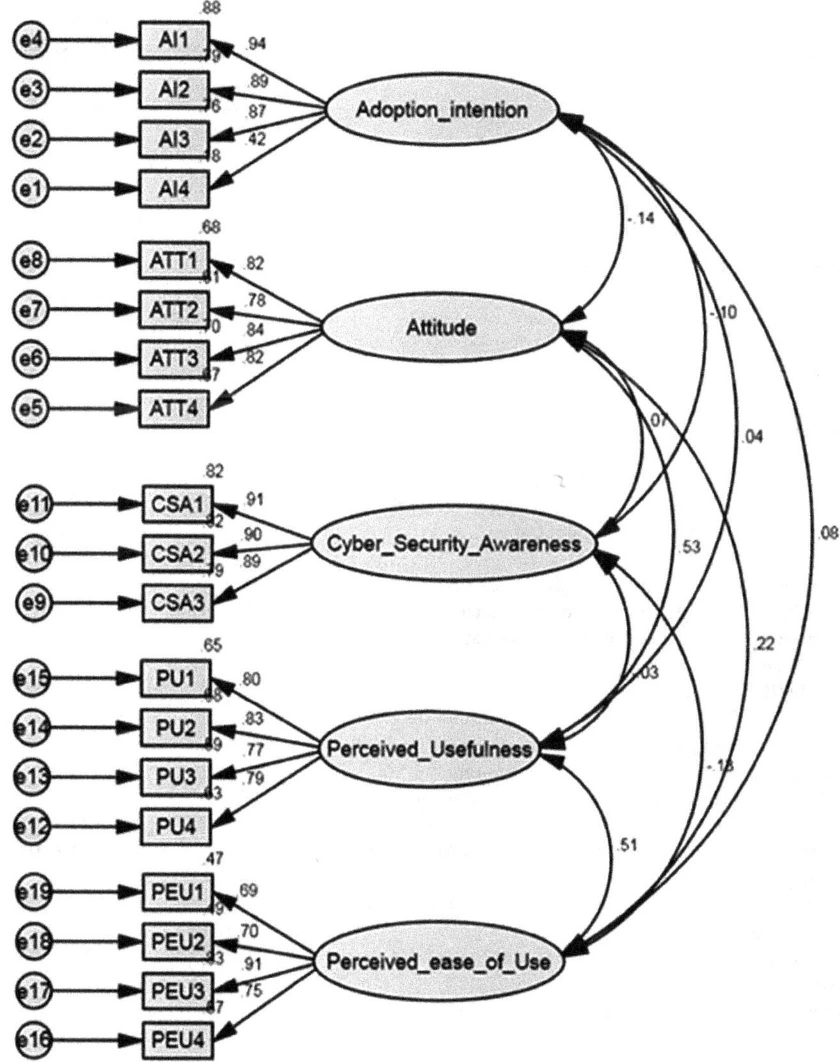

FIGURE 8.3 Confirmatory factor analysis.

- Construct reliability was analyzed using factor loadings for each component, which was > 0.7 (except for one component in the construct: perceived usefulness).
- Composite reliability (CR) should be more than 0.7.
- Convergent validity: average variance explained (AVE) should be > 0.5 for all constructs and CR should be greater than the AVE (CR > AVE).
- Discriminant validity: the AVE should be greater than the maximum shared variance (MSV), i.e., AVE > MSV. Also AVE > ASV (average shared variance) should be achieved, as presented in the correlation matrix Table 8.6 for all constructs.

From Tables 8.6 and 8.7 it can be seen that all the above validity aspects are achieved in the present model, therefore it can be further used for path analysis using SEM.

TABLE 8.6
Factor loading, CR, AVE, and ASV

Construct/items/source	Factor Loading	CR	AVE	ASV
Perceived usefulness (Davis, 1989)				
Use of cyber security measures will enhance my productivity	0.94	0.895	0.652	0.069
Use of cyber security will provide security to my data	0.89			
Use of cyber security will enhance my efficiency	0.87			
Use of cyber security will help in making my data less prone to attacks	0.42			
Perceived ease of use (Davis, 1989)				
Cyber security measures are easy to learn	0.82	0.888	0.665	0.025
Cyber security measures are easy to implement	0.78			
Cyber security measures are easy to integrate in my work	0.84			
Cyber security measures can be learnt fast	0.82			
Cyber security awareness				
I can achieve confidentiality by adoption of cyber security	0.91	0.928	0.810	0.010
I can achieve integrity by adoption of cyber security	0.90			
I can have uninterrupted availability of information by adoption of cyber security	0.89			
Attitude (Burnett et al., 2007)				
I feel positive about the cyber security measures for my LMS	0.80	0.875	0.636	0.107
I think cyber security measures will be favorable for my LMS	0.83			
It will be wise to use cyber security measures for my LMS	0.77			
It will be beneficial to use cyber security measures for my LMS	0.79			
Adoption intention				
I want to adopt cyber security measures for my LMS to help my teaching process	0.69	0.850	0.589	0.001
I plan to increase the intensity of using cyber security measures for my LMS	0.70			
For the future, I plan to use the cyber security measures for my LMS.	0.91			
For the future, I intend to use the cyber security measures for my LMS	0.75			

TABLE 8.7
Correlation Matrix and Roots of AVE

	MSV	MaxR(H)	Perceived Ease of Use	Adoption Intention	Attitude	Cyber Security	Perceived Usefulness
Perceived ease of use	0.264	0.888	**0.768**				
Adoption intention	0.023	0.936	0.079	**0.804**			
Attitude	0.283	0.890	0.217	–0.150	**0.816**		
Cyber security	0.016	0.927	–0.126	–0.110	0.067	**0.899**	
Perceived usefulness	0.283	0.877	0.514	0.037	0.532	–0.032	**0.798**

8.4.2 Data Analysis Using Structural Equation Modeling (SEM)

The model was entered in IBM AMOS for regression analysis (path analysis), as presented in Figure 8.4. The values of the regression coefficient (r^2) obtained led to hypothesis testing. From the values and significance it can be seen that perceived usefulness significantly affects attitude towards cyber security measures and that perceived ease of use affects the perceived usefulness. Therefore Hypothesis 1 and Hypothesis 3 are accepted. Hypothesis 2 is rejected due to the insignificant regression coefficient. Cyber security awareness significantly affects the attitude towards cyber security measures; however, it is very low (7%). Hypothesis 4 can be accepted. Attitude towards cyber security measures does not affect the intention to adopt them in the present case and hence Hypothesis 5 is also rejected.

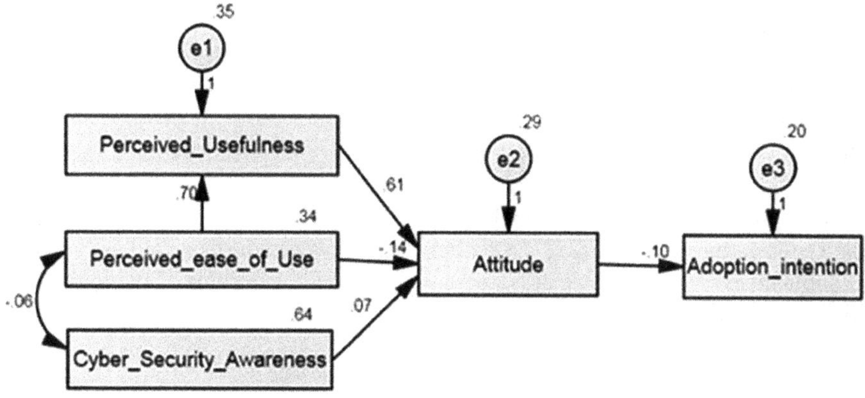

FIGURE 8.4 Regression path analysis.

8.5 RESULTS AND DISCUSSION

Cyber physical systems need to be successfully and securely implemented for education, delivery, and dissemination through the stakeholders. Amongst these stakeholders, the role of instructors is the most significant. This has been explored in the chapter, and the previous section highlighted the importance of security adoption mechanisms by instructors as participants. We can also spread the awareness of the vulnerabilities discussed in the literature review section to mitigate the respective exploitation of LMSs and thereby CPSs.

8.5.1 Theoretical Implications

This study has attempted to make a significant contribution to the existing literature of cyber security in LMSs. Based on the theory proposed, and the case study presented, the authors have developed an understanding about cyber security measures and the role of participants. With the given architecture of LMSs and the possible attacks, every participant is required to segment data and have a strict approach to sharing the information with other participants, thereby encouraging the adoption of and the intention for a given LMS.

8.5.2 Practical Implications

The findings of this study could be used to effectively implement the cyber security measures in LMSs. With the empirical investigation of the model shown in Figure 8.4, it can be seen that participants' perceived usefulness and cyber security awareness play a significant role in forming their positive attitude towards its adoption. In addition, the perceived ease of use affects the perceived usefulness of cyber security measures. Not all participants are well versed with the technology and cyber security. Therefore, there is a need for cyber security awareness among participants so that they can understand and mitigate the risk and threats associated with LMSs. Further, it may help in formulating the appropriate policies related to cyber security in an educational institute.

8.6 LIMITATIONS AND FUTURE SCOPE

Despite our efforts to present a holistic approach to cyber security measures, in no way do we claim it to be a complete one. The case study is confined to urban participants, therefore the findings cannot be generalized to rural areas. In the present study the focus is on the instructors as participants; however, further studies could be undertaken to explore the role of learners. Another interesting area of study for cyber security measures can be delivery of sessions via gamification.

REFERENCES

Al-Janabi, I., & Al-Shourbaji, J. (2016). A study of cyber security awareness in educational environment in the Middle East. *Journal of Information & Knowledge Management, 5*(1), 1650007-1–1650007-30.

Alkhalaf, S., Drew, S., & Alhussain T. (2012). Assessing the impact of e-learning systems on learners: A survey study in the KSA. *Procedia - Social and Behavioral Sciences, 47*, 98–104.

Ariu, D., Frumento, E., & Fumera, G. (2016). Social Engineering 2.0: A Foundational Work. In *Proceedings of the Computing Frontiers Conference, Combatting Cybercrime and Cyberterrorism - Challenges, Trends and Priorities,* 319–325.

Bachir S., & Abenia A. 2019. Internet of everything and educational cyber physical systems for university 4.0. In N. Nguyen, R. Chbeir, E. Exposito, P. Aniorté, B. Trawiński. (Eds) *Computational Collective Intelligence. ICCCI 2019. Lecture notes in computer science,* 11684. Springer. https://doi.org/10.1007/978-3-030-28374-2_50.

Bada, M., Sasse, A. M., & Nurse, J. R.C. (2015). Cyber security awareness campaigns: Why do they fail to change behaviour? In *Proceedings of Conference: International Conference on Cyber Security for Sustainable Society Projects.* 118–131.

Bandara, I., Loras F., & Maher, K. (2014). Cyber security concerns in e-learning education. In *Proceedings of International Conference of Education, Research and Innovation.* 0728–0734.

Bisson, D. (2020). Vulnerabilities in LMS plugins allow students to access records and edit data. Online from https://securityintelligence.com/news/vulnerabilities-in-lms-plugins-allow-students-to-access-records-edit-data/. Retrieved on January 12 2021.

Burnett, K., Bonnici, L.J., Miksa, S.D., & Kim, J. (2007). Frequency, intensity and topicality in online learning: an exploration of the interaction dimensions that contribute to student satisfaction in online learning. *Journal of Education for Library and Information Science, 48* (1), 21–35.

Burov O., Krylova-Grek Y., Lavrov E., Orliyk O., Lytvynova S., & Pinchuk O. (2021). Cyber safety in the digital educational environment: external and internal risks. In: D. Russo, T. Ahram, W. Karwowski, G. Di Bucchianico, R. Taiar. (eds) *Intelligent Human Systems Integration 2021. IHSI 2021. Advances in Intelligent Systems and Computing,* 1322. Springer.

Cuervo, M.C., Alarcón-Aldana, A.C., & López, A. B. (2016). Security evaluation model for virtual learning environments. In *Proceedings of XI Latin American Conference on Learning Objects and Technology (LACLO)*, San Carlos, 2016. 1–6.

Davis, F.D. (1989). Perceived usefulness, perceived ease of use and user acceptance of information technology. *MIS Quarterly, 3*(1), 319–340.

Elmaghraby, A.S., Losavlo, M. 2014. Cyber security challenges in smart cities: Safety, security and privacy. *Journal of Advanced Research, 5*(4), 491–497.

Glushkova, T., Todorov, J., Doychev, E., & Stoyanov, S. (2018). Implementing an internet of things e-learning ecosystem. In *Proceedings of the 44th International Conference on Applications of Mathematics in Engineering and Economics* AIP Conf. Proc. 2048, 020027-1-020027-8.

Gupta, M., & Sharman, R. (2008). *Social and human elements of information security: Emerging trends and countermeasures.* IGI Global.

Hadlington, L. (2017). Human factors in cybersecurity; examining the link between Internet addiction, impulsivity, attitudes towards cybersecurity, and risky cybersecurity behaviours. Online from http://dx.doi.org/10.1016/j.heliyon.2017. Retrieved on January 2 2021.

Hair, J.F., Black, W.C., Babin, B.J., & Anderson, R.E. (2010). *Multivariate data analysis.* Seventh Edition. New Jersey.

Jones, C. M., McCarthy, R.V., Halawi, L., & Mujtaba, B. (2010). Utilizing the technology acceptance model to assess the employee adoption of information systems security measures. *Issues in Information Systems, 11*(1), 9–16.

Khan, M., Naz, T., & Asir, M. (2019). A multi-layered security model for learning management system. *International Journal of Advanced Computer Science and Applications.* 10 (12), 207–211.

Kok, L. C., Oosting, D., & Spruit, M. (2020). The influence of knowledge and attitude on intention to adopt cybersecure behaviour. *Information and Security, 46*(3), 251–266.

Leia, C. Wanb, K., & Manb, K. L. (2013). Developing a smart learning environment in universities via cyber-physical systems. *Procedia Computer Science, 17*, 583–585.

Luminita, C.C., & Magdalena, C. N. (2012). E-learning security vulnerabilities. *Procedia - Social and Behavioral Sciences, 46*, 2297–2301.

Mourtzis, D., Vlachou, E., Dimitrackpoulos, G., Zogopoulos, V. 2018. Cyber-physical systems and education 4.0 – The teaching factory 4.0 concept. *Procedia Manufacturing, 23*, 129–134.

Nickolova, M., & Nickolov, E. (2007). Threat model for user security in e-learning systems. *International Journal Information Technologies and Knowledge*, 1, 341–347.

Rabai, l.B.A., Rjaibi, N., & Aissa, A. B. 2012. Quantifying security threats for e-learning systems. *International Conference on Education and e-Learning Innovations*, Sousse, 1–6.

Rahima, N., Othmanb, Z., Hamidc, F. Z., & Yeop, O. (2020). Cyber security and the higher education literature: A bibliometric analysis. *International Journal of Innovation, Creativity and Change, 12*(12), 852–870.

Salimovna, F.D., Yuldasheva, N., & Zokirugli, I.S. (2019). *Security issues in E-Learning system.* In *Proceedings of International Conference on Information Science and Communications Technologies (ICISCT)* 2019. 1–5.

Santos, R., Devincenzi, S., Bothelo, S., & Bichet, M. (2017). A model for implementation of educational cyber physical systems. *Espacios, 39*(10), 36–54.

Seuwou, P., Banissi, E., & Ubakanma, G. (2016). User acceptance of information technology: A critical review of technology acceptance models and the decision to invest in information technology. In Jahankhani, H., Carlile, A., Emm, D., Hosseinian-Far, A., Brown, G., Sexton, G, & Jamal, A. (Eds.), *Global Security, Safety, and Sustainability—The Security Challenges of the Connected World*, Springer International Publishing, 1–22.

Stankovic, J.A., Sturges, J.W., & Eisenberg, J. (2017). A 21st century cyber-physical systems education. *Computer, 50*(12), 82–85.

Torngren, M., Bensalim, S., McDermid, J., Passerone, R., Vincentelli, A.S., & Schatz, B. (2015). Education and training challenges in the era of Cyber-Physical Systems: Beyond traditional engineering. In *Proceedings of the WESE'15: Workshop on Embedded and Cyber-Physical Systems Education, 8*, 1–5. https://doi.org/10.1145/2832920.2832928.

Trustradious. (2021). Moodle vs Sakai. Online from https://www.trustradius.com/compare-products/moodle-vs-sakai. Retrieved on February 10 2021.

Wang, P.A. (2010). Information security knowledge and behavior: An adapted model of technology acceptance. In *Proceedings of 2nd International Conference on Education Technology and Computer.* Shanghai, 364–367.

9 A Security Model for Cloud-computing-based E-governance Applications

Aditya Makwe
Institute of Engineering and Technology DAVV, Indore, MP

Anand More
Computer Centre DAVV, Indore, MP

Priyesh Kanungo
SCSIT DAVV, Indore

Niranjan Shrivastava
IMS DAVV, Indore

CONTENTS

9.1 Introduction ... 133
9.2 Security of e-governance Applications ... 135
9.3 Related Work ... 136
9.4 Threats to the Security of e-governance Services ... 137
9.5 Security Model for E-governance ... 138
  9.5.1 Firewall and Access Control ... 139
  9.5.2 Intrusion Detection and Intrusion Prevention System (IDS/IPS) ... 139
  9.5.3 Encryption ... 140
9.6 Experiment and Results Analysis ... 141
9.7 Conclusion ... 142
9.8 Limitation and Future Scope ... 143
References ... 145

9.1 INTRODUCTION

In the present scenario, e-governance-based applications require high performance computation models for maintaining and executing the data generated from applications. With the rise of high performance computation technologies like cluster

DOI: 10.1201/9781003146711-9

computing, grid computing, and cloud computing, services can be provided to the user cost effectively and with a reduced overhead for maintenance. For implementing these models high security features must be considered. E-governance provides interactive online services to citizens and businesses in developing countries like India. This requires the use of various technologies like networking, the Internet, and mobile computing by the government. These services are required by various stakeholders – such as citizens, business, and industry, employees, and other parts of government – in a faster and more reliable manner. The term "e-governance" is generally used for Web services in various departments of government, where the citizens of the country are able to conveniently and rapidly access this Web enabled service for official processes. The users may be interested either in getting some information or performing activities or making related payments using these applications (Sharma, 2006). Various categories of e-governance services include (More & Kanungo, 2016):

1. **Government to citizens or G2C:** These are the activities where Web-based services are provided to citizens by the government. Government to citizen applications empower users to: raise queries about government activities and to receive replies to them; pay direct taxes (income tax, house tax, water tax, etc.); pay traffic tickets; make or renew driver licenses; update addresses and vehicle pollution control inspections; and make appointments for passports, driving licenses, and so on. Various services provided by government to its citizens include agriculture services, land records, rural services, and municipal services. An important advantage of G2C applications are 24/7 availability of government information and services through the Internet.
2. **Government to business or G2B:** G2B refers to two-way interactions and transactions, i.e. government to business (G2B) and business to government (B2G). Thus, the government is able to communicate with businesses using the Internet. B2G refers to businesses that are providing products and services to government. Government is responsible for policy enforcement, taxation, and entering into contracts with various businesses. Some common examples of G2B are online auctions and the e-procurement of stocks of various items needed by the government.
3. **Government to employees or G2E:** With the help of G2E, enterprises are able to increase their effectiveness and efficiency, by improving employee satisfaction and retention.
4. **Government to government or G2G:** This refers to the activities among various governments and among government departments. A number of such activities have been created to improve the effectiveness and working of overall government functions to serve the country. These applications provide interaction among various government departments and require a large number of messages to be passed between them. G2G enables government in solving many problems like controlling unemployment and crime while improving security.

E-governance is the real-time delivery of multiple government utilities and its services to users over the Internet. This availability requires security infrastructure so that the services can be made available all the time to the user in a secured easy-to-use manner. Cloud computing enables a convenient platform which provides the benefits of cost saving, improved work efficiencies, agile secured business, and quality of services (QoS) (Pokharel & Park, 2009).

E-governance services can be implemented using the software as a service (SaaS), platform as a service (PaaS), and infrastructure as a service (IaaS) models of the cloud. With a sharp rise in industries on digital platforms the demand for various services, particularly SaaS, has increased exponentially. For example, e-commerce industries and Internet banking use cloud computing for providing services to the user. Due to the increasing use of the cloud the necessity of providing services to the user in a secured way is also increasing. While providing e-governance services, the data center encounters many problems like deciding the priority of jobs, management of physical and virtual machines, secure scheduling of user jobs on virtual machines, and providing output to users in real time (Xiao et al., 2012).

While providing e-governance services, security and privacy becomes an important factor for users for ensuring effective data sharing and protection. Providing real-time secured services to the user requires a smart infrastructure. Designing such an environment is a very cumbersome process because the organization has to pay attention not only to security but also to the optimal execution of user tasks.

In this chapter, the problem of the real-time secured execution of e-governance jobs is addressed. The real-time output of execution depends on various factors like the time of applying security measures (e.g., firewalls, intrusion detection and intrusion prevention systems (IDS/IPS), encryption), deciding the priority of jobs (filtering high priority jobs), and execution time on virtual machines (VMs). Apart from that, the various factors affecting the security of e-governance applications is also discussed in the chapter. The proposed architecture involves three layers. The first layer describes the security of user jobs; the second layer decides the priority of user jobs (filtering out the e-governance jobs from other jobs in the cloud); and a third layer is responsible for the execution of sorted jobs on VMs.

9.2 SECURITY OF E-GOVERNANCE APPLICATIONS

The perception of computation as a utility based on a service provisioning model has provoked an enormous change in the computing industry by which computing services can be made available on demand to the user without any difficulty. According to the National Institute of Standards and Technology (NIST) the cloud provides convenient and on demand Internet-based access to cloud resources. These resources may include servers, storage, bandwidth, and applications. They can be released instantly and with minimum interaction (Mell and Grance, 2011).

Cloud computing provides services in a pay-as-you-use model to consumers. While providing services to users, all internal infrastructure and its functionalities are hidden. Users only experience the services provided via the Internet and a browser. Cloud computing provides various computational resources like storage, servers, and platforms on a rental basis.

For SME organizations, buying a huge infrastructure requires a lot of money and technical skill for maintaining them. Moreover, an optimal utilization of all the resources with less power is not possible for such industries. All such problems can be easily managed by using the cloud computing model. For accessing e-governance services, the user will acquire their login credentials when they start accessing the services. After login, they submit their jobs to the cloud which returns the results to the user.

Although the cloud offers various utilities to the user, there are few issues and challenges that need to be addressed before migrating to it. This is essential for both the users who are accessing services, and the organizations which are hosting their services on the cloud. Security threats have a serious impact on the growth of the cloud. Users don't consider the internal factors that the cloud does while providing services. They only see the benefits and convenience of use. Data in the cloud are stored in remote servers which may be geographically located in some other country from where the user is accessing the services. An attack on these remote servers can prove to be vulnerable to the privacy and security of user data. Therefore, cloud service providers need to address all these issues for users (either general users or organizational users).

9.3 RELATED WORK

While providing cloud services, security and privacy become important factors for ensuring that data are protected. Providing real-time secured services to the user requires a smart infrastructure. Designing such an environment is a cumbersome process as the organizations that adopt the cloud must pay attention not only to security but also to the optimal execution of user tasks.

Various scientific and business computation models, like climate modeling, weather forecasting, and e-commerce, require parallel processing of data on high performance computing platforms like cloud computing. Along with this, it is equally important to address the related security issues. Patra and Chakroborty have highlighted cloud-related challenges and security concerns regarding big data applications. With the help of encryption techniques like fully homomorphic encryption and functional encryption, one can securely deploy such applications in a cloud infrastructure (Patra & Chakraborty, 2014).

Cloud computing provides a platform for business models without an upfront investment. The cloud has various advantages, which are still growing. With the rise of various new dimensions the cloud, like multi-tenancy and elasticity, which provide security to the infrastructure, has become more challenging. Mohamed, Grundy, and Müller discussed cloud security problems from a characteristics perspective, such as service delivery and stack-holders, and determined the key problems and features that must be considered when providing security to the cloud.

A number of concerns related to privacy and security raise barriers when providing cloud services. Users must be familiar with these security problems that exist in the cloud. Various features, like virtualization and multi-tenancy, provide access to cloud resources present in data centers in different locations. A proper isolation between VMs needs to be maintained to preserve the security of the system. Singh & Chatterjee discussed various important cloud features of the security concerns caused by its distributed, virtualized, and public nature. Authors also discussed various solutions that are needed for privacy and security problems (Singh & Chatterjee, 2015).

The use of cloud computing reveals many security threats and access control issues for users. Cloud servers are used by the cloud service provider to store sensitive user data. Users don't have to worry about its trustiness, since that is a responsibility of the cloud service provider. Existing solutions like cryptography involves computational overheads such as key distribution and don't address performance, confidentiality, and scalability issues simultaneously. Various security techniques like lazy re-encryption, proxy re-encryption, and attribute-based encryption may be used to achieve security related goals. The user acquires secret key accountability and access privilege confidentiality using the proposed scheme (Yu et al., 2010).

Varadharajan & Tupakula discuss problems in implementing a secured cloud computing infrastructure against various threats and attacks. The authors also discussed different types of security service that a service provider must offer to its users. They proposed a security model which offers a basic security mechanism for the protection of the data centers of cloud service providers and provides options to users to satisfy their security needs (Varadharajan & Tupakula, 2014).

9.4 THREATS TO THE SECURITY OF E-GOVERNANCE SERVICES

State-of-the-art computing technologies like cloud computing, fog computing, and roof computing promise efficient, agile, cost effective, and secured delivery of services to organizational users for efficiently operating their businesses. These promises can be fulfilled only if the service provider does address security issues adequately. For the business adaptation of cloud computing, some security threats need to be addressed. To identify the most serious security threats in cloud computing, the Cloud Security Alliance conducted a survey of industrial experts to find out the weaknesses in the security of the cloud. In the recent release updates, the following security threats have been identified.

1. **Data breaches:** A data breach is a security issue in which information is accessed and modified without authorization. When the cloud stores vital information related to the user, it must ensure the highest level of security regarding authentication and authorization – for user data protection. If vulnerability in clients' applications persist, this allows attackers to access not only clients' precious data but also other data stored in the cloud.
2. **Denial of services:** This is the most general type of attack in which the attacker makes services unavailable for the end user by temporarily disrupting the resources which provide services to the user.
3. **Data loss:** It could be dangerous for the organization to lose user data. Data stored in the cloud can be lost due to various reasons like a malicious attack or a natural disaster, such as an earthquake or a flood. The cloud needs to take various counter-measures for protecting user data, like backup at the server at some other geographical location.

4. **Insecure application program interfaces (APIs):** Cloud service providers use API for providing services to the user. Authentication and authorization of user identity is performed using these APIs. Apart from this, API also helps in encryption and activity monitoring. The design of APIs must include these features.
5. **Account hijacking:** This is a type of activity in which attackers steal user's information for carrying out malicious or unauthorized activities. Various methods, like phishing and spoofing, are used for obtain the information.
6. **Malicious insiders:** The cloud stores all the vital user information for accessing its services. This information is maintained by an administrator, who has access to all the information by default which he or she can leak to attackers for personal interest or when under pressure.
7. **Insufficient due diligence:** Prior to migrating to the cloud, it is essential for an organization to understand the security measures used by a cloud service provider. Many of these organizations often don't fully understand how their services are going to be secured when making the decision to migrate.
8. **Shared technology issue:** The cloud provides services by sharing its infrastructure, platform, and application with the user. While providing these services either in IaaS, PaaS, or SaaS, complete security needs to be implemented. A single vulnerability may lead to comprising an entire cloud infrastructure.
9. **Abuse of services:** The cloud provides all type of resources to users on demand. An infinite availability of resources and elastic properties of the cloud allows even a small organization to access resources on rent. However, if the organization doesn't need these huge infrastructures and if attackers have access to the organization, then cloud infrastructure becomes vulnerable to attack.

9.5 SECURITY MODEL FOR E-GOVERNANCE

Security and privacy have been challenges in the implementation of e-governance applications and are of primary concern. Executing user jobs cost effectively while securing data is a major objective of all cloud service providers. The internal implementation and infrastructure are transparent to the user. Therefore, the selection of the service provider is a major issue in the implementation of e-governance.

To address the above problems, a security model is required that not only executes e-governance applications in a secured way but also supports an optimal execution. Figure 9.1 shows a model for the secured execution of user jobs. The proposed model consists of three layers, that is layer 1, layer 2, and layer 3. Each layer has its own functionality and helps in achieving optimal and secured execution of user jobs. When a user wants to access the resources, he or she has to go through three layers. Layer 1 provides security to user jobs and identifies the users by the service provider. Once the user is authenticated, the second layer decides on the priority of user jobs by assigning a higher priority to the e-governance jobs. After deciding the priority, layer three executes the e-governance jobs on the VMs using a scheduling policy. A complete description of each of these layers is given in the following.

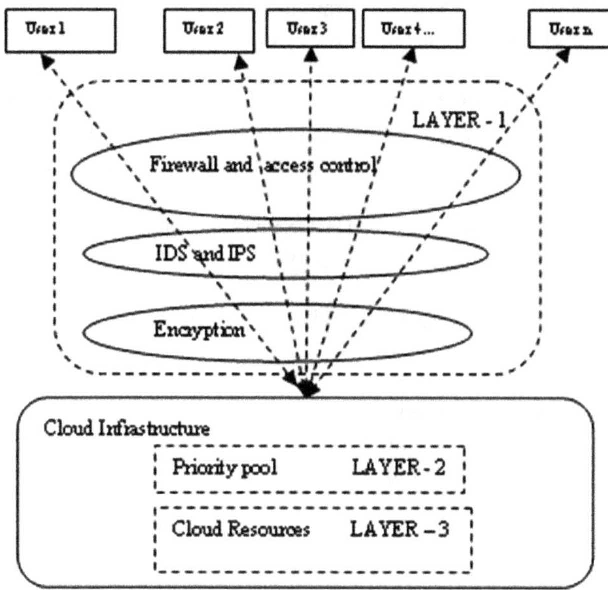

FIGURE 9.1 Security model for the execution of e-governance.

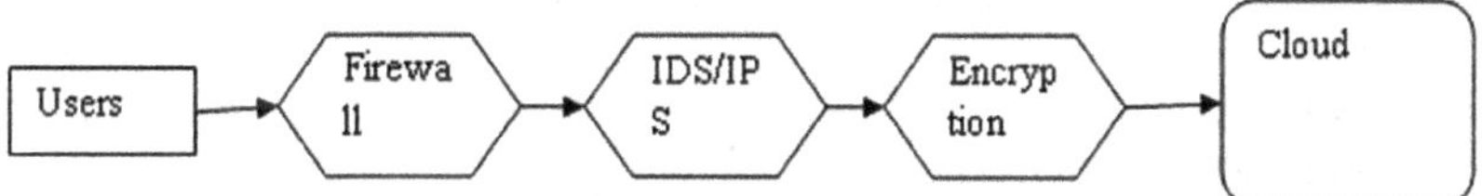

FIGURE 9.2 Layer 1 security implementation.

The layer 1 model provides the security aspects related to the user and consists of three stages as shown in Figure 9.2. The stage-wise description of layer 1 is as follows.

9.5.1 Firewall and Access Control

The first stage is the firewall and access control, which is restricted to members. The typical functionality of the firewall is deep packet inspection, which checks the header of the packet as a shallow inspection and the payload as a deep packet inspection. The traffic is processed and matched with an existing set of signatures. The processing of the packet depends upon the traffic in the network.

9.5.2 Intrusion Detection and Intrusion Prevention System (IDS/IPS)

The second stage is the IDS/IPS and the function here is to detect and prevent any possible attack or anomalous activity such as port scanning, denial of service (DoS), anti-spoofing, pattern-based attack, parameter tampering, and cookie poisoning in the cloud. A typical IDS has the following components:

1. Network intrusion detection system (NIDS) which matches the traffic that passes through the network with traffic with known attacks.
2. Network node intrusion detection system (NNIDS) which performs the monitoring of traffic on a single node.
3. Host intrusion detection system (HIDS) which monitors the activities of a particular host or machine within the network.

Besides the above components, IDS also has the following properties:

1. Monitoring users and system activities.
2. Auditing the system configuration and vulnerabilities.
3. Accessing critical system integrity and data files.
4. Statistical analysis of system activities.

9.5.3 Encryption

This feature monitors and provides end-to-end continuous assurance of the security of user data and can be achieved by using various cryptographic techniques. The cloud not only stores the data but also performs processing on stored data. Thus a scheme is needed that enables performing computation on the encrypted data. For processing of data, the traditional encryption algorithms can't be used. To support such an environment, a fully homomorphic encryption scheme is used which allows operations on the encrypted data (Gentry, 2009).

Layer 2 of the security model helps in deciding the priority of jobs in the cloud environment. Figure 9.3 depicts the cloud environment that receives user requests at

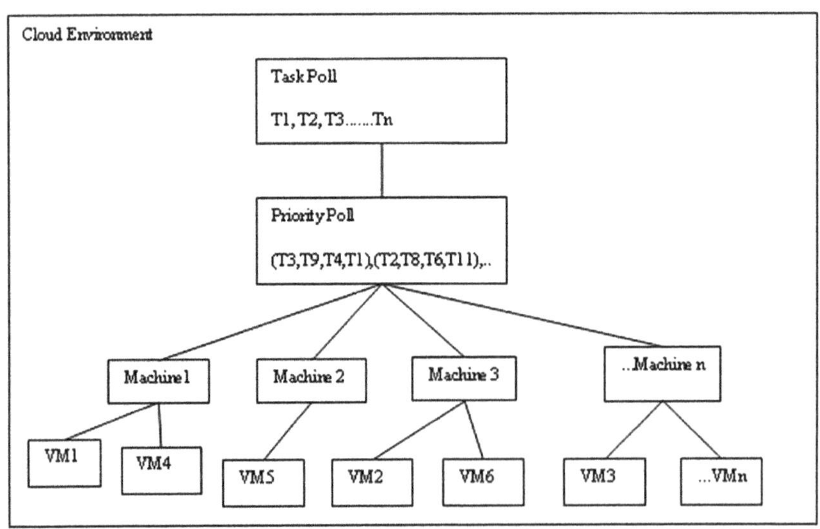

FIGURE 9.3 Layer 2 of e-governance security model.

a single instant of time for acquiring the resources. Providing resources to the user in real time is a trivial task for service providers. To solve this problem, the priority of jobs needs to be decided. The proposed model uses the concept of a multicriteria decision approach to solve the problem. An analytical hierarchy process (AHP) model is used for determining the sequence of user tasks to be scheduled on physical machines (Makwe and Kanungo, 2015). The comparison of tasks based on the definition and the sequence of tasks is decided by applying AHP. This method is appropriate for scheduling tasks of a different nature based on varying resource needs, for example, computing power, primary memory, and network speed.

Once the priority set of tasks to be scheduled on different physical machines is decided, layer 3 of the security model creates the VMs that correspond to the physical machines. Thereafter, tasks are scheduled on the VMs by using various types of scheduling schemes, like first come first serve (FCFS) and shortest job first (SJF).

9.6 EXPERIMENT AND RESULTS ANALYSIS

For real-time execution of user tasks, the processing time of all three layers needs to be calculated. The overall execution time of the proposed model is a function of three factors:

$$\text{Ft}(\text{model}) = \text{T1}(\text{layer}\,1) + \text{T2}(\text{layer}\,2) + \text{T3}(\text{layer}\,3)$$

The time complexity of the model can be reduced by minimizing the function defined in this equation. T1(layer1) defines the time taken by the firewall to process the incoming data packets and to determine their authenticity. T2(layer2) defines the time used for deciding the priority of the user task on the physical machine. T3(layer3) defines the time used for scheduling the task on a VM. The focus of the study was to reduce the time at layer 2 and layer 3, so that the security of the data is not compromised. Parameters that affect the time complexity of the firewall are the set of rules defined for the firewall and the time taken to search the rule while processing the packet. Various search algorithms like linear search, binary search, tuple space search (Srinivasan, Suri & Varghese, 1999), aggregated bit vector (Boescu & Varghese, 2001), and hierarchical space mapping (Xu, Jiang, and Li, 2005) can be used. The time complexity of layer 2 depends on the number of parameters for determining the priority of user tasks. An average time complexity of an AHP model is $O(n^2)$. As the value of the parameter increases, the time complexity also increases. At layer 3, the time complexity depends on the policy used for scheduling user tasks on the VM. In this study, a simulation tool kit called CloudSim (Calheiros et al., 2011) was used and user tasks were scheduled by using FCFS and SJF algorithms. The setup for the experiment included the data center having multiple physical machines with two types of specifications. Machine 1 had 4 GB of RAM, 100 GB of storage, and a Quad core processor. Machine 2 had 2 GB of RAM, 100 GB of storage, and a dual core processor. When the cloud has no request to process, these machines remain in an off state. When the request arrives, the cloud service manager starts these physical machines as per the requirement for creating VMs over these physical machines.

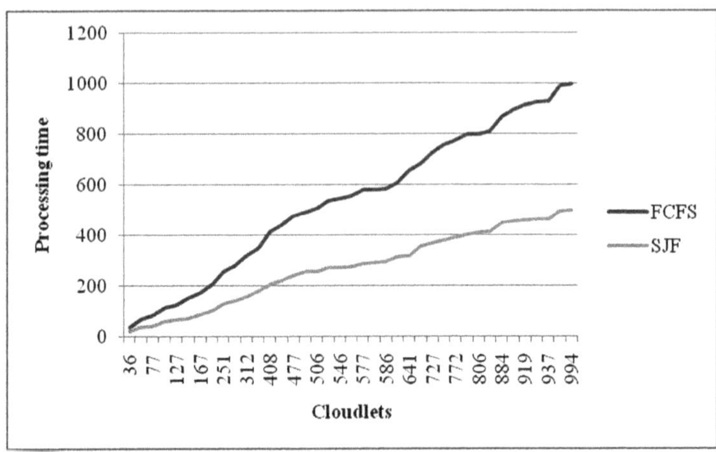

FIGURE 9.4 Comparison of processing time of FCFS and SJF under case 1.

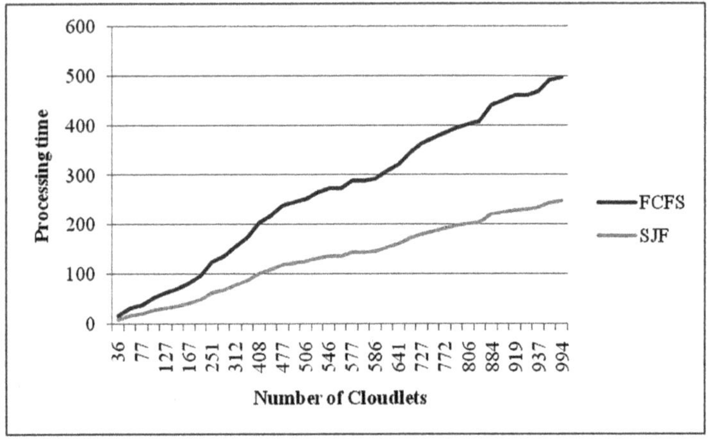

FIGURE 9.5 Comparison of processing time of FCFS and SJF under case 2.

In this study, the performance of FCFS and SJF with random scheduling were compared for the following two cases: (1) when user tasks compete for RAM; (2) when user tasks compete for processing power. Figures 9.4 and Figure 9.5 depict the performance of SJF and FCFS policy, and the performance of algorithms evaluated over 1000 cloudlets, corresponding to the execution time of jobs. The results of SJF policy gave better results as compared to FCFS scheduling policy.

9.7 CONCLUSION

In this chapter, the problem of the secured execution of e-governance applications was addressed. The real-time output of execution depends on various factors such as applying security measures (firewall, IDS/IPS, encryption), deciding the priority of jobs (filtering high priority jobs), and execution time on VMs. Moreover, various factors concerning the security of the cloud have also been discussed. The e-governance security model proposed involves three layers. Layer 1 provides security to user jobs and identifies the users by the service provider; the second layer decides the priority of user jobs; and the third layer executes the e-governance jobs on VMs.

The overall execution time of the security model can also be computed. This execution time is divided into three parts (i.e., on each layer). Layer 1 time depends upon the processing time of the firewall. The execution time of the second layer depends on the various parameters considered for comparing the tasks. In layer three, various scheduling policies are implemented depending upon the nature of the requests.

9.8 LIMITATION AND FUTURE SCOPE

E-governance is an important area of research. A number of state-of-the-art technologies and techniques can be used to improve the effectiveness of e-governance applications. High performance computing with Web technologies can provide a better infrastructure for the optimal execution of these applications at less cost and overheads. Still, there are some areas that are needed to be focus on in e-governance:

1. **Using big data in e-governance:** Conventional relational database management system (RDBMS) databases, such as SQL Server, Oracle, and MySQL, have several limitations for e-governance applications when compared with big data solutions developed over the last ten years. With the exponential growth of data in e-governance applications, the need for using machine learning and big data analysis is also increasing. The development of NoSQL databases recently has improved the big data analytics technique. In e-governance applications, the focus is on the processing of structured and unstructured data. Therefore, NoSQL is the best choice for moving towards big data in these applications. For the integration of NoSQL with e-governance applications, various NoSQL databases, like CouchDB, MongoDB, and Cassandra, are available in the open source community. Though all these databases have common features, their performance varies according to specific parameters. Our future plan is to explore the performance of these databases on e-governance applications (Rajagopalan & Vellaipandiyan, 2013).
2. **Automation of e-governance applications using artificial intelligence:** The use of artificial intelligence (AI) can make e-governance effective for a broad category of services by providing advanced and sophisticated solutions for preserving security and privacy and identifying lapses in policy regulations. AI systems can also provide transparency to optimize government functioning. E-governance can help in detecting anomalies using AI to prevent fraud and secure data to automate manual tasks. Machine learning together with AI helps in many e-governance activities (e.g., enforcement of law and order using facial recognition, speech recognition, drones, predictive analytics, and cyber defense). The use of these techniques provides knowledge about violations of the law. These systems can help in solving issues rapidly.

Another example could be disaster management with the help of AI automated systems and predictive data analytics. Weather forecasting might help with preparatory action regarding government response during natural calamities or major accidents and also by managing public infrastructure to help in effective operations. Maintaining an AI-driven public platform helps in solving public queries at faster speed. The public has 24/7 access to information regarding every government service.

3. **Application of machine learning in e-governance:** Machine learning can enable e-learning to handle complex data so that it can be used for strategic data-driven decision making. We need an exhaustive approach to focus on using machine learning technology in e-governance applications (Lachana et al., 2018).
4. **The Internet of Things and fog computing in e-governance:** E-governance systems can provide input for a number of services with the help of the Internet of Things (IoT). Devices connected with the IoT have a number of problems like network latency, privacy and security, network speed, and high speed computations with cloud computing. The challenges can be addressed by fog computing architecture for e-governance. In fog computing, the computations are performed at a data collection source to reduce the latency. Fog enhances network speeds among IoT devices and data centers. The use of the IoT and fog computing enables organizations to move towards smart governance (Tewari & Datt, 2020).
5. **Edge computing for e-governance:** Modern e-governance applications need to improve their quality of service and delivery speed. Governments are undergoing critical challenges to improve resource allocation, service response, storage of data, and cost effectiveness. Edge data centers have the potential to address these needs. In e-governance services, in various sectors including education, healthcare, agriculture, and other business functions, edge data centers may improve the response, availability, and utilization of resources.
6. **E-governance for smart cities:** Smart cities provide a high level of services in an efficient, effective, and transparent manner for balanced regional development and requires linkages among different government departments. Last mile connectivity can be enhanced by mobile governance which also enables the collection and analysis of data. Participation can be improved using e-governance. Cities can have not just a Web presence but also smart governance.
7. **Quantum Internet and e-governance:** Information technology is being used by almost all citizens for e-commerce, online banking, and social media applications like WhatsApp, Twitter, and Facebook. However, this also gives rise to many security and privacy concerns. The conventional algorithms for security based on cryptography have now become vulnerable due to the evolution of high speed computation techniques like quantum computing, causing security threats to existing e-governance applications. To address these issues, we need to consider new security mechanisms like quantum cryptography and the quantum Internet. Therefore, we need to further study e-governance challenges and solutions and its relation to the evolution of the quantum Internet (Thakkar & Vanzara, 2020).

REFERENCES

Almorsy, M., & Grundy, J., Müller, I. (2016). An analysis of the cloud computing security problem. *arXiv preprint arXiv:1609.01107.*

Calheiros, R. N., Ranjan, R., Beloglazov, A., De Rose, C. A., & Buyya, R. (2011). CloudSim: A toolkit for modeling and simulation of cloud computing environments and evaluation of resource provisioning algorithms. *Software: Practice and experience, 41*(1), 23–50.

Gentry, C. (2009). Fully homomorphic encryption using ideal lattices. *Proceedings of the Forty-first Annual ACM Symposium on the Theory of Computing*, 169–178.

Lachana, Z., Alexopoulos, C., Loukis, E., & Charalabidis, Y. (2018). Identifying the different generations of Egovernment: an analysis framework. *The 12th Mediterranean Conference on Information Systems,* 1–13.

Makwe, A., & Kanungo, P. (2015). Scheduling in cloud computing environment using analytic hierarchy process model. *International Conference on Computer, Communication and Control,* 1–4.

Mell, P. M., & Grance, T. (2011). Sp 800-145. The NIST definition of cloud computing.

More, A., & Kanungo, P. (2016). Various e-governance applications, computing architecture and implementation barriers. *Proceedings of the International Congress on Information and Communication Technology*, Springer, Singapore, 635–643.

Patra, G. K., & Chakraborty, N. (2014). Securing cloud infrastructure for high performance scientific computations using cryptographic techniques. *International Journal of Advanced Computer Research*, 4(1), 66–73.

Pokharel, M., & Park, J. S. (2009). Cloud computing: future solution for e-governance. *Proceedings of the 3rd International Conference on Theory and Practice of Electronic Governance*, 409–410.

Rajagopalan, M. R., & Vellaipandiyan, S. (2013). Big data framework for national e-governance plan. *Eleventh International Conference on ICT and Knowledge Engineering, IEEE,* 1–5.

Sharma, S. K. (2006). E-Government services framework. *Encyclopedia of E-commerce E-government and Mobile Commerce* IGI Global, 373–378.

Singh, A., Chatterjee, K. (2015). Cloud security issues and challenges: a survey. *Journal of Network and Computer Applications, 79*, 88–115.

Srinivasan, V., Suri, S., & Varghese, G. (1999). Packet classification using tuple space search. *Proceedings of the Conference on Applications, Technologies, Architectures, and Protocols for Computer Communication IEEE,* 135–146.

Tewari, N., & Datt, G. (2020). Towards FoT (Fog-of-Things) enabled architecture in governance: transforming e-governance to smart governance. *International Conference on Intelligent Engineering and Management IEEE,* 223–227.

Thakkar, M. D., & Vanzara, R. D. (2020). Quantum internet and e-governance: a futuristic perspective. *Quantum Cryptography and the Future of Cyber Security* IGI Global, 109–132.

Varadharajan, V., & Tupakula, U. (2014). Security as a service model for cloud environment. *IEEE Transactions on Network and Service Management, 11*(1), 60–75.

Xiao, Z., Song, W., & Chen, Q. (2012). Dynamic resource allocation using virtual machines for cloud computing environment. *IEEE Transactions on Parallel and Distributed Systems, 24*(6), 1107–1117.

Xu, B., Jiang, D., & Li, J. (2005). HSM: a fast packet classification algorithm. *19th International Conference on Advanced Information Networking and Applications*, 1, 987–992.

Yu, S., Wang, C., Ren, K., & Lou, W. (2010). Achieving secure scalable and fine-grained data access control in cloud computing. *Proceedings IEEE INFOCOM,* 1–9.

10 Automatic Time and Motion Study Using Deep Learning

Jefferson Hernandez

Industrial Artificial Intelligence (INARI) Research Lab at ESPOL, Guayaquil, Ecuador

Sofia Lopez and Gabriela Valarezo

Escuela Superior Politecnica del Litoral (ESPOL), Guayaquil, Ecuador

Andres G. Abad

Industrial Artificial Intelligence (INARI) Research Lab at ESPOL, Guayaquil, Ecuador

CONTENTS

10.1 Introduction 147
10.1.1 Research Contribution 148
10.2 Literature Review 149
10.3 Methodology 150
10.3.1 Micro-action Recognition 151
10.3.2 Macro-action Recognition 151
10.4 Case Study: Order Preparation in a Distribution Center 152
10.4.1 Action-effectiveness Evaluation 153
10.4.2 Labor-productivity Metrics 154
10.5 Conclusions 159
Acknowledgment 160
References 160

10.1 INTRODUCTION

Measuring manual-labor performance has been a key element of work scheduling and resource management, with particular relevance in areas such as manufacturing, construction, and logistics, where human labor can account for up to 50% of the total cost of a project (Gouett, Haas, Goodrum, & Caldas, 2011). It has been proposed that the biggest inefficiency losses in the workplace are due to human-effort waste (Taylor, 1911). To overcome these inefficiencies a work measurement system may be implemented during process planning to allow the performance of a wide variety of

DOI: 10.1201/9781003146711-10

operations with standard times. This procedure is called a predetermined time system (PTS) and was developed by Frederick Winslow Taylor in 1881 (Aft, 2000; Freivalds, 2009); it was later modified into a technique called a time and motion study (TMS) by Maynard, Stegemerten, and Schwab (1948) – who focused on the time aspect of these studies – and Frank and Lilian Gilbreth (Baumgart & Neuhauser, 2009) – who focused on the motion aspect of these studies.

Implementing the TMS technique in workplaces can improve the efficiency and effectiveness of manual tasks by: characterizing and simplifying their design; assessing and measuring productivity; and assisting in ergonomic evaluations and calculations of the distribution of work tasks (Meyers & Stewart, 2002).

Attempts to perform TMS usually rely on defining a hierarchy of worker tasks, allowing the abstraction of them into levels that result in a more organized observation. One such of these abstractions consists of hierarchically differentiating between *micro* and *macro* actions (Freivalds, 2009): the former are defined as the movement of a specific body part, while the latter are defined as the combination of successive micro-actions. A worker or a group of workers is selected to collect data, with post-processing often being required. Variations of this methodology have been applied to manufacturing (Al-Saleh, 2011; Chatzis, 1999), health (Finkler, Knickman, Hendrickson, Lipkin Jr, & Thompson, 1993; Harewood, Chrysostomou, Himy, & Leong, 2008), and other fields (Gunasekaran, Forker, & Kobu, 2000).

Many industries still rely on human effort to perform the TMS technique, which is usually performed by direct and detailed observation using a stopwatch and a sheet of paper to determine the time and motion utilized to perform specific tasks. Shehata and El-Gohary (2011) show that a single TMS study usually needs observations in the hundreds, taken multiple times during the day, and spanning more than half a month. Clearly, this can be time-consuming, error-prone, and expensive overall.

Specifically, the *human-in-the-loop* approach to TMS has three problems: (P1) the duration of the study is usually not enough to capture the complexity of activities commonly seen in industries (Kanawaty, 1992); (P2) it may be difficult for analysts to distinguish between micro- and macro-actions (Meyers & Stewart, 2002); and (P3) it is a costly effort that prohibit continuous measurement. In order to tackle these issues, human observers may be replaced with automatic procedures which provide a continuous stream of information (solving P1) and a finer understanding of worker activities (mitigating P2) at lower costs and increased speed (alleviating P3).

10.1.1 Research Contribution

In this chapter, we propose an automatic TMS approach to measure manual-labor performance *in situ* at two levels of abstraction: micro- and macro-actions. We present a system (Section 10.2) that uses deep neural networks to infer a 3D skeleton model of the workers from RGB videos performing various tasks. Likewise, object detection is used to extract features of objects held in the workers' hands. This information is combined and fed to a neural network to classify the micro-actions (see Figure 10.1). These micro-actions are used in a continuous-time hidden Markov model (CT-HMM) to identify macro-actions. We show that the proposed system can

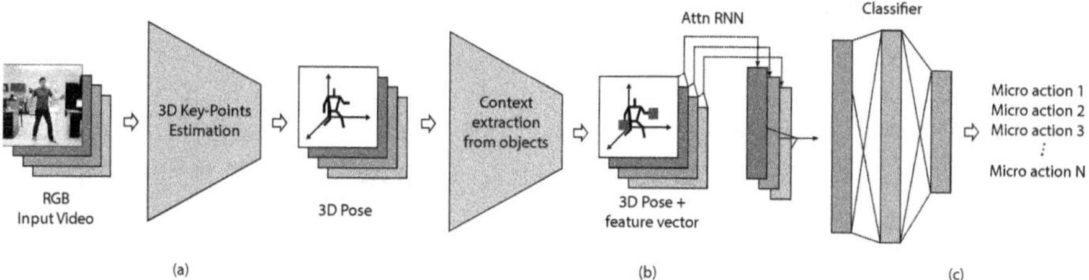

FIGURE 10.1 Micro-action recognition steps. (a) 3D human pose estimation; (b) object identification for context feature extraction; (c) micro-action classification.

be used to measure the performance of manual labor (Section 10.3), becoming a suitable replacement to the human-in-the-loop approach to TMS.

The contribution of this chapter is threefold: (1) we present an automatic replacement to the TMS methodology that works at two levels of abstraction (micro- and macro-actions); (2) we integrate context from objects, differentiating our methodology from other approaches which have primarily focused on skeletal or sensor data; and (3) we show how the proposed system can be used to make operational engineering decisions by estimating *standard times*, and calculating performance indicators – such as *worker availability*, *worker performance*, and *overall labor effectiveness* – as well as hand-usage indicators – such as *hand speed* and *handedness*.

10.2 LITERATURE REVIEW

The TMS technique is one of the most commonly used to measure manual labor performance; a detail discussion of the procedure can be found in Thomas and Daily (1983). Research in this area has focused on improving data collection using software applications based on single-frame analysis, like the multimedia video task analysis (MVTA) tool (Yen & Radwin, 2006) and Simdata (Createasoft.com, 2012), although they still require a human to record the observations. Work on automating the TMS technique have been primarily focused on the use of automatic sensor- and vision-based solutions (Joshua & Varghese, 2010; Teizer, 2015). Since human micro- and macro-actions produce particular patterns, machine learning algorithms are used to learn and identify worker tasks.

Among the sensor-based solutions, some research has used global positioning system (GPS) and radio-frequency identification (RFID) technologies as a proxy to measure labor productivity and to manage worker safety (Cheng, Migliaccio, Teizer, & Gatti, 2012; Jaselskis & El-Misalami, 2003). Likewise, body-worn sensors such as motion-capture markers (Tinoco, Ovalle, Vargas, & Cardona, 2015) and accelerometers (Joshua & Varghese, 2010, 2014) have been used. Such devices, however, have major drawbacks as accuracy is affected by their location in the body (Bao & Intille, 2004); they also provide sparse data, making detailed activity monitoring unavailable, resulting in low micro-action recognition accuracy.

Vision-based approaches have recently gained interest over sensor-based ones, since they provide data with a less intrusive collection methodology (Seo, Han, Lee, & Kim, 2015; Yang, Park, Vela, & Golparvar-Fard, 2015). These methods have been predominantly based on the use of depth sensors (e.g., Kinect™) to build a 3D skeleton model (Han, Achar, Lee, & Peña-Mora, 2013; Van Blommestein, Van Der Merwe, Matope, & Swart, 2012). However, usage of such devices may admit only minimal disturbance which can result in high implementation costs and noisy measurements. These methods have only been shown to be successful in controlled-laboratory environments and, therefore, may not be suited to tackle the intricacies of industrial workplaces.

The work of Yang, Shi, and Wu (2016) and Wang, Qin, Yan, and Guo (2019) are the most similar to our own. Yang et al. (2016) propose an automatic system to monitor construction activities; they perform action recognition using dense trajectories and achieve an accuracy of 59% but their method is computationally intensive and might not scale to real-life scenarios. Wang et al. (2019) propose an automatic system to monitor piece assembling activities; they perform action recognition using a hierarchical-clustering-based convolutional neural network (CNN) model which achieves an accuracy of 56%, though the dataset in which they evaluate their method is very limiting since it only contains hands and their method requires extensive preprocessing, limiting its real-life use. The existing literature does not incorporate context from the tools the worker is using which can help distinguish action with a high inter-class similarity. We may note that, to the best of our knowledge, our work is the first to measure manual labor performance using two levels of abstractions which enable the calculation of performance and hand-usage indicators.

10.3 METHODOLOGY

In this section we introduce our methodology, in which micro-actions are segmented and learned by a recurrent neural network (RNN) classifier, while macro-actions are learned by a CT-HMM classifier which models their relationship with micro-actions.

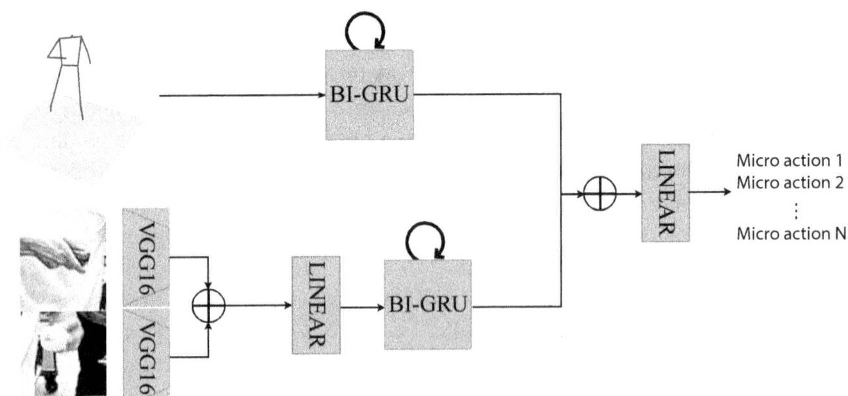

FIGURE 10.2 Proposed worker micro-action recognition system consisting of two RNNs: one is composed of a bidirectional gated recurrent unit (Bi-GRU) network that processes skeletal data; the other one is composed of a VGG16 network to extract features around the left and right hands, which are concatenated and passed through a linear layer before being processed by a Bi-GRU network. The vectors extracted from both RNNs are concatenated and fed to a linear classifier.

10.3.1 Micro-action Recognition

Micro-action recognition is comprised of three steps (see Figure 10.2). (1) *3D human pose estimation*: a human pose consists of 2D or 3D coordinates of human joints or keypoints – such as elbows, wrists, and shoulders – resulting in a compact and lightweight representation of the human body. We estimate 2D pose data for every video frame using the OpenPose framework (Cao, Simon, Wei, & Sheikh, 2017) and infer 3D poses from 2D poses to obtain a richer representation, using a residual feedforward network, based on the work in Martinez, Hossain, Romero, and Little (2017). (2) *Object identification for context feature extraction*: skeleton-based recognition usually struggles to classify actions related to the use of objects. To overcome this issue, we use the pose to locate the worker's hands and extract features that represent objects the worker is currently handling. (3) *Micro-action classification*: we use an RNN classifier based on the *encoder–decoder* framework (Cho et al., 2014; Sutskever, Vinyals, & Le, 2014). For the encoder we use two RNNs: one for the human pose data and another for the extracted hand features. Each RNN is augmented with the self-attention mechanism devised in Vaswani et al. (2017) that allows the RNNs to decide to which parts of the sequence it pays attention to. The decoder is trained to predict the class y_t given the output of the encoder c, which is modeled as $p(y_t|c) = g(c)$; here g is a feedforward neural network that outputs the conditional distribution of y_t.

10.3.2 Macro-action Recognition

The system concurrently learns hierarchical-structure information and how macro-actions are composed by the sequencing of the micro-actions extracted from video data. We model the higher level structure of macro-actions by using a CT-HMM: an HMM in which both observations and transitions between hidden states can occur at arbitrary instants (see Figure 10.3).

A CT-HMM is defined by the set $\lambda = \{\mathbf{b}, \pi, \mathbf{Q}\}$, where $\mathbf{b}$ is the observation model $p(o\,|\,s)$ that relates hidden states s to observations o, π is the initial hidden state distribution, and $\mathbf{Q}$ is a state transition rate matrix whose elements q_{ij} describe the rate at

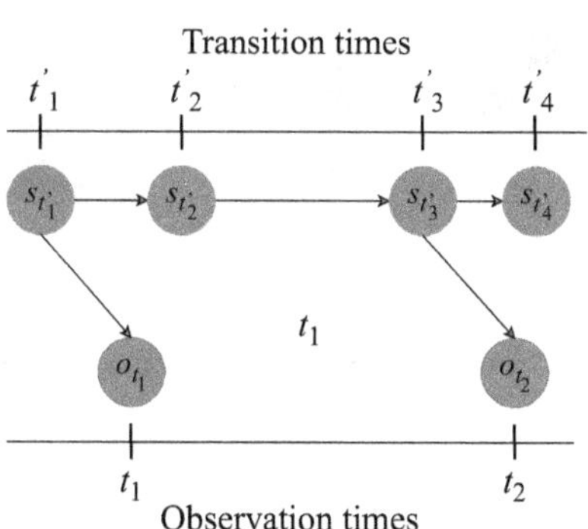

FIGURE 10.3 In a continuous-time hidden Markov model both the hidden states and the transition times are unobserved. Furthermore, multiple hidden state transitions can occur before an observation is obtained.

which the process transitions from hidden states i to j. The sufficient statistics of the model are the cumulative amount of time τ_i the model is in hidden state i and the number of times n_{ij} the model transitions from hidden states i to j. Given a current estimate of the model parameters of λ and observations O, the model can be trained using the expectation maximization (EM) algorithm (Liu, Li, Li, Song, & Rehg, 2015; Nodelman, Shelton, & Koller, 2012). Classification of a macro-action is done using an ensemble methodology by training a CT-HMM for each category and computing the optimal weights for the ensemble using a held-out dataset.

10.4 CASE STUDY: ORDER PREPARATION IN A DISTRIBUTION CENTER

Order preparation is the process associated with various operational tasks for packaging and shipping orders in a distribution center; it can be done in a variety of ways (Richards, 2017). For our purposes, we chose a system consisting of rack aisles in which workers place items into shelves following instructions from light indicators. Order preparation commonly involves whole-body and upper-limb motions such as

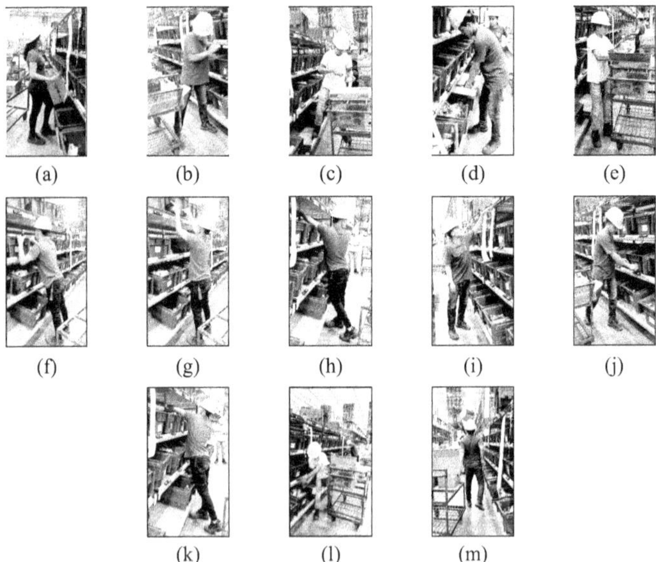

FIGURE 10.4 Samples from the selected micro-actions for the sorting activities in the distribution center. The micro-actions are: **(a) search**, person searches where to put product; **(b) find**, person finds product to process; **(c) select**, person selects product to scan; **(d) grasp**, person takes product with hand; **(e) hold**, person holds product in hand for a duration of time; **(f) transport loaded**, person moves a product from or to location; **(g) transport empty**, person moves him or herself or empty box; **(h) position**, person places product in final location; **(i) use**, person uses his or her equipment; **(j) inspect**, person inspects whether position, labels, or anything else is incorrect; **(k) preposition**, person moves product to make space; **(l) release load**, person releases product from hand; and **(m) delays**, any micro-actions or movement not related to the process.

repetitive bending and material or tool handling. We recorded RGB video data of these complex actions and used it to test our proposed methodology. This case study considers micro-actions – predefined in Freivalds (2009) and shown in Figure 10.4 – and macro-actions – namely label, scan, search, put, and confirm.

Our case study considered six workers with different levels of proficiency. Each subject was recorded for 40–60 min with a relative viewpoint variation of 60°, 120°, 240°, and 300°. RGB videos were recorded at a resolution of 1280 × 720 pixels at 30 frames per second. The collected data was manually labeled. Actions irrelevant to the activities were considered as delays and labeled together in one category. These actions include taking irregular rest times, walking to grab a box, and disposing boxes. Other actions not taken into account included communicating with coworkers and using the restroom. The final dataset was partitioned into training and validation sets, discarding 10 min between sets to ensure data independence. Data was partitioned again using a moving window approach with a size of 1 s. All experiments were run on an Intel i9 3.3 GHz processor with 20 cores, 48 GB of RAM memory, and endowed with an Nvidia GTX 1080 Ti with 12 GB of vRAM using the PyTorch deep-learning framework (Paszke et al., 2017).

For micro-action classification, we evaluated the model using a cross-view protocol – designed to test if the models can generalize to view variation – corresponding to angles 60°, 120° for training and 240°, 300° for testing.

For classifying activities, the protocol was to compare the classical end-to-end classification framework with the CT-HMM model framework. In the end-to-end framework, we used a variable frameskip to reduce the duration of activities to 64 frames; no information of the primitives was provided. In the CT-HMM model framework, we kept the duration of each video and used the logits produced by the models in the cross-view primitives classification protocol as input to the model.

All models were trained for 100 epochs, with a learning rate of 10^{-3}, reducing it by a factor of 10 in the epochs 60 and 80 as a learning rate schedule. We used standard data augmentation techniques for human action recognition in videos.

The proposed system achieved an average accuracy of 78.36% for the task of micro-action recognition. The micro-action categories with the highest accuracy were position, inspect, and preposition, while the categories with the lowest accuracy were the grasp, hold, and release (see Figure 10.6) task. The ROC curves for the cross-view task are shown in Figure 10.5 and our model achieves an area under curve (AUC) of 0.963. Likewise, the proposed system achieved an average accuracy of 85.57% for the task of macro-action recognition. The macro-action categories with the highest accuracy were put and confirm, while the categories with the lowest accuracy were label, scan, and search (see Figure 10.8). The effectiveness of the CT-HMM model is seen in Figure 10.7 which shows an improvement on performance over the end-to-end baseline which obtains an AUC of 0.9563, while the CT-HMM obtains an AUC of 0.989.

10.4.1 Action-effectiveness Evaluation

According to Aft (2000), micro-actions can be divided into effective – such as transport loaded, grasp, release load, use, preposition, inspect – and ineffective – such as hold, position, search, transport empty, find, delays – depending on whether these

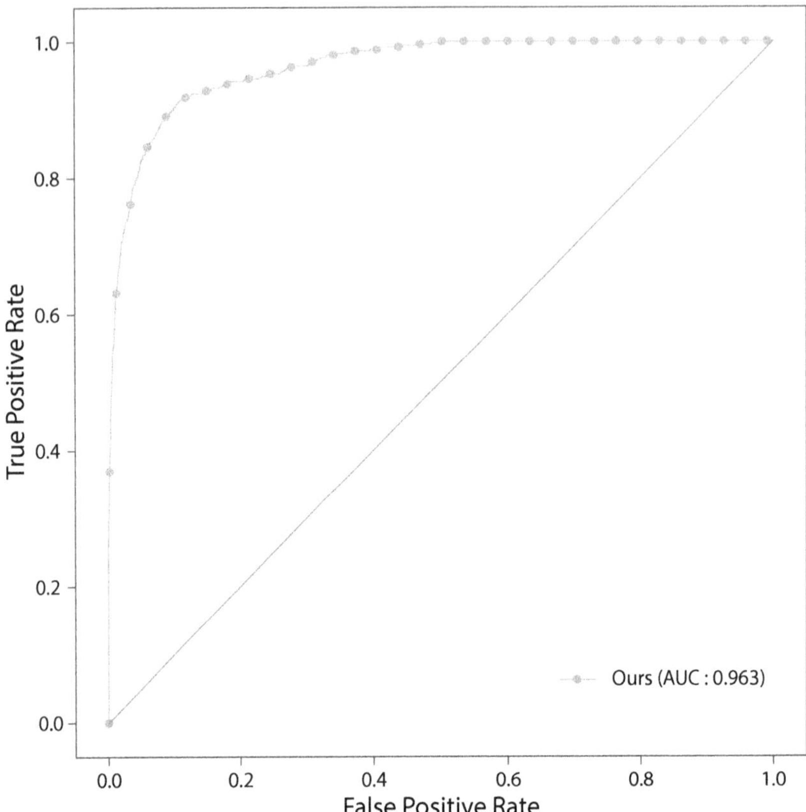

FIGURE 10.5 ROC curve for micro-action recognition.

micro-actions add value to activities. In our case study, worker 4 spends 58% of his time in delays and 17% holding the product. Likewise, worker 1 spends 21% of his time in releasing the product, while worker 3 prepositions the product 29% of his available time. However, both operators (1 and 3) spend 27% of their time holding or taking the product in their hands (as shown in Figure 10.9). All operators except worker 4 spend more than 50% of their time performing movements that contribute effectively to the process of order preparation. While worker 3 spends 76% of the time performing effective micro-actions, worker 4 spends only 37% of his time on these movements. In this way it is possible to know the percentage of time the operator makes effective and ineffective movements (see Table 10.1).

10.4.2 Labor-productivity Metrics

We used the procedure described in Freivalds (2009) to calculate standard times (see Table 10.2). This procedure considers personal, fatigue, and delay (PFD) allowances. We have considered a base allowance of 5% for the five macro-actions, which represents the personal needs allowance. For the search activity we considered an extra

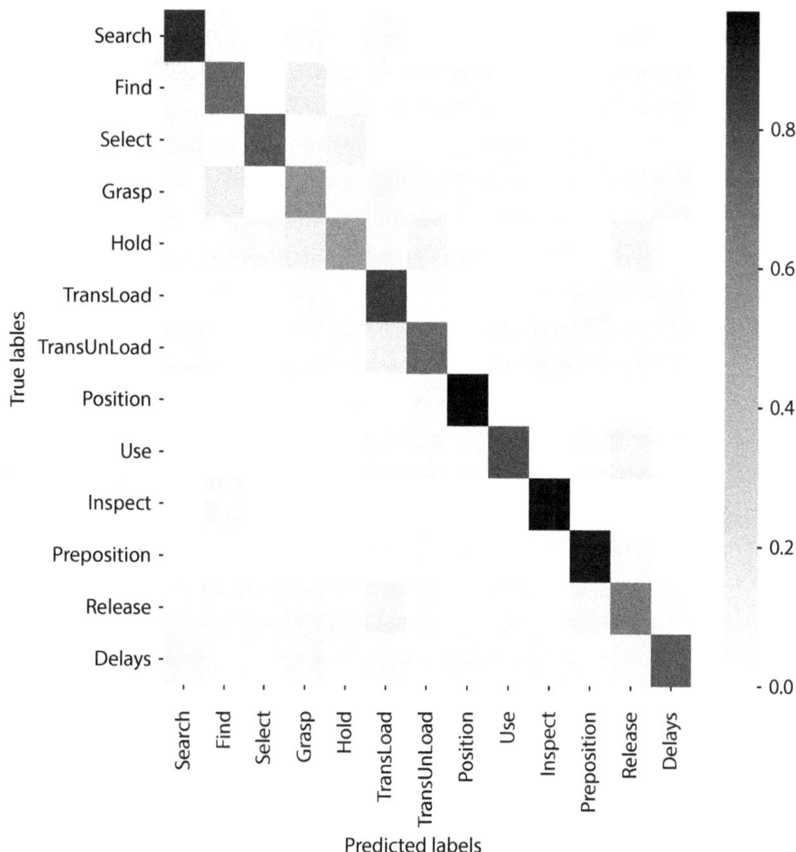

FIGURE 10.6 Confusion matrix of micro-action recognition.

allowance of 5% since it requires a high level of attention. Finally, for the putting activity we considered an extra allowance of 12% since it generates mental stress and requires a high level of attention.

We used the proposed system to calculate labor productivity metrics such as worker availability, performance, and overall labor effectiveness (OLE). These metrics allow managers to assess operational decisions since they provide the required information to analyze the combined effect of these decisions over activities (Gordon, 2011). These metrics can help locate areas in which providing an optimal work schedule can be critical to the number of productive hours. Figure 10.10 details the obtained metrics for each worker.

Worker availability measures the percentage of time spent doing productive (value-adding) activities and allows us to understand the origin of time losses caused by delays and their impact on the activities. Likewise, performance compares execution times of workers to standard times and allows us to determine opportunities to increase performance at the individual level. OLE is calculated as the product of

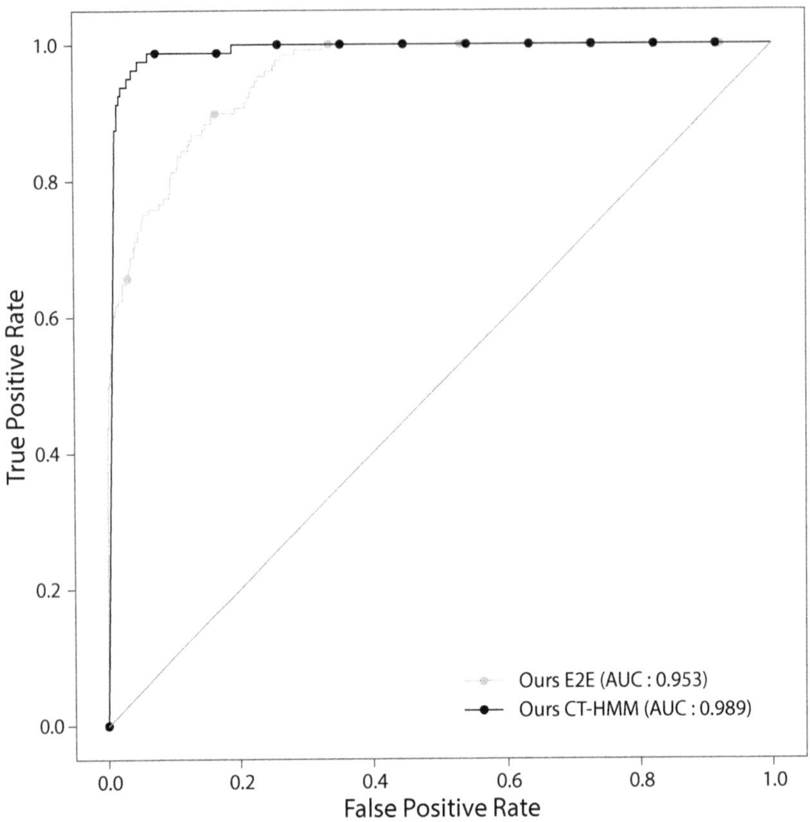

FIGURE 10.7 ROC curve for macro-action recognition with the CT-HMM model and without.

these two metrics and the quality of work obtained from historical data, and allows us to determine which worker is more productive as well as identify corrective actions so that all operations are up to standard. These metrics can help to diagnose whether proposed solutions to a problem improved overall productivity.

For the hand-usage measurements, the operator's hand speed – when he or she makes and/or breaks contact with the workpiece – was calculated. As shown in Figure 10.11, four of the six operators showed a preferred handedness: worker 2 and worker 3 moved the left hand faster than the right and for longer periods of time; worker 4 and worker 5 moved the right hand faster and for longer periods of time, while worker 0 and worker 1 showed mixed-handedness movements. Worker 3 showed a disproportionate use of his dominant hand, which may create the risk of developing a musculoskeletal disorder such as carpal tunnel syndrome. In addition, it is worth mentioning that this operator spent 27% of the time holding an object in his hands, as shown by our system.

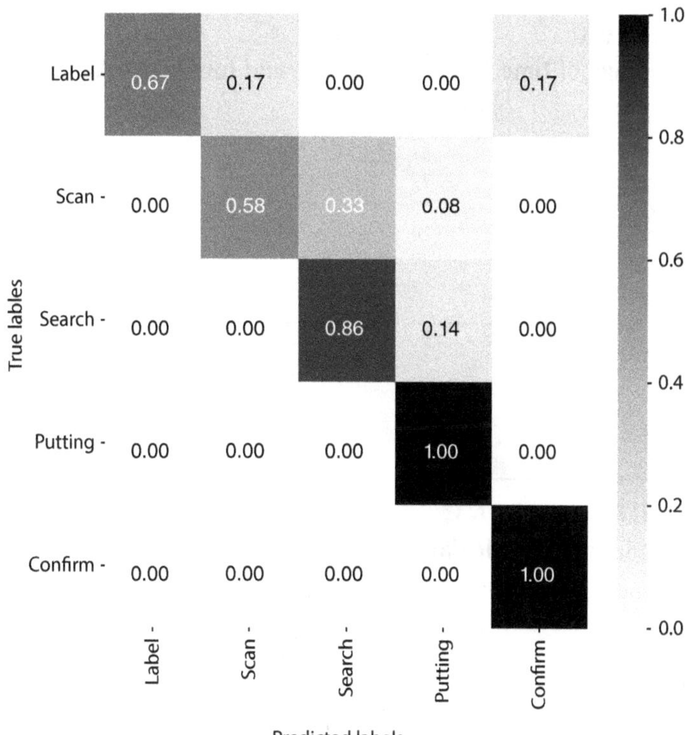

FIGURE 10.8 Confusion matrix of macro-action recognition.

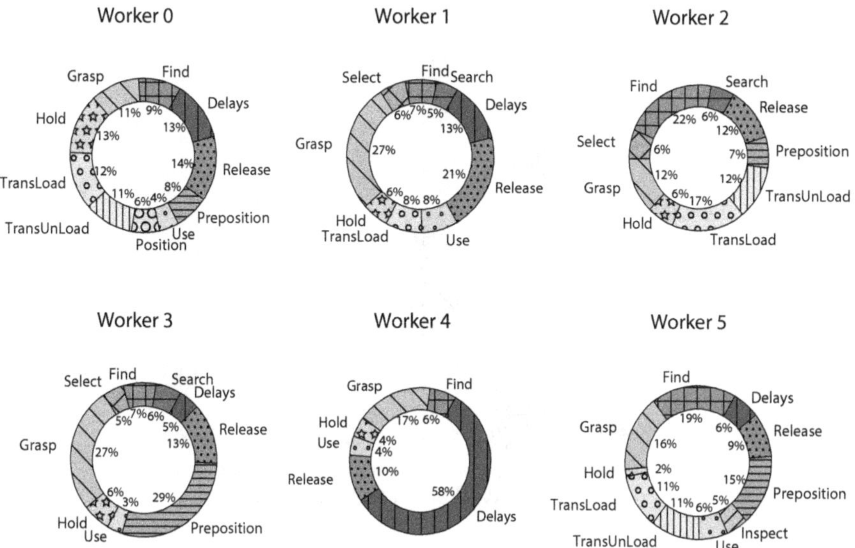

FIGURE 10.9 Percentage of time spent by operators performing each micro-action.

TABLE 10.1
Percentage of Time Spent on Effective and Ineffective Actions

	Effective Actions (%)	Ineffective Actions (%)
Worker 0	51	49
Worker 1	72	28
Worker 2	63	37
Worker 3	76	24
Worker 4	37	63
Worker 5	63	37

TABLE 10.2
Standard Time Calculation

Activity	Normal Times(s)	PFD Allowance (%)	Standard Time(s)
Label	2.45	105	2.57
Scan	1.34	105	1.41
Search	1.18	110	1.30
Putting	3.47	117	4.06
Confirm	1.17	105	1.23

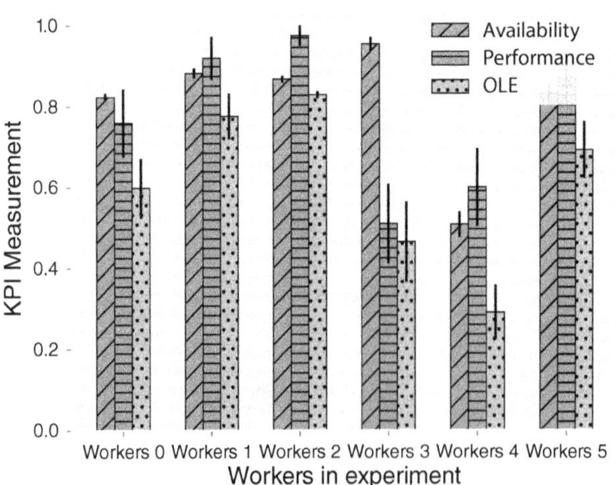

FIGURE 10.10 Worker availability, performance, and overall labor effectiveness measured by the proposed system.

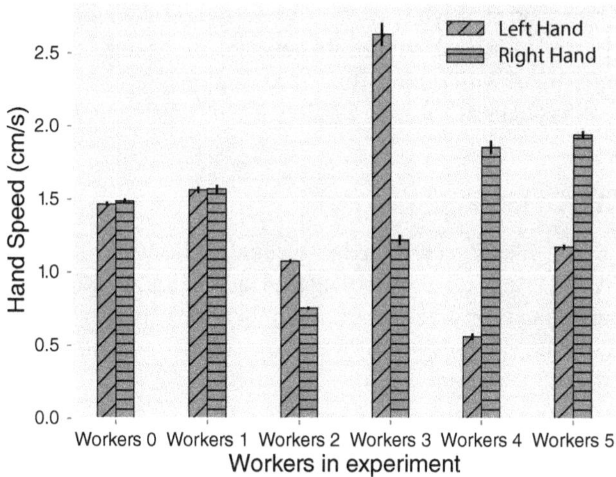

FIGURE 10.11 Hand speed measured by the proposed system.

10.5 CONCLUSIONS

In this chapter we have presented an automatic replacement for the human-in-the-loop TMS methodology to measure the performance of manual labor using skeletal data and features extracted from images of objects around the hands. We focused on the TMS methodology at two levels of abstraction: micro- and macro-actions. As a case study, we collected RGB video data from order preparation tasks in a distribution center and labeled it, selecting from 13 predefined micro-actions and five macro-actions. This dataset had several challenging characteristics including illumination and view angle change, and interrupted work flow. The achieved accuracy was 78.36% and 85.57% for micro- and macro-action recognition, respectively.

These results demonstrate that video data along with deep-learning technology can take advantage of particular patterns according to the type of micro- and macro-actions performed, and that our methodology is capable of reaching human-level proficiency in measuring the performance of manual labor. Since we have used standard and predefined micro-actions, the proposed approach can be adapted to other industrial settings. We used the proposed system to calculate labor productivity metrics – which allows us to determine which worker is more productive as well as identify corrective actions so that all operations are up to standards – and hand-usage metrics – which can be used in ergonomic and health evaluations.

Current limitations of our system include: our methodology requires that actions are recorded with little angle variation; there should be very little intra-class variability, meaning that micro-actions must be performed in a very similar way; objects are identified using pretrained networks which could confuse the system when similar-looking objects are used; and the need for dense annotations at the micro-action level.

Directions for future work include: testing the proposed methodology in other areas; using context-free grammars instead of the CT-HMM which may allow us to represent a higher level of abstractions (Kuehne, Arslan, & Serre, 2014); performing unsupervised learning of the micro- and macro-action taxonomy, incorporating a multi-object tracker to include actions that need worker interactions; and including more ergonomic metrics like the lifting load National Institute for Occupational Safety and Health (NIOSH) equation (Waters, Putz-Anderson, Garg, & Fine, 1993).

ACKNOWLEDGMENT

The authors would like to acknowledge Tiendas Industriales Asociadas Sociedad Anonima (TIA S.A.), a leading grocery retailer in Ecuador, for providing funding for this research effort and for granting access to their Distribution Center, particularly to their put-to-light system, which provided a rich environment for developing our case study.

REFERENCES

Aft, L. S. (2000). *Work Measurement and Methods Improvement*. (Vol. 9). John Wiley & Sons.

Al-Saleh, K. S. (2011). Productivity improvement of a motor vehicle inspection station using motion and time study techniques. *Journal of King Saud University-Engineering Sciences, 23* (1), 33–41.

Bao, L., & Intille, S. S. (2004). Activity recognition from user-annotated acceleration data. In *International Conference on Pervasive Computing* (pp. 1–17).

Baumgart, A., & Neuhauser, D. (2009). *Frank and Lillian Gilbreth: Scientific Management in the Operating Room*. BMJ Publishing Group Ltd.

Cao, Z., Simon, T., Wei, S.-E., & Sheikh, Y. (2017). Realtime multi-person 2D pose estimation using part affinity fields. In *Proceedings of the IEEE Conference on Computer Vision and Pattern Recognition* (pp. 7291–7299).

Chatzis, K. (1999). Searching for standards: French engineers and time and motion studies of industrial operations in the 1950s. *History and Technology, an International Journal, 15* (3), 233–261.

Cheng, T., Migliaccio, G. C., Teizer, J., & Gatti, U. C. (2012). Data fusion of real-time location sensing and physiological status monitoring for ergonomics analysis of construction workers. *Journal of Computing in Civil engineering, 27* (3), 320–335.

Cho, K., Van Merriënboer, B., Gulcehre, C., Bahdanau, D., Bougares, F., Schwenk, H., & Bengio, Y. (2014). Learning phrase representations using RNN encoder-decoder for statistical machine translation. *arXiv preprint arXiv:1406.1078*.

Createasoft.com. (2012). *Time and Motion Studies Software - SimData Time Motion Software*. Retrieved 2021-02-12, from https://www.createasoft.com/Products/Time-and-Motion-Studies-Software

Finkler, S. A., Knickman, J. R., Hendrickson, G., Lipkin Jr, M., & Thompson, W. G. (1993). A comparison of work-sampling and time-and-motion techniques for studies in health services research. *Health Services Research, 28* (5), 577.

Freivalds, A. (2009). *Niebel's Methods, Standards, and Work Design* (Vol. 700). Mcgraw-Hill higher education Boston.

Gordon, G. (2011). *Lean Labor: A Survival Guide for Companies Facing Global Competition*. Kronos Publishing.

Gouett, M. C., Haas, C. T., Goodrum, P. M., & Caldas, C. H. (2011). Activity analysis for direct-work rate improvement in construction. *Journal of Construction Engineering and Management, 137* (12), 1117–1124.

Gunasekaran, A., Forker, L., & Kobu, B. (2000). Improving operations performance in a small company: a case study. *International Journal of Operations & Production Management, 20* (3), 316–336.

Han, S., Achar, M., Lee, S., & Peña-Mora, F. (2013). Empirical assessment of a rgb-d sensor on motion capture and action recognition for construction worker monitoring. *Visualization in Engineering, 1* (1), 6.

Harewood, G. C., Chrysostomou, K., Himy, N., & Leong, W. L. (2008). A "time-and-motion" study of endoscopic practice: strategies to enhance efficiency. *Gastrointestinal Endoscopy, 68* (6), 1043–1050.

Jaselskis, E. J., & El-Misalami, T. (2003). Implementing radio frequency identification in the construction process. *Journal of Construction Engineering and Management, 129* (6), 680–688.

Joshua, L., & Varghese, K. (2010). Accelerometer-based activity recognition in construction. *Journal of Computing in Civil Engineering, 25* (5), 370–379.

Joshua, L., & Varghese, K. (2014). Automated recognition of construction labour activity using accelerometers in field situations. *International Journal of Productivity and Performance Management, 63* (7), 841–862.

Kanawaty, G. (1992). *Introduction to Work Study*. International Labour Organization.

Kuehne, H., Arslan, A., & Serre, T. (2014). The language of actions: Recovering the syntax and semantics of goal-directed human activities. In *Proceedings of the IEEE Conference on Computer Vision and Pattern Recognition* (pp. 780–787).

Liu, Y.-Y., Li, S., Li, F., Song, L., & Rehg, J. M. (2015). Efficient learning of continuous-time hidden markov models for disease progression. In *Advances in Neural Information Processing Systems* (pp. 3600–3608).

Martinez, J., Hossain, R., Romero, J., & Little, J. J. (2017). A simple yet effective baseline for 3D human pose estimation. In *Proceedings of the IEEE International Conference on Computer Vision* (pp. 2640–2649).

Maynard, H. B., Stegemerten, G. J., & Schwab, J. L. (1948). *Methods: Time Measurement*, American Psychological Association (APA).

Meyers, F., & Stewart, F. (2002). *Motion and Time Study for Lean Manufacturing*. Prentice Hall.

Nodelman, U., Shelton, C. R., & Koller, D. (2012). Expectation maximization and complex duration distributions for continuous time bayesian networks. *arXiv preprint arXiv:1207.1402.*

Paszke, A., Gross, S., Chintala, S., Chanan, G., Yang, E., DeVito, Z., … Lerer, A. (2017). Automatic Differentiation in Pytorch. In *Nips-w.*

Richards, G. (2017). *Warehouse Management: A Complete Guide to Improving Efficiency and Minimizing Costs in the Modern Warehouse*. Kogan Page Publishers.

Seo, J., Han, S., Lee, S., & Kim, H. (2015). Computer vision techniques for construction safety and health monitoring. *Advanced Engineering Informatics, 29* (2), 239–251.

Shehata, M. E., & El-Gohary, K. M. (2011). Towards improving construction labor productivity and projects' performance. *Alexandria Engineering Journal, 50* (4), 321–330.

Sutskever, I., Vinyals, O., & Le, Q. V. (2014). Sequence to sequence learning with neural networks. *Advances in Neural Information Processing Systems,* 3104–3112).

Taylor, F. W. (1911). The principles of scientific management. *New York,* 202.

Teizer, J. (2015). Status quo and open challenges in vision-based sensing and tracking of temporary resources on infrastructure construction sites. *Advanced Engineering Informatics, 29* (2), 225–238.

Thomas, H. R., & Daily, J. (1983). Crew performance measurement via activity sampling. *Journal of Construction Engineering and Management, 109* (3), 309–320.

Tinoco, H. A., Ovalle, A. M., Vargas, C. A., & Cardona, M. J. (2015). An automated time and hand motion analysis based on planar motion capture extended to a virtual environment. *Journal of Industrial Engineering International, 11* (3), 391–402.

Van Blommestein, D., Van Der Merwe, A., Matope, S., & Swart, A. (2012). Automation of work studies: An evaluation of methods for a computer based system. In 42nd International Conference on Computers & Industrial Engineering 2012.

Vaswani, A., Shazeer, N., Parmar, N., Uszkoreit, J., Jones, L., Gomez, A. N., … Polosukhin, I. (2017). Attention is all you need. In *Advances in Neural Information Processing Systems* (pp. 5998–6008).

Wang, Z., Qin, R., Yan, J., & Guo, C. (2019). Vision sensor based action recognition for improving efficiency and quality under the environment of industry 4.0. *Procedia CIRP, 80,* 711–716.

Waters, T. R., Putz-Anderson, V., Garg, A., & Fine, L. J. (1993). Revised niosh equation for the design and evaluation of manual lifting tasks. *Ergonomics*, *36* (7), 749–776.

Yang, J., Park, M.-W., Vela, P. A., & Golparvar-Fard, M. (2015). Construction performance monitoring via still images, time-lapse photos, and video streams: Now, tomorrow, and the future. *Advanced Engineering Informatics*, *29* (2), 211–224.

Yang, J., Shi, Z., & Wu, Z. (2016). Vision-based action recognition of construction workers using dense trajectories. *Advanced Engineering Informatics*, *30* (3), 327–336.

Yen, T. Y., & Radwin, R. G. (2006). Usability testing by multimedia video task analysis. *Medical Instrumentation: Accessibility and Usability Considerations*, *159*, 159–171.

11 Applications of IoT Based Frameworks in Industry 4.0

Applications of IoT Based Frameworks

Kaushik Ghosh and Sugandha Sharma
UPES, Dehradun

Piyush Bagla and Kuldeep Kumar
Dr. B.R. Ambedkar NIT, Jalandhar

CONTENTS

11.1 Introduction 163
11.2 Literature Review 164
11.3 Proposed Framework 166
  11.3.1 Radio Model 167
    11.3.1.1 Degree of Proximity 169
    11.3.1.2 Working Phases of the Proposed Framework 169
11.4 Applications 170
  11.4.1 Application in Agricultural Industry 170
  11.4.2 Application in Healthcare 171
  11.4.3 Application in the Oil and Gas Industry 172
11.5 Results 174
11.6 Limitations 176
11.7 Conclusion and Future Work 176
References 176

11.1 INTRODUCTION

Industry 5.0 is going to be the reality of a soon-to-be future, one equipped with a host of cutting edge technologies. Big data, blockchain, the Internet of Things (IoT), cloud computing, and so on are some of the prominent areas of research today, finding their application in multi-domain and cross-domain industries. The IoT for that matter has found its application in different industries. Be it healthcare, agriculture,

DOI: 10.1201/9781003146711-11

military applications, or manufacturing industry – the application of the IoT has not only enriched these industries but has also made them move a step further towards Industry 5.0 standards. Moreover, the cross-domain application of the IoT is also becoming popular under the current scenario where a proposed framework may be used in different platforms with nominal customization. This cross-platform usage of a single framework other than being inclusive is sure to bring down the development cost. Therefore, a cost effective solution to address the needs of industry today is to have a 1 to *n* mapping between the framework and the industry domain.

In this chapter we propose a single framework that may be used for three different industries: agriculture, healthcare, and the oil industry. The proposed framework will address the specific issues of these industries with some changes in the application layer.

11.2 LITERATURE REVIEW

In this section, we will discuss the usage of wireless sensor networks (WSN) in different industries like precision agriculture, healthcare, and the oil industry for the specific domains related to this chapter (e.g., smart irrigation, telemonitoring/telemedicine, and intrusion detection).

An IoT-based drip technique of irrigation has been discussed in Parameswaran and Sivaprasath (2016). This is an automated irrigation scheme that waters agriculture land depending upon the level of humidity in the soil, which is measured with humidity sensors and, based on the readings, solenoid valves are operated by microcontroller drives. For measuring soil moisture, SY-HS220 was used. The authors infer that the setup will reduce the level of human interference in the system along with reducing the total energy requirement.

Singh and Saikia (2016) discussed an Arduino-based smart irrigation technique. They used three different kinds of sensors: water flow, soil moisture, and temperature sensors. Data collection was done by Arduino and was linked to an interactive website where the real time values of different soil parameters are shown along with their standard values. The website is interactive in the sense that actuators like sprinklers and pumps can be controlled through the website. The proposed system is Arduino based. Here the data is sent and received remotely by an ESP8266 Wi-Fi module. The authors point out the ease of implementation and low maintenance cost of the system.

Another IoT-based smart irrigation technique was proposed by Saraf and Gawali (2017). Here, remote monitoring of farm parameters and controlling of actuators are made possible using android phones. The users can view farm parameters in real time over a GUI, designed using MATLAB, on the phone. This work is different from the previous ones in the sense that it introduces the use of cloud technology for the seamless access of remote data, reduction in hardware requirement along with overall cost, and increasing data security. The authors used DHT11 for sensing the ambient temperature and humidity. Moreover, an LM 393 sensor was used for measuring soil moisture and M116 sensors were used for measuring the water level. Sensors transmitted data once the value of the farm parameter was beyond the threshold window set. The same data was processed by AtMega 328 microcontrollers and transmitted to Zigbee. The data is finally transmitted to cloud servers.

A novel method of intelligent irrigation was discussed in Roopaei, Rad, and Choo (2017) through the cloud of things technology. Here, irrigation monitoring was done through thermal imaging of the agricultural land, where it was possible to figure out the patches of land that required hydration.

Sahu and Behera (2015) propose a low cost smart irrigation system that is affordable by a middle-class farmer. The working principle of the prototype proposed in this paper is similar to that of the ones proposed in Parameswaran and Sivaprasath (2016) and Singh and Saikia (2016) due to the fact that all three works are Arduino based and use humidity sensors for finding out the soil moisture level. However, it is different from the previous research as it uses Raspberry Pi for sending messages over the Internet. Actually, the data sensed by the sensors is sent to the microcontroller node wirelessly. Thereafter, if the soil moisture level in a given patch is below the standard value, then the controller node will switch on the motor to irrigate it. The Raspberry Pi is for processing data. It sends notifications to the mobile phone registered with it. LM393 is the soil moisture sensor used for this work.

Sales, Remédios, and Arsenio (2015) discuss the use of WSN and the role of actuators in IoT-based smart irrigation. The paper proposes the deployment of a cloud based network, comprising sensor nodes and actuators. As an add on, a Web service is also employed to empower the system with weather information.

Much work has been done on the application of the IoT on the healthcare system as well, particularly on remote monitoring and reporting of vital parameters. Sathya, Madhan, and Jayanthi (2018) discuss the challenges faced in IoT-based health monitoring using sensors that are either wearable or can be embedded in the body of the patient. The authors also discuss different WSN and IoT-based approaches to remote healthcare monitoring.

Baker, Xiang, and Atkinson (2017) discuss different technologies adopted in healthcare along with the challenges and opportunities. Here, a detailed discussion concerns wearable sensors and the future of the IoT in healthcare. The communication of data from the sensors was subdivided into two parts: short range and long range. In short range communication, the patient data was collected by the sensors and transmitted to a central node. A comparison between Bluetooth and Zigbee was also made for short range communication. From the central node onwards, data was transmitted to a database using long range communication. In data thus collected, the database is available for authenticated users in a secure manner. The option of cloud storage of the collected data was also proposed in the paper. However, security measures must be taken in order to keep the cloud data secure and safe from malicious parties.

Another IoT-based vital parameter-monitoring prototype was proposed by Banka, Madan, and Saranya (2018) using Raspberry Pi. Here the authors proposed one Web-based and a mobile-based application for monitoring patient conditions along with reporting of the same to doctors and relatives. The patient information here is stored in the cloud. The system has three distinct modules: health monitoring, emergency alert, and health status prediction. Different sensors were used for measuring the temperature, heartbeat, and blood pressure of the patients.

Sharma and Sunanda (2017) present a survey on the application of the IoT in smart healthcare. Along with discussing the application of the IoT in the field, the paper presents a discussion on IoT security, open issues, and challenges.

Another comprehensive survey was presented by Islam, Kwak, Kabir, Hossain, and Kwak (2015) regarding the use of the IoT for healthcare. Along with discussing the applications of the IoT in healthcare services, they also consider different such as security. They also propose a layered networking model for a secured IoT. They segregate issues in IoT healthcare under three broad heads: topology, architecture, and platform. They subdivide IoT services and applications under many smaller heads (e.g., community healthcare and child healthcare). Not only this, the authors should be praised for listing the different possible attacks on IoT healthcare. The different technologies required (wearables, cloud, etc.) are dealt with along with the policies adopted by different countries towards it. Open issues and challenges present in IoT healthcare were also discussed.

The use of WSN for intrusion detection within given premises is a very prominent security feature, particularly in military applications. In this chapter we have extended the concept so that it can be used in the oil industry, where security is no less important as far as sabotage is concerned. Ghosh, Neogy, Das, and Mehta (2018) propose a WSN-based two-dimensional network for intrusion detection, using proximity sensors. They take both grid and random deployment under consideration. The results show the combined effect of the coverage area and node density on the lifetime of the deployed network. Similar work is proposed by Naz, Hengy, and Hamery (2012) for soldier detection, though the sensors used here were acoustic and seismic.

In perimeter protection schemes, day/night cameras are used along with infrared cameras and radars for detecting radiation from intruders. This requires sensor nodes and a proper algorithm to fuse the data (Dulski et al., 2011; He, Fallahi, Norwood, & Peyghambarian, 2011).

11.3 PROPOSED FRAMEWORK

The sensor nodes may be viewed as deployed over a two-dimensional Cartesian plane. As seen in Figure 11.1, the shape of the sensor field is taken as either square or rectangular. Three sinks are placed and assumed to be at the three corners of the field. It is always better to have multiple sinks in a WSN for the following reasons:

1. It reduces the load on individual sinks by distributing the load of data collection to multiple sinks.
2. It reduces the sink hole problem.
3. It reduces the total number of hops a packet takes on its way to the destination.
4. It provides the required infrastructure support at multiple interfaces.
5. It increases the overall network lifetime.
6. It eliminates single point failure.

Node deployment is assumed to be in grid fashion, where the grid size may be changed, depending upon the nature of the application and requirement for node density. However, it is to be noted that the size of the sensor field remains the same. A change in grid size changes the *degree of proximity* of the sensor field. The effects of a changed degree of proximity on the lifetime of the deployed network is

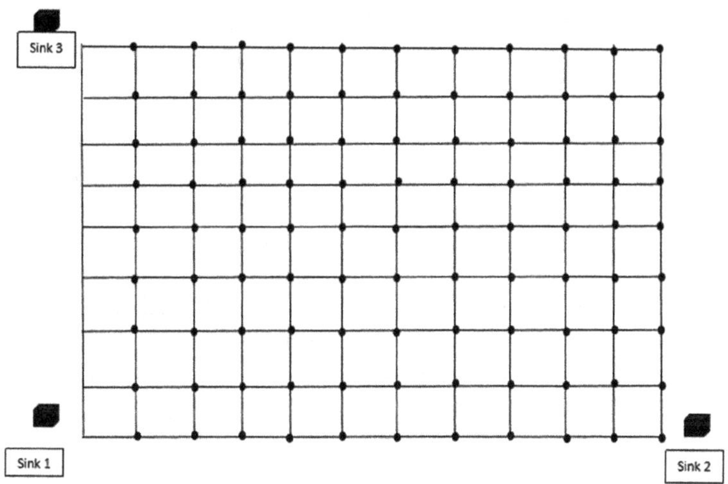

FIGURE 11.1 Common deployment setup of the sensor nodes.

discussed later in Section 11.4. The nodes keep measuring the required parameter at the sites they are deployed to, and report to the sinks when the value of that parameter goes above/below the permissible threshold for the application. Network lifetime is defined as "n-out-of-n" (Dietrich & Dressler, 2009). Here, "n" denotes the number of sensors deployed, which implies that a deployed network is considered "dead" as soon as the very first node's battery power becomes zero. The justification behind this definition is that as soon as the first node fails, the network suffers the first topology change.If data from all the nodes are equally important, then this definition may reasonably be used (Dietrich & Dressler, 2009). Sink nodes were excluded from the scope of this definition as sinks were rightly assumed to be plugged nodes (Madan, Cui, Lall, & Goldsmith, 2006).

11.3.1 Radio Model

Assumptions:

1. All the nodes are homogeneous in terms of initial energy and transmission range.
2. The transmission ranges of the nodes can be altered by changing the transmission power of their antennas.

The proposed framework considers possible energy consumption sources under the following heads: (i) sensing ($E_{Sensing}$), (ii) computation ($E_{Computation}$), (iii) forwarding ($E_{Forwarding}$), (iv) receiving ($E_{Receiving}$), and (v) listening ($E_{Listening}$). Before a source node can transmit, first of all it has to sense the physical parameter (sensing). Then, it has to do the necessary computation in order to convert the physical parameter under consideration into transmissible form (computation). Finally, it forwards the ready packet to the next hop forwarder (forwarding). Energy required for transmission (E_{TX}) therefore will comprise three components: $E_{Sensing}$, $E_{Computation}$, and $E_{Forwarding}$.

The intermediate nodes on the other hand consume energy for the following tasks: (i) energy consumed for receiving a packet ($E_{Receiving}$) and (ii) forwarding the received packet ahead to the next hop forwarder ($E_{Forwarding}$). Even if a sensor node is not performing its duty as a source or a relay, then it too has a certain energy expenditure as it has to listen to the transmissions of its neighboring nodes ($E_{Listening}$). We have considered a node as "on" when it serves either as a source node or as a relay node. A node's state on the other hand is "off" when it is simply listening to the communications of its neighboring nodes. So, t_{on} indicates the "on" period of a node and is defined as the time during which the node either transmits (T_{tx}) or forwards (T_{fwd}) data. On the other hand, its "off" period (t_{off}) is the time for listening to the communications among its neighboring nodes.

With t_{on} and t_{off} we have expressed the duty cycle (D) of any node as

$$D = \frac{t_{on}}{t_{on} + t_{off}} \tag{11.1}$$

The more the number of neighbors, the more the chance of selecting a different neighbor for an upcoming transmission. This will ensure that the same node will not be selected for subsequent data forwarding actions. As a result, the same set of nodes are never overburdened by relaying data, which increases the network lifetime. $E_{Computation}$ here is taken as 117 nJ/bit from Min and Chandrakasan (2001). The value of $E_{Sensing}$ on the other hand is taken as 1.7 μJ/bit (Min et al., 2001). $E_{Listening}$ again is taken as 570 μJ/s as per the findings in Anastasi, Falchi, Passarella, Conti, and Gregori (2004).

The energy model of this framework is thus finally expressed as:

$$E_{TX} = m * 117 * 10^{-9} + m * 1.7 * 10^{-6} + D * m * \varepsilon * d^{n} \tag{11.2}$$

$$E_{forwarding} = D\left(m * E + m * \varepsilon * d^{n}\right) \tag{11.3}$$

$$E_{listening} = \left(1 - D\right) * 570 * 10^{-6} \tag{11.4}$$

where,

m = number of bits per packet
n = path loss exponent
D = duty cycle
E = 50 nJ/bit
ε = 8.854 pJ/bit/m^2
d = distance between two immediate nodes.

The values of E and ε for the present work are taken from Heinzelman, Chandrakasan, and Balakrishnan (2002). ε records the "per bit energy expenditure per meter square," and E is the "minimum start-up energy" required for any communication.

11.3.1.1 Degree of Proximity

Now that the energy model of the framework has been proposed, let us introduce a novel network parameter called the degree of proximity (DoP). DoP gives the density of neighbors for a node. For two and three-dimensional scenarios, DoP may be defined by Equations 11.5 and 11.6 respectively:

$$DoP = \sqrt[2]{ROUND\left(TXR/X\right) * ROUND\left(TXR/Y\right)} \tag{11.5}$$

$$DoP = \sqrt[3]{ROUND\left(TXR/X\right) * ROUND\left(TXR/Y\right) * ROUND\left(TXR/Z\right)} \tag{11.6}$$

where,

TXR = transmission range of the nodes
X = number of neighbors within transmission range along x-axis
Y = number of neighbors within transmission range along y-axis
Z = number of neighbors within transmission range along z-axis.

It is to be noted that for grid deployment with a fixed grid size, and assuming the same *TXR* for all the nodes, the value of DoP will be constant for all the nodes in a network. We have used DoP to regulate the lifetime of a given network. In the results section we will see how changing the value of DoP changes the lifetime (L) of a network, when all other parameters are kept constant. We have shown that the lifetime of a network can be maximized for an optimum value of DoP.

11.3.1.2 Working Phases of the Proposed Framework

1. **Initialization:** In this phase, after deployment, individual nodes will theoretically discover the coordinates of the Fermat point of an imaginary quadrilateral, formed by the node itself and the three sinks at its vertex.

   Definition of the Fermat point: A point within the bounds of a triangle or polygon, such that the sum of the distances of all the vertices from that point comes out to be the minimum as compared to that of any other point within that triangle or polygon.

   Our proposed scheme exploits the concept of the Fermat point for data forwarding. The primary reason for choosing such a scheme is the fact that it ensures lifetime enhancement by minimizing energy consumption (Ghosh, Das, & Neogy, 2016). This happens as the total distance traveled from the source to the sinks can be minimized if data transmission takes place through the Fermat point (Lee & Ko, 2006).

   The node located nearest to the Fermat point thus found is identified as the Fermat node (FN), for a given source node. So, the transmission phase is sub-divided into two sub-phases. At first, data is transmitted to the FN. During the second part of transmission, data received at the FN is aggregated into packets which are further forwarded to the three sinks.

2. **Transmission phase:** As discussed, transmission is sub-divided into two sub-phases. In the first one, a node, on sensing the data, transmits it to its FN in multiple hops. A node, on having a data packet, first checks if the FN is within its transmission range (TXR). If so, it transmits the packet directly to the FN. Otherwise, it has to select a next hop forwarder, which could relay the packet to the FN. In order to select the next hop neighbor, the scheme calculates the forwarding potential (κ) of a node's one-hop neighbors. The data packet is then forwarded to the neighbor having the greatest value of κ, which is calculated as per Equation 11.7:

$$\kappa_{ij} = \frac{res\_energy_i}{dist_j} \tag{11.7}$$

where,

res_energy_i = residual battery power of node i in millijoules
$dist_j$ = distance of a node from a destination j in meters.

The second sub-phase deals with aggregating data and forwarding aggregated data packets from the FN to the sinks. If a particular sink is within the transmission range of the FN, it transmits the packet directly to that sink. Otherwise, the same technique as the previous sub-phase is used for finding the next hop forwarder from the FN to the sinks. One of the main tasks of sub-phase 2 is data aggregation, which ensures a reduction in energy consumption and thereby enhances network lifetime (Ghosh, Das, & Neogy, 2015). In this framework, we propose a data aggregation scheme with flexibility in choosing the degree of aggregation, which is indicated by the parameter AGFACT. AGFACT = n indicates that n data packets are to be packed together and forwarded further to the sinks through multi-hop transmissions. This reduces the number of transmissions and thereby prolongs network lifetime. If no aggregation is required, the value of AGFACT should remain as 1.

11.4 APPLICATIONS

11.4.1 Application in Agricultural Industry

The proposed common framework can be used in agricultural industry for the automated watering of patches of land, where the soil moisture level has gone down below a specified threshold (Ghosh, Neogy, Das, & Mehta, 2017). Of the different application areas where the IoT and WSN may be used, precision agriculture (PA) is a prominent one (Parameswaran & Sivaprasath, 2016). Introducing WSN in agricultural activities has reduced the running cost of the setup due to an 80% reduction in the cost incurred due to wiring (Wang, Zhang, & Wang, 2006). In PA, the sensor nodes record farm parameters and then report these to one or many sinks. In certain other cases, a sensor node may also be used for aggregating data. After aggregation,

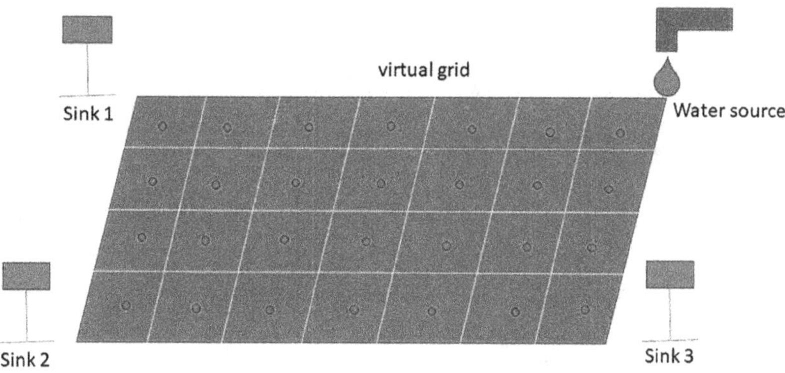

FIGURE 11.2 Nodes deployed over a rectangle of agricultural land.

the resultant packet is forwarded to the respective sinks (Wark et al., 2007). In either case, the presence of multiple sinks is advocated due to the points already discussed in Section 11.2.

IoT-based nodes keep measuring the soil moisture level of the sites they are deployed at and report to the sinks when the value of the soil moisture has gone down below the permissible threshold for the crop. As a result, only the affected patches of the land are watered and not the entire land. This way, we can optimize the watering of the agricultural field and thereby save water. If in addition to having three sinks in the three corners of the agricultural land, we have a water source in the fourth corner, then a node for sensing the soil moisture level below the threshold can transmit the information to the sinks. Moreover, we consider each grid (see Figure 11.1) to have a sprinkler along with a node. The setup therefore would look somewhat as shown in Figure 11.2.

The sink nearest to the patch of land would activate the sprinkler for that particular grid only, and not for all the grids. When the said patch of land is adequately watered, the sprinkler will stop.

11.4.2 Application in Healthcare

The proposed framework may also be used in healthcare. Particularly during the time of a global pandemic, like COVID-19, it may be used by health workers for the remote monitoring of the vital parameters of patients. This will reduce the probability of frontline corona workers from becoming infected. Here, every patient admitted will have a wearable sensor node that will record his or her vital parameters, such as pulse rate, blood pressure, and temperature, and will report these to multiple sinks. From the sinks, the recorded data may be transmitted to the relevant doctors for appropriate action. The number of transmissions may be reduced by making them event driven. That is, a node worn by a patient will only transmit data when any of the vital parameters is above/below the acceptable threshold.

For the healthcare industry, we propose a three-dimensional deployment scenario. Here, we assume the patients to be on different floors of a hospital building, with

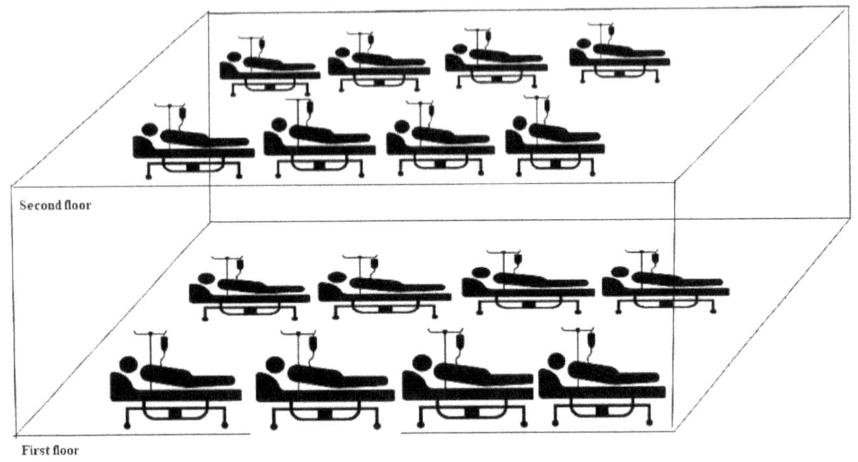

FIGURE 11.3(A) Physical placement of beds on different floors of a hospital building.

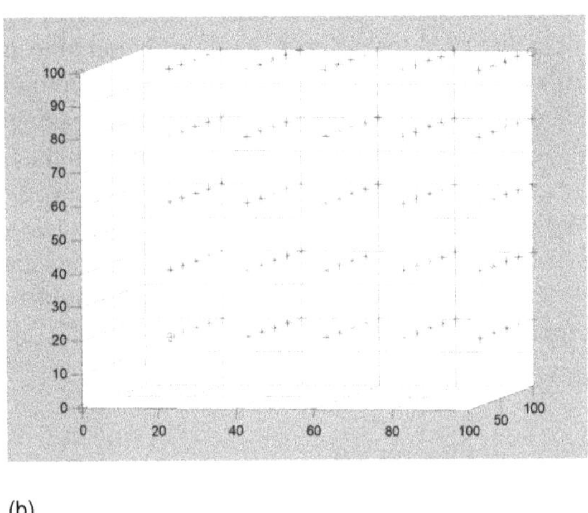

FIGURE 11.3(B) Logical location of the sensor nodes on different floors of a hospital building.

limited mobility, as depicted by Figure 11.3(a). Figure 11.3(b) depicts the three-dimensional-grid deployment scenario.

11.4.3 Application in the Oil and Gas Industry

By adopting IoT applications, the oil and gas (O&G) sector could simultaneously improve safety and increase profits. Asset tracking and predictive maintenance solutions are becoming more and more popular. It is predicted that IoT applications may even become necessary for a competitive edge among escalating geopolitical tensions related to climate change.

The IoT will enable the industry to digitize, improve, and automate the various processes in order to save time, money, and increase safety. The different sub-areas in the O&G industry, where the IoT plays an important part are:

1. **Security of O&G fields:** One of the security issues faced in the O&G industry is sabotage and igniting inflammables. If the proper security measures are not taken the result could be a huge loss of infrastructure, money, and lives. WSN and IoT-based systems are therefore deployed in O&G fields for remote surveillance and monitoring.
2. **Remote operations:** Non-productive time (NPT) leads to the loss of billions of dollars in upstream industries (the initial discovery and extraction of oil). In order to decrease NPT, the IoT can be leveraged using near-real-time data to predict breakdowns and plan preventive schedules. This can further avoid accidents and optimize processes (AbdElminaam, Alenezi, & Ali, 2018). For example, the voluminous amount of reservoir data can be integrated with near-real-time field data to plan the establishment of wells.
3. **Predictive and preventive maintenance:** Most O&G facilities (tank levels and pressure, flow rates, machinery and equipment involved) require monitoring on a regular basis (Abdul-Elminaam & Alenezi, 2017). For example, if a coil is running too hot, its failure is unavoidable and the cause needs diagnosing. Predictive maintenance is performed by relying on the present condition of equipment. It allows O&G companies to perform the remote monitoring of equipment through sensors to make strategic decisions related to whether something needs to be shut down, fixed, or replaced. Preventive maintenance allows an increased chance of diagnosing a failure in a remote manner by tracking the deterioration of equipment in advance. The collected data can prevent an entire process from failing or shutting down.
4. **Health and safety:** O&G sites are usually located in remote and dangerous geographical locations that are not conducive to the health as well as safety of workers. Remote sensors and drones can allow the monitoring of and detecting of leakages; smart sensors can be integrated into aworker's clothing as an intelligent wearable device. By deploying IoT-based safety devices O&G companies will not only keep their workforce safe but further benefit from a reduction in corporate liability.
5. **Asset tracking and monitoring:** IoT sensors are capable of monitoring and surveying equipment more accurately and at reduced costs. They also permit oil companies to monitor processes along with inventory supplies and O&G shipments in order to maximize asset utilization. For example, if an item is found missing, an IoT-based application can be used to track shipments with the precise location of each O&G asset.
6. **Data management:** A massive amount of data collected with numerous connected devices like sensors or video cameras allows O&G companies to implement safe, efficient, and effective practices for operational benefits using data analytic techniques that require big data for accurate predictions, forecasting,

and decision making (Hossein Motlagh, Mohammadrezaei, Hunt, & Zakeri, 2020). Large refineries generate up to 1 TB of raw data per day.

Of the different sub-areas discussed here, we are considering the security of O&G fields. Although sensitive sites are well protected and guarded, the chance of an unforeseen accident or sabotage cannot be ruled out. In order to deal with situations like this and to monitor the security of an oil field, we propose creating a sensor field, similar to the one depicted in Figure 11.1. Here, the nodes detect a human presence and send alarm signals to different security points. The sensors used might be thermal or proximity sensors. Even ultrasonic sensors may be combined with the above mentioned sensors for better precision in intrusion detection.

11.5 RESULTS

Tools such as MATLAB, ThingSpeak, and Simulink help to design, prototype, install, and test IoT-based systems and applications for real life scenarios. Arduino-based systems are capable of communicating with MATLAB through serial ports present in Arduino boards (via a USB connection) or through buses like SPI, I2C, or Bluetooth.

Keeping these facts in mind, the proposed layout was emulated in a MATLAB environment. It was seen that the lifetime (L) of the network is maximized when a DoP value of 3 is reached by changing the transmission range (TXR) of the nodes. This finding is shown in Figure 11.4(a) and 11.4(b) for two and three-dimensional deployment scenarios respectively.

The network parameters taken for two and three-dimensional deployments are given in Tables 11.1 and 11.2 respectively.

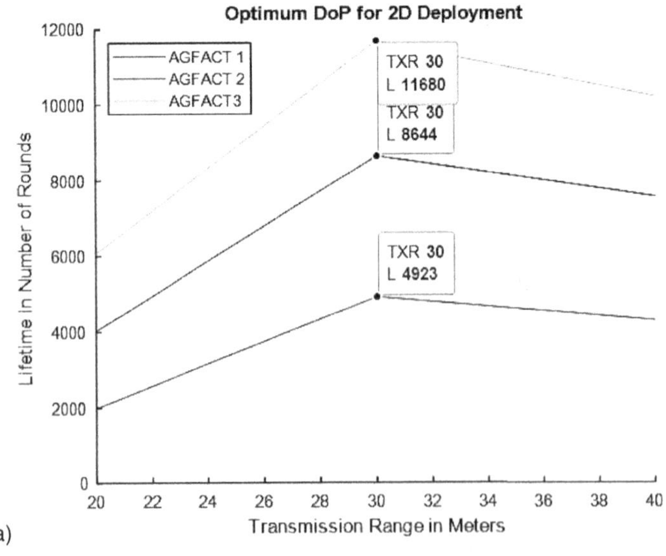

FIGURE 11.4(A) Optimum degree of proximity in a two-dimensional deployment for maximum lifetime.

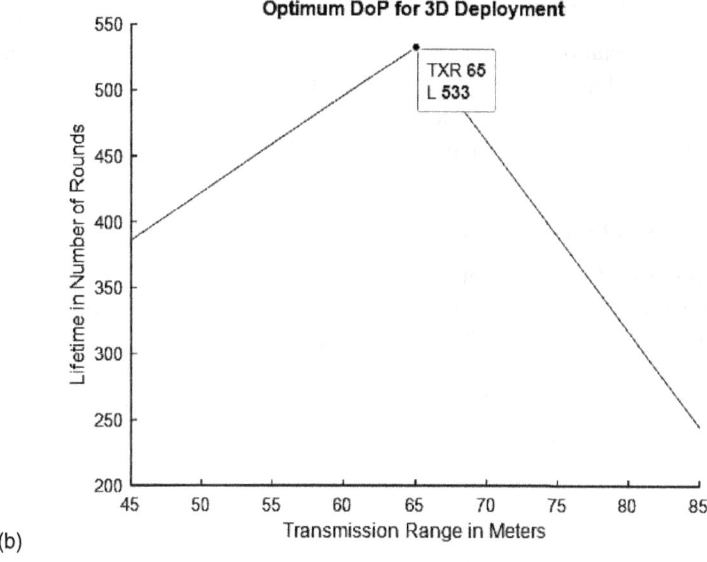

FIGURE 11.4(B) Optimum degree of proximity in a three-dimensional deployment for maximum lifetime.

TABLE 11.1
Network Parameters for a Two-dimensional Deployment

Network Parameter	Value
Nodes	100
Initial energy of nodes	1 J
Area of coverage	100 m × 100 m
Number of sinks	3
Grid size	10 m × 10 m
Path loss exponent (PLE)	3

TABLE 11.2
Network Parameters for a Three-dimensional Deployment

Network Parameter	Value
Nodes	125
Initial energy of nodes	1 J
Area of coverage	100 m × 100 m×100 m
Number of sinks	3
Grid size	20 m × 20 m
Path loss exponent (PLE)	3

11.6 LIMITATIONS

One primary limitation of the proposed framework is that it is solely sensor based. For accurate readings and for doing away with false alarms (e.g., intrusion detection), we may need to incorporate and integrate cameras with the framework proposed in this chapter. This we have decided to include in our future work.

11.7 CONCLUSION AND FUTURE WORK

The IoT is one of the foundation pillars of the Industry 4.0 revolution. A holistic view about the subject is therefore required for determining the application areas, where the IoT can fit in. Today, the IoT has found its application in different industries having altogether different customer bases. It is therefore the need of the hour to propose a common IoT-based framework that might be used for different applications in different industries, requiring minimum improvisation. In this chapter we proposed such a framework that might be used for different applications in three completely different industries.

As sensor nodes are the backbone of any IoT application, the framework proposed here is also based upon their grid deployment. Our framework may be used for both two and three-dimensional deployment scenarios. It provides effective data communication in an energy efficient manner, between the deployment site and the sinks. Although the industries discussed here are agriculture, healthcare, and oil, the same framework may be used for a range of other industries as well, for the purpose of data collection and/or generating security alerts.

For future work, along with the incorporation of cameras and image processing, we have included some other industries as well, where the same framework may fit for some relevant applications. For example education, military, and habitat monitoring.

REFERENCES

AbdElminaam, D. S., Alenezi, T. M. M., & Ali, M. A. (2018). Smartsepog: IoT based system for enhancement of the performance of KJO oil and gas fields in Kuwait. *Far East Journal of Electronics and Communications*, *18*(6), 915–944.

Abdul-Elminaam, D. S., & Alenezi, T. M. M. (2017). Building smart oil and gas field using IOT. *International Journal of Advancements in Computing Technology (IJACT)*, *9*(3), 43–56.

Anastasi, G., Falchi, A., Passarella, A., Conti, M., & Gregori, E. (2004, October). Performance measurements of motes sensor networks. In *Proceedings of the 7th ACM International Symposium on Modeling, Analysis and Simulation of Wireless and Mobile Systems* (pp. 174–181).

Baker, S. B., Xiang, W., & Atkinson, I. (2017). Internet of things for smart healthcare: Technologies, challenges, and opportunities. *IEEE Access*, *5*, 26521–26544.

Banka, S., Madan, I., & Saranya, S. S. (2018). Smart healthcare monitoring using IoT. *International Journal of Applied Engineering Research*, *13*(15), 11984–11989.

Dietrich, I., & Dressler, F. (2009). On the lifetime of wireless sensor networks. *ACM Transactions on Sensor Networks (TOSN)*, *5*(1), 1–39.

Dulski, R., Kastek, M., Trzaskawka, P., Piątkowski, T., Szustakowski, M., & Życzkowski, M. (2011, June). Concept of data processing in multisensor system for perimeter protection.

In *Sensors, and Command, Control, Communications, and Intelligence (C3I) Technologies for Homeland Security and Homeland Defense X* (Vol. 8019, p. 80190X). International Society for Optics and Photonics.

Ghosh, K., Das, P. K., & Neogy, S. (2015). Effect of source selection, deployment pattern, and data forwarding technique on the lifetime of data aggregating multi-sink wireless sensor network. In Rituparna Chaki, Khalid Saeed, Sankhayan Choudhury, and Nabendu Chaki (eds.), *Applied Computation and Security Systems* (pp. 137–152). Springer.

Ghosh, K., Das, P. K., & Neogy, S. (2016). Kps: A fermat point based energy efficient data aggregating routing protocol for multi-sink wireless sensor networks. In Rituparna Chaki, Khalid Saeed, Sankhayan Choudhury, and Nabendu Chaki (eds.),*Advanced Computing and Systems for Security* (pp. 203–221). Springer.

Ghosh, K., Neogy, S., Das, P. K., & Mehta, M. (2017). On regulating lifetime of a 3-sink wireless sensor network deployed for precision agriculture. *International Journal of Next-Generation Computing*, *8*(3), 153–170.

Ghosh, K., Neogy, S., Das, P. K., & Mehta, M. (2018). Intrusion detection at international borders and large military barracks with multi-sink wireless sensor networks: An energy efficient solution. *Wireless Personal Communications*, *98*(1), 1083–1101.

He, J., Fallahi, M., Norwood, R. A., & Peyghambarian, N. (2011, June). Smart border: ad-hoc wireless sensor networks for border surveillance. In *Sensors, and Command, Control, Communications, and Intelligence (C3I) Technologies for Homeland Security and Homeland Defense X* (Vol. 8019, p. 80190Z). International Society for Optics and Photonics.

Heinzelman, W. B., Chandrakasan, A. P., & Balakrishnan, H. (2002). An application-specific protocol architecture for wireless microsensor networks. *IEEE Transactions on Wireless Communications*, *1*(4), 660–670.

Hossein Motlagh, N., Mohammadrezaei, M., Hunt, J., & Zakeri, B. (2020). Internet of Things (IoT) and the energy sector. *Energies, 13*(2), 494.

Islam, S. R., Kwak, D., Kabir, M. H., Hossain, M., & Kwak, K. S. (2015). The internet of things for health care: A comprehensive survey. *IEEE Access*, *3*, 678–708.

Lee, S. H., & Ko, Y. B. (2006, May). Geometry-driven scheme for geocast routing in mobile ad hoc networks. In *2006 IEEE 63rd Vehicular Technology Conference* (Vol. 2, pp. 638–642). IEEE.

Madan, R., Cui, S., Lall, S., & Goldsmith, A. (2006). Cross-layer design for lifetime maximization in interference-limited wireless sensor networks. *IEEE Transactions on Wireless Communications*, *5*(11), 3142–3152.

Min, R., & Chandrakasan, A. (2001, November). Energy-efficient communication for ad-hoc wireless sensor networks. In *Conference Record of Thirty-Fifth Asilomar Conference on Signals, Systems and Computers (Cat. No. 01CH37256)* (Vol. 1, pp. 139–143). IEEE.

Min, R., Bhardwaj, M., Cho, S. H., Shih, E., Sinha, A., Wang, A., & Chandrakasan, A. (2001, January). *Low-power wireless sensor networks.* In *VLSI Design 2001. Fourteenth International Conference on VLSI Design* (pp. 205–210). IEEE.

Naz, P., Hengy, S., & Hamery, P. (2012, May). Soldier detection using unattended acoustic and seismic sensors. In *Ground/Air Multisensor Interoperability, Integration, and Networking for Persistent ISR III* (Vol. 8389, p. 83890T). International Society for Optics and Photonics.

Parameswaran, G., & Sivaprasath, K. (2016). Arduino based smart drip irrigation system using Internet of Things. *International Journal of Engineering Science and Computing*, *6*(5), 5518–5521.

Roopaei, M., Rad, P., & Choo, K. K. R. (2017). Cloud of things in smart agriculture: Intelligent irrigation monitoring by thermal imaging. *IEEE Cloud computing*, *4*(1), 10–15.

Sahu, C. K., & Behera, P. (2015, February). A low cost smart irrigation control system. In *2015 2nd International Conference on Electronics and Communication Systems (ICECS).* (pp. 1146–1152). IEEE.

Sales, N., Remédios, O., & Arsenio, A. (2015, December). Wireless sensor and actuator system for smart irrigation on the cloud. In *2015 IEEE 2nd World Forum on Internet of Things (WF-IoT)* (pp. 693–698). IEEE.

Saraf, S. B., & Gawali, D. H. (2017, May). IoT based smart irrigation monitoring and controlling system. In *2017 2nd IEEE International Conference on Recent Trends in Electronics, Information & Communication Technology (RTEICT)* (pp. 815–819). IEEE.

Sathya, M., Madhan, S., & Jayanthi, K. (2018). Internet of things (IoT) based health monitoring system and challenges. *International Journal of Engineering & Technology*, *7*(1.7), 175–178.

Sharma, C., & Sunanda, D. (2017). Survey on smart healthcare: An application of IoT. *International Journal on Emerging Technologies (Special Issue NCETST-2017)*, *8*(1), 330–333.

Singh, P., & Saikia, S. (2016, December). Arduino-based smart irrigation using water flow sensor, soil moisture sensor, temperature sensor and ESP8266 WiFi module. In *2016 IEEE Region 10 Humanitarian Technology Conference (R10-HTC)* (pp. 1–4). IEEE.

Wang, N., Zhang, N., & Wang, M. (2006). Wireless sensors in agriculture and food industry—Recent development and future perspective. *Computers and Electronics in Agriculture*, *50*(1), 1–14.

Wark, T., Corke, P., Sikka, P., Klingbeil, L., Guo, Y., Crossman, C., Valencia, P., Swain, D. & Bishop-Hurley, G. (2007). Transforming agriculture through pervasive wireless sensor networks. *IEEE Pervasive Computing*, *6*(2), 50–57.

12 Impact of Deep Learning and Machine Learning in Industry 4.0

Impact of Deep Learning

Umesh Kumar Lilhore, Sarita Simaiya, Amandeep Kaur, Devendra Prasad, Meenu Khurana
Chitkara University Institute of Engineering and Technology, Chitkara University, Punjab, India

Deepak Kumar Verma
Chhatrapati Shahu Ji Maharaj University, Kalyanpur, Kanpur Uttar Pradesh, India

Afsan Hassan
Chitkara University Institute of Engineering and Technology, Chitkara University, Punjab, India

CONTENTS

12.1 Introduction 180
12.2 Industry 4.0 and Its Key Players 181
12.3 Literature Survey 183
12.4 Challenges and Role of Machine Learning and Deep Learning in IR 4.0 183
 12.4.1 Challenges in IR 4.0 183
 12.4.2 Role of Machine Learning and Deep Learning in IR 4.0 186
  12.4.2.1 Machine Learning Methods in IR 4.0 186
  12.4.2.2 Deep Learning Methods in IR 4.0 188
 12.4.3 Advantages of Machine Learning and Deep Learning in IR 4.0 190
 12.4.4 Applications of Machine Learning and Deep Learning in IR 4.0 190
12.5 Proposed Model for IR 4.0 192
12.6 Conclusions and Future Work 193
References 194

DOI: 10.1201/9781003146711-12

12.1 INTRODUCTION

Intelligent production devices require creative technologies to optimize the effectiveness and productivity of production processes and yet at the same time to lower expenses. In another framework, artificial-intelligence (AI)-driven innovations – collateralized through IR 4.0 main transforming methods such as the Internet of Things (IoT), cloud computing, advanced embedded systems, virtual reality, cognitive systems, big data, and augmented reality – are capable of creating modern manufacturing frameworks. IR 4.0 may not be the future, but perhaps the truth. This has been around for more than a few decades and will continue to grow throughout the coming years, adding creativity and productivity through development. It has become essential for everyone who focuses on that stage of industrialization as it is much more interesting, purpose-driven, as well as profitable (Peters et al., 2020).

The phrase "IR 4.0," or the industrial revolution, refers to the implementation of digital technologies, mostly in munitions factories. However, this terminology is now commonly used to refer to organizational change inside the industrial sector. It is a general term that can be used to refer to the employment of cyber-physical devices, including the IoT, machine learning (ML), deep learning (DL), cloud computing, and robotics. A few other applications concerning the use of AI in IR 4.0 will also be discussed in detail (Çınar et al., 2020).

ML and DL generate frameworks through sample training as well as testing, and extend the learning to new applications. DL is just a subfield of ML, and unlike simple ML, DL networks frequently depend upon layers of neural networks called "artificial neural networks," which correspond to new representations of a database. Whether or not specific ML algorithms still need guidance from the developer for the forecast to be successful, a deep neural network will decide by itself (Villalba-Díez et al., 2020).

- **IR 1.0:** An invention of the 21st century and a consequence of technological developments and industrialization.
- **IR 2.0:** At the start of the 20th century, a factor that contributes to IR 2.0 is the emergence of energy produced, which improves the effectiveness of manufacturing operations.
- **IR 3.0:** The next technological revolution is planned for the electronics industry and the production sector, including business technology.
- **IR 4.0:** The fourth phase strengthens reproduced corporeal systems and human interface communication, and communicates, measures, as well as coordinates smarter activities aimed at various industrial activities so as to automate the technology (Khan et al., 2020).

The foundation of IR 4.0 can be described mostly by digital technologies as well as the automation of modern integrated logistics operations with the use of mobile and Internet devices in machines and products. Virtual and physical environments converge through automation products, which have made popular international systems of social devices. Business intelligence challenges have been increasing in difficulty, although there is a clear necessity to incorporate advanced technologies that seem worthwhile (Romeo et al., 2020).

12.2 INDUSTRY 4.0 AND ITS KEY PLAYERS

Industrial activities have a significant impact on economic growth and social development. More as a widely accepted framework for research organizations including institutions, the IR 4.0 framework has gained great interest from both the industrial and technological environments. While the concept is not fresh and has been on the agenda for intellectual research for many years and through different interpretations, the phrase "Industry 4.0" has only just been coined and is becoming very well recognized not just in intellectual circles but also in economic ones. Strengthening IR 4.0 involves integrating the IoT, data analysis, AI, expert systems, ML, DL, and the cloud: a technology that facilitates highly automated start-ups (Büchi et al., 2020). Now we have a clear understanding of what IR 4.0 is, let us take a glance at what else is out there. Each module that facilitates IR 4.0 can be described under nine important aspects:

1. **System integration:** In IR 4.0 the systems are highly advanced around their services, and yet they attempt to connect with several other applications. Requirements through external analysis promote the simple transmission of learning, mostly to the enterprise but also to the client and the individual consumer. This may include identifying popular network communication environments, such as those established in accordance with the need for information about the job, source images, and a hard copy of documents.
2. **Big data and data analytics:**IR 4.0 structures are gradually authenticated as well as linked; a significant amount of information can be collected and evaluated. However, the volume of information has become an obstacle. The level of detail makes it extremely difficult to determine whether information is necessary, including patterns that can contribute towards wise as well as intelligent assessments. That is where "big data" analytics is growing in IR 4.0. Big data analytics makes it possible to define the performance of an individual factor and its functional deficiencies, through a sufficient quantity and preventing potential manufacturing problems and finding proactive measures (Pokhrel & Garg, 2020).
3. **Virtualization and simulation:**The virtualization of ecosystems, including simulation processes in IR 4.0, make it more difficult to test various situations. If a technology is assessed, cost-effective methods can also be established, tested, and implemented much faster, ultimately due to the low consumer costs and delays. The benefits of the simulation would be color monitoring and control, where in-line changes have been used to decrease the defined hours and also to enhance the subsequent interaction process.
4. **The IoT:**This is a central player in IR 4.0. The IoT connects physical devices over the infrastructure to collect information for all the final selection phases. This integrated computing strengthens the efficiency as well as the likelihood of its devices getting manufactured (Demertzis et al., 2020).
5. **Cloud computing:**In IR 4.0 the cloud plays a key role. Cloud computing is used for applications including service, display performance plans, stress testing, as well as its function in many other industry sectors which continue

growing. With consistent technological improvements, digital information including usability will indeed begin to move to cloud computing services. Cloud computing makes it possible to carry out enhancements and vitality utilization, including distribution solutions, significantly quicker than any other stand-alone framework. The manufacturing sector has shown a dramatic change in the utilization of cloud services.

6. **Cyber security:** As it moves towards multiple routes (to expanded communication through cloud services), security is essential. Data security and integrity allow the practical application of such a real modern system and digitally creates an enabling environment, taking full advantage of the benefits of a globalized economy.
7. **Automated robots:** While automation is in its early stages in the visual communication industry, researchers have seen a growth in usage across specialized production processes; operations including the transport of products (as in the Cox Focus Advertising site) and product inventory control. Researchers expect to boost the levels of effectiveness through communication between individuals and technologies, and they foresee expanded usage of automation among innovative industries. Econ et al., 2020).
8. **Augmented reality:**IR 4.0 is progressively used for delivering real-time services in an effective way that helps people to properly integrate as well as communicate. Descriptions may involve the communication of maintenance feedback to a component as seen in portable platforms or indeed in the retention of employees utilizing animations, including Create 3d of a manufacturing plant or machinery.
9. **Additive manufacturing:** IR 4.0 continues to be particularly important for small quantity implementation and the manufacture of individual components or quality services. It can be used effectively with customers by manufacturers to enhance models through efficiency improvements, versatility, and cost optimization.

Compelling factors for other businesses to move towards IR 4.0 (Ghobakhloo et al., 2020) and automation include:

- Boosting productivity.
- Reducing manual interference.
- Improving efficiency.
- Reducing cost.
- Focusing personal effort towards non-overlapping activities to increase productivity.

Businesses compete against daunting obstacles in the implementation of such technological developments. To create and maintain leadership in the quest for full adoption, companies need to extend further their technical experience of emerging technologies and the relevant usage applications, and afterward design and evaluate personalized marketing production methods.

12.3 LITERATURE SURVEY

ML and DL methods play an important role in Industry 4.0. IR 4.0 is mainly related to smart industry management. This section presents an overview of the related studies of Industry 4.0 and ML, as well as the ML techniques. Various researchers have suggested numerous methods in the IR4.0 field, a few of them are listed in Table 12.1.

Industry 4.0 is mainly based on automation of industry work. ML techniques in manufacturing processes can be recognized and enhanced intelligently. This is accomplished with the examination of the data obtained through the manufacturing process. Throughout this assessment, existing innovations are acquired which can adapt to growing companies constantly. Thus the various personal procedures are not only stronger linked but can be improved.

12.4 CHALLENGES AND ROLE OF MACHINE LEARNING AND DEEP LEARNING IN IR 4.0

Automation and robotic manufacturing processes move hand in hand. Their convergence is accomplished through the correct computer linking. Further focus needs to be paid to the difficulties in the Industry 4.0 revolution across the world (Das et al., 2020).

12.4.1 CHALLENGES IN IR 4.0

Based on the literature survey, the following key challenges need to be addressed in researching IR 4.0:

- **Security:** The integration of production materials increases the risks to the safety of the whole processing facility. When the entire network is breached via an attack from malicious users, the development cycle may be changed. In instances where a robotics merging process directly impacts humans, then they may be exposed to risk if the robotics are corrupted (Chauhan et al., 2021).
- **The complexity of the system:** Currently, the automotive sector is faced with emerging demands in terms of difficulty and external disturbances, although maintenance of the manufacturing base is influenced by instability. The relentless growth of big data, combined with its accessibility, poses a significant challenge to the industrial climate because information cannot be removed.
- **Misuse of data:** There is sometimes widespread concern about the storage of big data regarding confidentiality, financial advantage, and protection because most companies process transactions on digital cloud platforms. Manipulation of information throughout the manufacturing industry is already on the rise, as most of the tools involved in data collection are remotely operated. Additional studies are also needed to enhance storage space protection safety.
- **Integration:** A manufacturing industry becomes primarily distinguished by either a tradition of replacement or one of repair of machinery after it has been damaged. Many devices are used to the point where it becomes difficult to get replacement pieces. Industries recommend using them but repairing machines will be

TABLE 12.1
Analysis of Various Existing Work in the Field of Industry 4.0

Reference	Key Method Used	Data Set/Outcomes	Challenges/Limitations
A. Angelopoulos et al., 2020	➢ Deep learning method	➢ Data collected from 4600 respondents. ➢ Tested against the research model by using structural equation modeling.	➢ Integration of data takes more time. ➢ Takes more time for decision making.
A. Essien and C. Giannetti, 2020	➢ Machine learning method	➢ A comprehensive review of the recent advancements of ML techniques. ➢ ML category, machinery, and equipment used, the device used in data acquisition, classification of data, size, and type, and highlights the key contributions.	➢ Poor identification of required data. ➢ Consumes more resources.
J. P. Usuga Cadavid et al., 2019	➢ Geometric deep learning	➢ Big data. The acquisition of data associated with the cyber-physical systems of Industry 4.0.	➢ Open security system. ➢ Individual privacy. ➢ Terminal-security function.
A. Diez-Olivan et al., 2020	➢ IoT-block chain, ➢ advanced deep learning	➢ Achieved percentile latency, response time, and resource utilization by the proposed method.	➢ The result can be enhanced by investigating more complex business networks.
J. A. Carvajal Soto et al., 2020	➢ Machine learning-based design support system	➢ Achieved better precision, recall, and accuracy.	➢ Data security needs improvement.
L. Ma et al., 2020	➢ Machine learning	➢ Builds an operation of the concepts of openness to Industry 4.0 and performance.	➢ Set-up cost is higher. ➢ Static data.
M. Carletti et al., 2019	➢ Deep Q-network	➢ The proposed scheme significantly outperforms the baseline schemes. ➢ Highly dynamic and multidimensional network environments, i.e., mobile nodes and time-variant settings.	➢ As long as the rate of change into the system is in dynamic mode, speed becomes slower. ➢ Data security issues.
D. A. Rossit et al., 2019	➢ Blockchain deep learning	➢ Enhances the capabilities of the network, ➢ Blockchain security architecture improves security.	➢ Data integrity is challenging. ➢ Takes more time.
S. Mittal et al., 2019	➢ Deep learning	➢ Enhances response of IR 4.0 devices.	➢ Data processing time is greater.
J. Francis et al., 2019	➢ Geometric deep learning	➢ Provides improved results for IR 4.0.	➢ Data security is challenging

A. G. Frank et al., 2019	➢ Information and digital technologies	➢ Perceived benefits and management support. ➢ Explains the complex precedence relationships that exist among determinants of smart manufacturing IDT adoption.	➢ Digitalization in Industry 4.0 needs data security.
D. P. Penumuru et al., 2020	➢ Machine-learning solutions	➢ Industry 4.0 ecosystem has the role of human operators and workers. ➢ Allowing humans to be in the loop of the manufacturing processes in a symbiotic manner with minimal errors.	➢ High error rate. ➢ Security is not covered.
G. Aceto et al., 2019	➢ Deep learning, model ➢ supervised learning using a sliding-window approach	➢ Empirical analyses based on real-world data of metal packaging plant collected from the United Kingdom. ➢ Improved results for accuracy over existing methods.	➢ Machine speed can be used in time production to improve speed. ➢ Multivariate time series based on machine states as well as external sensor data that can be applied.
D. B. Durocheret et al., 2021	➢ Machine learning	➢ IoT for data collection. ➢ ML model to adapt the production system changes.	➢ More real-time data and methods can be applied.
A. Majumdar et al., 2021	➢ Data fusion ➢ ML model	➢ Apply data fusion and data analysis for industrial prognosis. ➢ Determine predictive monitoring. ➢ Minimize impact of prescription on the industry.	➢ More methods can be applied for comparison.
R. Zhou et al., 2021	➢ Machine learning	➢ Mainly describe how to build a scalable and flexible system for real-time applications. ➢ Better accuracy and production performance.	➢ Security can be applied. ➢ More data/parameters can be applied.
V. J. Ramírez-Durán et al., 2021	➢ Deep learning model	➢ Improves average prediction error. ➢ Improves the prediction accuracy.	➢ The real-time dataset can be applied. ➢ More ML methods can be used.
C. Schockaert et al., 2020	➢ Machine learning	➢ Threefold analysis, hardware and software implications. ➢ Analyze what caused the failure in the industry.	➢ More parameters can be applied. ➢ The real-time dataset can be used.

expensive, and light manufacturing companies are unable to satisfy the burden whenever the original equipment is functioning properly (O'Donovan et al., 2019).

- **Insufficient skills:** Adding intelligent manufacturing processes to something like a corporation is of high significance as having the capacity to realize the innovation. When an organization is preparing to introduce a smart factory, this must also have the necessary expertise to manage the machinery involved. Since this is high technology, most factories are confronted with the task of recruiting additional staff with the right knowledge or with training current staff in an intelligent production climate (Misra et al., 2021).

12.4.2 Role of Machine Learning and Deep Learning in IR 4.0

It is perhaps necessary for businesses to acknowledge, first and foremost, the functionality, as well as the quality, of IR 4.0, and its future transition through machine, controlled production to smart technologies. To accomplish an effective transition, participants must explicitly evaluate their functions, including the corresponding potential outcomes, against the basic criteria set out in the IR 4.0 norm.

This might enable others to produce a more well-defined blueprint. A few strategies including debates have taken place on this, as well as plans that are already being developed. ML is being used successfully throughout fields of massive information, including object recognition. But on the other hand, all the DL strategies train continuously by changing the behavior of continuous processes to maximize the required outcome. AI procedures can also be used in combination for specific processes or by implementing several methods sequentially or simultaneously. DL models will be used to enhance information gathering as well as management, contributing to the detection of potentially faulty items across several manufacturing plants. These same results were obtained by ML as well as deep ML approaches across the area of AI. Throughout recent decades, a leading division of AI has turned out to beML, including DL (Wang et al., 2021).

12.4.2.1 Machine Learning Methods in IR 4.0

The importance of ML is apparent in production and data-driven analysis through data processing. In comparison to basic computer guidance, ML is advanced in evaluation. Essentially, the topic is closely linked to computational statistics and percentages. It also assumes a significant function in the evaluation of insight participation in management services. Authors refer to this subject as mathematical programming, including heuristic classification algorithms (Teoh et al., 2021). The model building, like that at the center of all operating conditions, is optimized by utilizing ML technologies.

The major forms of ML are differentiated as:

- **Supervised learning:** The system is trained in supervised learning using input that has been tagged. The tags assign each dataset to one or more classifications. Therefore the model neither learns why this learning information is represented nor uses it through parameters to identify the new correct output. An ultimate objective of the accomplished learning algorithm is that the outputs are similar to, or can even be used with, all design ranges.

- **Unsupervised learning:** This is efficient training where an assessment of the activity isn't based, given, or supervised, since there is no competent specialist. An objective of unsupervised learning is just to explore the set of related input feature findings, especially in the form of clusters of data items.
- **Reinforcement learning:** This learning introduces a further ML plan that relies upon learning from past mistakes. It is a mixture of unsupervised and supervised teaching strategies, in which an incredible opportunity lies within the dataset. It is balanced through supervised and unsupervised learning. This learning system seeks feedback when communicating through industrial operations as well as implementing incremental judgments to enhance the possible benefits. In reinforcement learning, rather than training datasets that just represent the successful result for just a particular item, information technology is thought to provide an indicator of whether or not the behavior is appropriate. Unless the response is not right, the issue of selecting the right response persists (Faridi et al., 2020).

Some important ML methods which are widely used in IR 4.0 are:

- **Support vector machine (SVM):** The supervised training algorithm is often used for either linear or nonlinear challenges, including classifiers. SVM was established based on the concept of constructing a flat hyperplane and perhaps a series of the hyperplane (Pankova et al., 2020). A hyperplane mainly splits the high-dimensional as well as limitless-dimensional vector space across different sections, with only a decision boundary gap between the two closest data points, including its categories. The objective of its SVM is to generate a model, based on training examples, which determines the evidence test point at which a subsection of training data becomes required. Following the objective(i.e., to obtain the optimum precision forecast), SVM involves a strong representative sample (Lilhore et al., 2020).
- **Decision tree:** This is a machine learning algorithm that can be effectively recognized by humans for its visual analysis. The issue is to identify an appropriate form of a decision tree using training data. Hinton et al. (2012) distinguishes between two types of decision trees. The first form is a classification tree that provides a categorical outcome, while the second type is a regression model that provides a numerical performance. A further disadvantage of a decision tree is the failure to overcome nonlinear equations, unlike with the SVM classifier; however, the frequency of training is quite strong. The decision tree provides widespread use in discovery, including forecasting challenges, owing to its capacity to perform so strongly on the critical features of data collection (Tiwari et al., 2018).
- **Rule-based learners:** This is often recognized as an expert system and considered to be one of the most relevant uses of AI in conjunction with data analysis. Often, rule-based learners have been used to acquire data based on demographic importance, with the aid of "if-then" principles. While they are applicable across both supervised and unsupervised scenarios, individuals are more commonly used during unsupervised teaching practices for IR 4.0 because of their comprehensibility. The major advantage of rule-based learning seems to

be that the framework can fully understand how well the outcome was created, even where supervised learning always systematically controls all data sources (Lilhore et al., 2020).

- **K-nearest neighbor:** A k-NN method is the ML technique used in nonlinear equations (i.e., classification as well as pattern detection), which involves the measurement of ranges among test data in which each input image is labeled by its nearest k test. A K-NN algorithm is indeed very responsive towards data points that are incomplete, fuzzy, noisy, meaningless, as well as inconsistent, or where the rate of identification is weak. In particular, this contributes to precision, which is also very poor. A k-NN method is also difficult to detect due to its unstructured array of training wide datasets. In comparison, k-NN has a strong learning rate which allows for one of the quickest training ML techniques (Lilhore et al., 2020).
- **Naive Bayesian:** NB represents the ML algorithm, which is defined as a very simple model composed of acyclic graphs. Acyclic graphs of the NB model have only one parent and multiple children. The parent represents the unobserved node, while the children are observed nodes. NB algorithms are usually less precise than some other ML algorithms that are more sophisticated (e.g., artificial neural network (ANN)) due to the supposition of independence among children. NB is useful for problems related to classification, regression, clustering, and others. Also, it requires little storage space during both stages: training and classification. But the major advantages of the NB classifier are its short computational time for training, fast process of learning, and ability to work with big, fuzzy, noisy, and incomplete data (Tiwari et al., 2018).
- **Artificial neural network:** An ANN is a method of ML that is popular across nonlinear classification and regression difficulties, implemented in different manufacturing sectors because it performs a significant function in today's parallel processing crisis, in which it replicates the decentralized "computation" of the actual computational structure. An ANN helps the virtual users to manage supervised, unsupervised, as well as enhanced learning activities. A challenge for more than just ANN is to achieve a high level of accuracy in which big data is considered necessary. There are other problems related to coping with over-fitting, lacking numeric value, the pace of training, as well as the nature of the prototypes they develop (Feng et al., 2021).

12.4.2.2 Deep Learning Methods in IR 4.0

This is emerging among the main practices across fields such as manufacturing, tourism, smart devices, the IoT, and automobiles. Also with the growing utilization of ML, businesses are exploiting their software to become a portion of IR 4.0. Throughout this section, we discuss different approaches to DL in IR 4.0, including their commercial application (Mat Lazim et al., 2020).

12.4.2.2.1 Convolutional Neural Networks

The approach which has introduced remarkable success to deep language is based on the convolutional neural network(CNN), initially proposed by Yann Le Cun (1998). Since then, developers have improved on this framework and started to create networks with much better biological complexities. Designs with these acceptable results are able to work inIR 4.0 in various automation sectors.

The significance of CNN lies in all the tasks that it is capable of performing:

- Semantic segmentation and classification.
- Classification and localization.
- Object detection.
- Detection of literary objects.
- Segmentation of cases.
- Approximate human exposure estimate.

12.4.2.2.2 Auto-encoders

This is a neural network that passes the input to its auto-encoder (AE) output value. Usually, the AEs consist of two main parts: the encoder as well as the decoder. The method mainly attempts to extract useful data attributes; the latter attempt to reproduce the inputs as long as they satisfy the data. However, a network that knows how to transfer input data to its target is not very effective. The purpose of AEs is to generate feature vectors (i.e. learning similarities between input data). To achieve non-trivial learning outcomes of the AE, developers need to incorporate another kind of information constraint to the core network. The first approach is to limit the set of invisible patterns which contain increasing neurons in a convolutional layer, as is the situation with a full AE (Khan et al., 2020).

12.4.2.2.3 Recurrent Neural Networks

The traditional neural network seems to be uncertain regarding primary states, which is a critical limitation in trying to deal with time-dependent problems. Solving speech recognition, machine translation, channels offered, and emotional interpretation require the infrastructure which can create decisions on behalf of the given frame as well as the details of its prior groups combined. This is the feature of recurrent neural networks (RNNs). The distinctive alteration is a circle that allows data to flow from one phase of the system to the next.

An integrated treatment theory including useful illustrations is set out below. Unlike many other channels, RNNs also seem to be qualified to work through adjustable output variables (Ting et al., 2019). A further feature of RNNs is the activity of a series of variables across time as well as a non-fixed number of equations. They may view the design of the RNN of many iterations of the very same version, for each system forwards a response to the replacement. The RNN may use the current method to direct the assigned tasks, and sometimes there is a large disparity between pertinent data and the location required (i.e., situations in which a further background is required to accomplish the task). In these cases, RNNs will not be able to attribute the details.

12.4.2.2.4 Deep Reinforcement Learning

In 2016, another method was performed against the iconic Go player Lee Sedol, recipient of 18 major championships overall, and generally known as the biggest player of the century. A core premise behind the whole approach is the use of an individual who carries out any action defined by his policy. The actions carried out in the world, as well as the reaction, is then detected by the investigator that discovers them in such a new nation. The progress and weakness of the entity's behavior are calculated by compensation (Yew et al., 2019).

12.4.2.2.5 Generative Adversarial Networks

An exciting DL method began to emerge when Ian Goodfellow published his article in 2014. Generative adversarial networks (GANs) consist of two independent generators as well as classifier systems. A generator produces new data instances, and the discriminator tests them for validity. The generator's ambition is just to create pretty decent data which is distinct from actual statistics. The purpose of the classifier is to correctly predict data: either true (information from training data) or false (generator produces) (Juhary,2019).

12.4.3 Advantages of Machine Learning and Deep Learning in IR 4.0

ML and DL approaches play an important role in IR 4.0, which is a major change among manufacturing companies that can open up new business markets and deliver benefits such as efficiency increases (Juhary, 2019).

The benefits are:

- **Cost reduction:** by predictive analytics leading to far less maintenance work, which implies cheaper labor, and decreased stock and wastage of resources.
- **Predicting future life span:** Heaping praise on the actions of machinery and appliances contributes to environments that boost efficiency while preserving the health of the system. Forecasting future life span, decreases the circumstances that cause unnecessary costs.
- **Enhanced supply-chain planning:** by better inventory management and also well-monitored and coordinated external factors.
- **Autonomous appliances as well as equipment:** use of stand-alone excavators including equipment to streamline operations by accepting containers through passenger cars, ships, trucks, and other HGVs.
- **Better quality management:** implementable knowledge to continuously increase customer satisfaction.
- **Enhanced human–machine cooperation:** thus enhancing the safety of workers and increasing performance.
- **Budget friendly:** Consumer-focused production can adapt rapidly to shifts in market performance.

12.4.4 Applications of Machine Learning and Deep Learning in IR 4.0

The main applications of ML and DL in IR 4.0 (Krima et al., 2020) are:

- **Smart process technology transformation:** The application of ML approaches through manufacturing processes, which can be recognized and enhanced smartly. This is done by analyzing the data obtained through development.

Existing innovations are acquired through such an assessment, which can cause errors and changes in production. Thus, the various individual procedures can not only be improved, but they can also be streamlined. This same automated, as well as real-time, integration of these improvements is regarded as "smart manufacturing" in IR4.0 (Wang et al., 2021).

- **Autonomous vehicles and machines:** Autonomous vehicles, as well as machines, are a particular case of implementation of AI-based techniques (i.e.,ML). In the industry, its use makes it possible to replace workers involved in repetitive activities or actual hazards.
- **Quality control:** Typically, the habits of consumers are measured mostly at the correct stage. The use of detectors, along with AI, allows continuous performance evaluation for each level of development.
- **Predictive maintenance:** In recent decades, miniaturization, as well as affordable sensors, has enabled the calculation of the condition of devices and obtain valuable data on the progress of each one of them. It makes machine management more accessible. Therefore, the real-time perception of the whole can indeed be achieved by multiplying hundreds or even thousands of measuring points in the devices. The above sets of data will be used later to teach machine supervised learning which can reduce the likelihood of inappropriate activity, including defects in the parts (Feng et al., 2021).
- **Demand prediction:** The sensitivity of production to requests is among the basic complaints in the business, particularly when an inventory is essential for life and therefore cannot be preserved at a time of heightened competition. The production of energy is also an illustration of this. Deficits cannot be recovered if the circumstances for growth are beneficial. Also, the introduction of alternative energy sources, where output varies, makes the system increasingly difficult. In this area, the emergence of ML allows everyone to meet energy consumption in an ideal manner. Even so, the expected demand can be estimated based on any past patterns in the usage of resources. On the other hand, however, the production of renewable energy can indeed be assessed, based on precipitation variables (Madireddy et al., 2019).
- **Chatbots:** Chatbots are programs where only consumers have a message or perhaps even audio communications. This can then be used to obtain data or conduct tasks that reduce the obstacles to obtaining solutions. Individuals are also used for client services as a backing track to consumers and during buying decisions or as channels for consulting with skilled human beings. Indeed they can decrease expenses as well as provide a 24/7 service, yet they are also used for learning from the most basic things.
- **Prototyping and outlining AI:** Existing plug-and-play learning models are mostly built to fix only particular issues. In certain cases, therefore, businesses need to apply further experiential and adaptive methods for applying AI software. Rapid application developments through small-scale implementation become helpful in analyzing an appropriate strategy before operational AI applications are rolled out. The prototyping also represents the position of smaller companies and their potential in IR 4.0 for sustainable development (Wu et al.,2020).

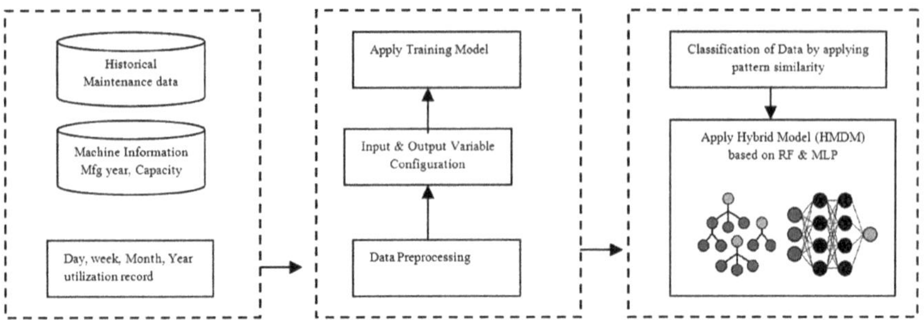

FIGURE 12.1 Proposed HDML model.

12.5 PROPOSED MODEL FOR IR 4.0

This section proposes an early fault and maintenance detection method using ML and DL (an HMDL model) for IR 4.0. The proposed HMDL model is based on the random forest ML method and multilayer perceptron networks (DL). Figure 12.1 shows the various modules in the proposed HMDL model.

The proposed HMDL model utilizes the following methods:

- **Random forest classifier (RFC):** An integration of classification models of the decision tree as well as the ensemble of trees which further considers the voting of the most powerful class. RFC seems to work well for validating, is not prone to over-fitting, and offers extra details such as particle significance. The massive dataset size can, however, result in high memory usage.
- **Multilayer perceptrons (MLPs):** A category of neural network that represents the brain properties of humans. This is a structure of neurons interlinked and supposed to represent a nonlinear mapping, with an input layer as well as a vector of the outcome. For small issues, the methodology performs excellently. It demonstrates one advantage: that it does not require either predictions based on the allocation of training samples or even a judgment on the perceived significance of iterated functions. The expense of determining the optimal number of layers in the network of nodes within these frames isn't insignificant, and there is no single way to do just that.

To overcome the difficulty of machine failure and maintenance monitoring, this research has proposed an HMDL model, which provides the following better results to improve performance in IR 4.0 (Al-Shabandar et al., 2021):

- **Classification accuracy:** For any modeling method, higher accuracy is always desirable. Higher values show better performance.
- **Mean square error (MSE):** This denotes the standard deviations of the variations between expected and actual (measured) values. With each sample group, we anticipated the mapping value and analyzed the quality against the real mapping value in terms of an MSE as per Equation 12.1.

$$MSE = \sum_{i=0}^{m} \left(yi - yi'\right) * 1^2 / m \tag{12.1}$$

where m = test length, yi' = prediction value for mapping, yi = actual mapping value.

- **Modeling and predicting time:** Observing the execution time of the ML model to generate the classifier model and to create new forecasts through using test cases combined.

12.6 CONCLUSIONS AND FUTURE WORK

ML and DL are now a foundation for technological development which automates the development process throughout manufacturing industry. The use of AI in IR 4.0 is indeed a development that will radically change the industry shortly. The transition will not only impact major corporations but also small and medium-sized enterprises. A cheaper equipment design for data processing allows innovation to be ever more available. Recent inventions in manufacturing industries can reduce equipment malfunction, improve production cost, and enhance reliability. When improvements are put together and distributed across a large sector, a business may dramatically maximize returns.

The implementation of ML is strongly connected with data analytics, and without it, it would not be feasible to design and execute methodological approaches because extracting useful information has been the most important action to ameliorate deficiency and fix processes. Further enhancement of ML implementation also allows the learning process to be improved. The most significant system is influenced by AI technology. Nowadays, ML algorithms are commonly used in numerous industrial plants, like modeling, control, debugging, safety, and authentication, where even the increase in the public accountability of the whole production system is advantageous for reducing costs without influencing quality performance.

To overcome the difficulty of machine failure and maintenance monitoring this research has proposed an HMDL model, which utilizes the random forest and multilayer perceptron method. In previous workDL and ML models also are implemented in such a heuristic way as to achieve better performance in IR4.0. The future of business can be seen as a mix of advanced computers and innovative apps. All of these important components are wired into a network or device. Machine challenges and future performance training are still strongly linked. The most difficult application concerns the collection of data analytics from production settings so as to have sufficient knowledge and extracted data for the development of ML algorithms capable of automatically learning from patterns and prior conduct. Big data is sometimes irrelevant, fuzzy, noisy, and redundant, which further complicates the process of learning. Also, because the same enormous amount of information is gathered, the space for storage is the other difficult problem that tends to lead to confidentiality, trade, and stability protection issues.

REFERENCES

Aceto, G., Persico, V., & Pescape, A. (2019). A survey on information and communication technologies for Industry 4.0: state-of-the-art, taxonomies, perspectives, and challenges. *IEEE Communications Surveys & Tutorials, 21*(4), 3467–3501. https://doi.org/10.1109/COMST.2019.2938259

Al-Shabandar, R., Jaddoa, A., Liatsis, P., & Hussain, A. J. (2021). A deep gated recurrent neural network for petroleum production forecasting. *Machine Learning with Applications, 3*, 100013. https://doi.org/10.1016/j.mlwa.2020.100013

Angelopoulos, A., Michailidis, E. T., Nomikos, N., Trakadas, P., Hatziefremidis, A., Voliotis, S., & Zahariadis, T. (2020). Tackling faults in the industry 4.0 era—a survey of machine-learning solutions and key aspects. *Sensors, 20*(1), 109. https://doi.org/10.3390/s20010109

Büchi, G., Cugno, M., & Castagnoli, R. (2020). Smart factory performance and Industry 4.0. *Technological Forecasting and Social Change, 150*, 119790. https://doi.org/10.1016/j.techfore.2019.119790

Cadavid, J. P. U., Lamouri, S., Grabot, B., Pellerin, R., & Fortin, A. (2020). Machine learning applied in production planning and control: a state-of-the-art in the era of industry 4.0. *Journal of Intelligent Manufacturing,* 1–28. https://doi.org/10.1007/s10845-019-01531-7

Carletti, M., Masiero, C., Beghi, A., & Susto, G. A. (2019, October). Explainable machine learning in industry 4.0: Evaluating feature importance in anomaly detection to enable root cause analysis. In *2019 IEEE International Conference on Systems, Man and Cybernetics (SMC)* (pp. 21–26). IEEE. https://doi.org/10.1109/SMC.2019.8913901

Carvajal Soto, J. A., Tavakolizadeh, F., & Gyulai, D. (2019). An online machine learning framework for early detection of product failures in an Industry 4.0 context. *International Journal of Computer Integrated Manufacturing,32*(4–5), 452–465. https://doi.org/10.1080/0951192X.2019.1571238

Chang, F. J., Chang, L. C., Kang, C. C., Wang, Y. S., & Huang, A. (2020). Explore spatio-temporal PM2. 5 features in northern Taiwan using machine learning techniques. *Science of the Total Environment, 736*, 139656.

Chang, H. S., Vembu, S., Mohan, S., Uppaal, R., & McCallum, A. (2020). Using error decay prediction to overcome practical issues of deep active learning for named entity recognition. *Machine Learning, 109*(9), 1749–1778 .https://doi.org/10.1007/s10994-020-05897-1

Chauhan, H., Kumar, D., Gupta, D., Gupta, S., & Verma, V. (2021). Blockchain and IoT based vehicle tracking system for industry 4.0 applications. In *IOP Conference Series: Materials Science and Engineering* (Vol. 1022, No. 1, p. 012051). IOP Publishing. https://doi.org/10.1088/1757-899X/1022/1/012051

Çınar, Z. M., Abdussalam Nuhu, A., Zeeshan, Q., Korhan, O., Asmael, M., & Safaei, B. (2020). Machine learning in predictive maintenance towards sustainable smart manufacturing in industry 4.0. *Sustainability,12*(19), 8211. https://doi.org/10.3390/su12198211

Das, L. B., Lijiya, A., Jagadanand, G., Aadith, A., Gautham, S., Mohan, V., & George, G. (2020, July). Human target search and detection using autonomous UAV and deep learning. In *2020 IEEE International Conference on Industry 4.0, Artificial Intelligence, and Communications Technology (IAICT)* (pp. 55–61). IEEE. https://doi.org/10.1109/IAICT50021.2020.9172031

Demertzis, K., Iliadis, L., Tziritas, N., & Kikiras, P. (2020). Anomaly detection via blockchained deep learning smart contracts in industry 4.0. *Neural Computing and Applications,32*(23), 17361–17378. https://doi.org/10.1007/s00521-020-05189-8

Diez-Olivan, A., Del Ser, J., Galar, D., & Sierra, B. (2019). Data fusion and machine learning for industrial prognosis: Trends and perspectives towards Industry 4.0. *Information Fusion, 50*, 92–111. https://doi.org/10.1016/j.inffus.2018.10.005

dos Santos, J. X., Vieira, H. C., Souza, D. V., de Menezes, M. C., de Muniz, G. I. B., Soffiatti, P., & Nisgoski, S. (2021). Discrimination of "Louros" wood from the Brazilian Amazon by near-infrared spectroscopy and machine learning techniques. *European Journal of Wood and Wood Products*, 1–10.

Durocher, D. B., & Sprinkle, L. (2020). The experiences of a global electrical manufacturing enterprise: The journey to become industry 4.0 ready. *IEEE Industry Applications Magazine, 27*(2), 67–75. https://doi.org/10.1109/MIAS.2020.3024476

Essien, A., & Giannetti, C. (2020). A deep learning model for smart manufacturing using convolutional LSTM neural network autoencoders. *IEEE Transactions on Industrial Informatics, 16*(9), 6069–6078. https://doi.org/10.1109/TII.2020.2967556

Faridi, M. S., Ali, S., Duan, G., & Wang, G. (2020, December). Blockchain and IoT based textile manufacturing traceability system in industry 4.0. In *International Conference on Security, Privacy and Anonymity in Computation, Communication and Storage* (pp. 331–344). Springer. http://dx.doi.org/10 10.1007/978-3-030-68851-6_24

Feng, S., Song, K., Wang, D., Gao, W., & Zhang, Y. (2021). InterSentiment: combining deep neural models on interaction and sentiment for review rating prediction. *International Journal of Machine Learning and Cybernetics, 12*(2), 477–488. https://doi.org/10.1007/s13042-020-01181-9

Francis, J., & Bian, L. (2019). Deep learning for distortion prediction in laser-based additive manufacturing using big data. *Manufacturing Letters, 20*, 10–14. https://doi.org/10.1016/j.mfglet.2019.02.001

Frank, A. G., Dalenogare, L. S., & Ayala, N. F. (2019). Industry 4.0 technologies: Implementation patterns in manufacturing companies. *International Journal of Production Economics, 210*, 15–26. https://doi.org/10.1016/j.ijpe.2019.01.004

Ghobakhloo, M. (2020). Determinants of information and digital technology implementation for smart manufacturing. *International Journal of Production Research, 58*(8), 2384–2405. https://doi.org/10.1080/00207543.2019.1630775

Hinton, G., Deng, L., Yu, D., Dahl, G. E., Mohamed, A. R., Jaitly, N., ... & Kingsbury, B. (2012). Deep neural networks for acoustic modeling in speech recognition: The shared views of four research groups. *IEEE Signal processing magazine*, *29*(6), 82–97. https://doi.org/10.1109/MSP.2012.2205597

Juhary, J. (2019). Perceptions of Students: Blended Learning for IR4. 0. *International Journal of Information and Education Technology, 9*(12). https://doi.org/10.18178/ijiet.2019.9.12.1322

Khan, F. (2020). Exploring IR4. 0 Strategy in Post Covid-19 era through Insights from FDI Inflows. Available at SSRN3690636. http://dx.doi.org/10.2139/ssrn.3690636

Khan, P. W., Byun, Y. C., & Park, N. (2020). IoT-blockchain enabled optimized provenance system for food industry 4.0 using advanced deep learning. *Sensors, 20*(10), 2990. https://doi.org/10.3390/s20102990

Kliestik, T., Nica, E., Musa, H., Poliak, M., & Mihai, E. A. (2020). Networked, Smart, and Responsive Devices in Industry 4.0 Manufacturing Systems. *Economics, Management and Financial Markets, 15*(3), 23–29. https://www.ceeol.com/search/article-detail?id=895600

Krima, S., Toussaint, M., & Feeney, A. B. (2020). Toward Model-Based Integration Specifications to Secure the Extended Enterprise. https://doi.org/10.1520/SSMS20200022

Lazim, R. M., Nawi, N. M., Masroon, M. H., Abdullah, N., & Iskandar, M. C. M. (2020). Adoption of IR4. 0 into agricultural sector in Malaysia: Potential and challenges. *Advances in Agricultural and Food Research Journal, 1*(2). https://doi.org/10.36877/aafrj.a0000140

Lilhore, U. K., Simaiya, S., Guleria, K., & Prasad, D. (2020). An efficient load balancing method by using machine learning-based VM distribution and dynamic resource mapping. *Journal of Computational and Theoretical Nanoscience, 17*(6), 2545–2551. https://doi.org/10.1166/jctn.2020.8928

Ma, L., Wu, J., Zhang, J., Wu, Z., Jeon, G., Zhang, Y., & Wu, T. (2019). Research on sea clutter reflectivity using deep learning model in industry 4.0. *IEEE Transactions on Industrial Informatics, 16*(9), 5929–5937. https://doi.org/10.1109/TII.2019.2957379

Madireddy, S., Li, N., Ramachandra, N., Butler, J., Balaprakash, P., Habib, S., & Heitmann, K. (2019). A Modular Deep Learning Pipeline for Galaxy-Scale Strong Gravitational Lens Detection and Modeling. *arXiv preprint arXiv:1911.* 03867

Madireddy, S., Park, J. H., Lee, S., Balaprakash, P., Yoo, S., Liao, W. K., Hauck, C.D., Laiu, M.P.&Archibald, R. (2020). In situ compression artifact removal in scientific data using deep transfer learning and experience replay. *Machine Learning: Science and Technology, 2*(2), 025010. https://doi.org/10.1088/2632-2153/abc326

Majumdar, A., Garg, H., & Jain, R. (2021). Managing the barriers of Industry 4.0 adoption and implementation in textile and clothing industry: Interpretive structural model and triple helix framework. *Computers in Industry, 125,*103372. https://doi.org/10.1016/j.compind.2020.103372

de Menezes, L. N., deAlencar Lira, M. C., & Neiva, L. S. (2021). IoT and knowledge economy: Two strong pillars of industry 4.0. *Scientia cum Industria, 9*(1), 10–15. http://dx.doi.org/10.18226/23185279.v9iss1p010

Mittal, S., Khan, M. A., Romero, D., & Wuest, T. (2019). Smart manufacturing: characteristics, technologies and enabling factors. *Proceedings of the Institution of Mechanical Engineers,Part B: Journal of Engineering Manufacture, 233*(5), 1342–1361. https://doi.org/10.1177%2F0954405417736547

O'Donovan, P., Gallagher, C., Leahy, K., & O'Sullivan, D. T. (2019). A comparison of fog and cloud computing cyber-physical interfaces for industry 4.0 real-time embedded machine learning engineering applications. *Computers in Industry, 110*, 12–35. https://doi.org/10.1016/j.compind.2019.04.016

Pankova, O. V., Ishchenko, O. V., & Kasperovich, O. Y. (2020). Labour and emloyment in a digital transformation: priorities for Ukraine in the context of global trends and formation of industry 4.0. *Economy of Industry, 2*(90), 133–160. http://www.ojs.econindustry.org/index.php/ep/article/view/203

Penumuru, D. P., Muthuswamy, S., & Karumbu, P. (2019). Identification and classification of materials using machine vision and machine learning in the context of industry 4.0. *Journal of Intelligent Manufacturing,* 1–13. https://doi.org/10.1007/s10845-019-01508-6

Peters, E., Kliestik, T., Musa, H., & Durana, P. (2020). Product decision-making information systems, real-time big data analytics, and deep learning-enabled smart process planning in sustainable industry 4.0.*Journal of Self-Governance and Management Economics, 8*(3), 16–22. https://www.ceeol.com/search/article-detail?id=895938

Pokhrel, S. R., & Garg, S. (2020). Multipath communication with deep Q-Network for industry 4.0 automation and orchestration. *IEEE Transactions on Industrial Informatics.* https://doi.org/10.1109/TII.2020.3000502

Pradeep, N., Paramasivam, K., Rajesh, T., Purusothamanan, V. S., & Iyahraja, S. (2021). Silver nanoparticles for enhanced thermal energy storage of phase change materials. *Materials Today: Proceedings, 45*, 607–611.

Rajesh, G., Raajini, X. M., & Dang, H.(Eds.). (2021). *Industry 4.0 interoperability, analytics, security, and case studies.* CRC Press.

Ramírez-Durán, V. J., Berges, I., & Illarramendi, A. (2021). Towards the implementation of Industry 4.0: A methodology-based approach oriented to the customer life cycle. *Computers in Industry, 126*, 103403. https://doi.org/10.1016/j.compind.2021.103403

Romeo, L., Loncarski, J., Paolanti, M., Bocchini, G., Mancini, A., & Frontoni, E. (2020). Machine learning-based design support system for the prediction of heterogeneous machine parameters in industry 4.0. *Expert Systems with Applications, 140*, 112869. https://doi.org/10.1016/j.eswa.2019.112869

Rossit, D. A., Tohmé, F., & Frutos, M. (2019). A data-driven scheduling approach to smart manufacturing. *Journal of Industrial Information Integration, 15*, 69–79. https://doi.org/10.1016/j.jii.2019.04.003

Schockaert, C. (2020). A causal-based framework for multimodal multivariate time series validation enhanced by unsupervised deep learning as an enabler for industry 4.0. arXiv preprint arXiv: 2008.02171. https://doi.org/2008.02171

Teoh, Y. K., Gill, S. S., & Parlikad, A. K. (2021). IoT and fog computing based predictive maintenance model for effective asset management in industry 4.0 using machine learning. *IEEE Internet of Things Journal*. https://doi.org/10.1109/JIOT.2021.3050441

Ting, A., Chieng, D., Sebastiampi, C. V., & Khalid, P. S. (2019). High precision location tracking technology in IR4. 0.https://doi.org/10.20474/jater-5.3.4

Tiwari, S., Lilhore, U., & Singh, A. (2018). Artificial neural network and genetic clustering based robust intrusion detection system. *International Journal of Computer Applications, 179*(36), 36–40. https://doi.org/10.116/ii.2018.00.03

Villalba-Díez, J., Molina, M., Ordieres-Meré, J., Sun, S., Schmidt, D., & Wellbrock, W. (2020). Geometric deep lean learning: Deep learning in industry 4.0 cyber–physical complex networks. *Sensors, 20*(3), 763. https://doi.org/10.3390/s20030763

Wang, H., Zheng, Z., Ji, C., & Guo, L. J. (2021). Automated multi-layer optical design via deep reinforcement learning. *Machine Learning: Science and Technology, 2*(2), 025013. https://doi.org/10.1088/2632-2153/abc327

Wu, C., Nguyen, D., Xing, Y., Barragan, A., Schuemann, J., Shang, H., … & Jiang, S. B. (2020). Improving proton dose calculation accuracy by using deep learning. *Machine Learning: Science and Technology*. https://doi.org/10.1088/2632-2153/abb6d5

Yew, K. H., Foong, O. M., & Sivarajan, T. P. (2019). Knowledge-based improvement of machine downtime management for IR4. 0. In *2019 9th IEEE International Conference on Control System, Computing and Engineering (ICCSCE)* (pp. 94–98). IEEE. https://doi.org/10.1109/ICCSCE47578.2019.9068584

Zhou, R., Awasthi, A., & Stal-Le Cardinal, J. (2020). The main trends for multi-tier supply chain in Industry 4.0 based on Natural Language Processing. *Computers in Industry*, 103369. https://doi.org/10.1016/j.compind.2020.103369 LeCun, Y., Bottou, L., Bengio, Y. and Haffner, P., 1998. Gradient-based learning applied to document recognition. *Proceedings of the IEEE, 86*(11), pp. 2278–2324.

Zhou, Z., Ren, L., Zhang, L., Zhong, J., Xiao, Y., Jia, Z., ... & Wang, J. (2020). Heightened innate immune responses in the respiratory tract of COVID-19 patients. *Cell Host & Microbe*, *27*(6), 883–890.

13 IoT Applications and Recent Advances

T. Mohana Naga Vamsi and Pratibha Lanka

Gayatri Vidya Parishad College for Degree and PG Courses(A), Visakhapatnam, Andhra Pradesh, India

CONTENTS

13.1 Introduction 200
13.2 IoT Services and Applications 201
13.2.1 Classification of IoT Services 201
13.2.1.1 Identity-related Services 201
13.2.1.2 Information Aggregation Services 201
13.2.1.3 Collaborative-aware Services 202
13.2.1.4 Ubiquitous Services 202
13.2.2 Prominent Applications of the IoT 202
13.2.2.1 The IIoT 202
13.2.2.2 The Internet of Medical Things (IoMT) 202
13.2.2.3 Smart Cities 204
13.3 Smart Environment Systems 204
13.3.1 Water Contamination Monitoring System 204
13.3.2 Air Pollution Monitoring System 205
13.4 Smart Intelligent Healthcare Systems 206
13.4.1 Research Frontiers in Smart Healthcare 206
13.4.2 The Role of Big Data Analytics in Smart Healthcare 207
13.4.3 The Role of Security and Privacy in Smart Healthcare Systems 207
13.5 IoT Communication and Network Protocols 207
13.5.1 Network Connectivity Protocols 208
13.5.1.1 Bluetooth 209
13.5.1.2 Wi-Fi 209
13.5.1.3 Zigbee 210
13.5.1.4 Z-Waves 210
13.5.1.5 Near Field Communication (NFC) 210
13.5.1.6 LoRa 211
13.5.1.7 6LowPAN 211
13.5.2 IoT Communication Protocols 211
13.5.2.1 MQTT 212
13.5.2.2 SMQTT 212
13.5.2.3 AMQP 213

DOI: 10.1201/9781003146711-13

13.5.2.4 CoAP 213
13.6 IoT Industrial Applications and Network Edges 214
13.7 Summary 217
13.8 Conclusions and Future Scope 218
References 218

13.1 INTRODUCTION

The term "Internet of Things" (IoT) was first coined by Kevin Ashton, the co-founder of the Auto-ID Center at MIT (Madakam et al., 2015). As per the International Organization for Standardization/International Electrotechnical Commission (ISO/IEC), the definition of the IoT is "an infrastructure of interconnected objects, people, systems and information resources together with intelligent services to allow them to process information of the physical and the virtual world and react" (ISO/IEC JTC1, 2015). The IoT is the revolutionary technology that is rapidly being applied to different fields with the support of contemporary wireless sensor networks (WSNs) in which the physical objects are treated as "things" that are connected via the Internet (Whitmore et al., 2015). This disruptive technology will become the next "Industrial Revolution" of this century. As the IoT is the combination of different technologies as well as being operated under different heterogeneous components this technology is considered to be disruptive.

During this current decade, with the advent of its role in many domains, the IoT is becoming the most familiar technology and a benchmark in the technology world. The IoT is the interconnectivity of embedded microcontrollers, mobile gadgets, buildings, and other things within the network that includes sensors and actuators, such that data is transferred between these things via the Internet (Khanna & Kaur, 2020). The IoT is a cross-discipline research technology that involves computer science and electronic engineering principles. The major research focus is on the design and development of intelligent and secured IoT-enabled hardware and software networking systems. This is made possible by incorporating artificial intelligence, machine learning, and Internet concepts while developing intelligent IoT systems (Agrawal & Vieira, 2013).

Since the research in IoT applications involves the amalgamation of embedded electronics, sensor technology, networking, cloud computing, artificial intelligence, data analytics, and other technologies, it is becoming a good area for interdisciplinary research for both academics and industrial people. At the outset there were a lot of new methodologies and findings that evolved with this IoT research, and these outcomes provided better solutions in the areas of smart environments, IoT services, intelligent systems, smart health and smart homes, smart agriculture, networking protocols, communication protocols, and autonomous industrial applications (Atzori et al., 2010).

Hence the research direction in IoT applications is definitely a challenge for academic researchers and industrial developers. The major challenges are in identification technology, choosing an IoT architecture technology, and establishing network and communication technologies for the specific research problem. In this chapter we will try to depict all such challenges by focusing on these IoT enabled topics and throw a light out for new researchers who are entering this domain.

13.2 IOT SERVICES AND APPLICATIONS

Today the main problem faced by IoT application development is interoperability. This is due to there being no specific uniform architecture and computing technology present in IoT systems (Gigli & Koo, 2011). The combination of communication technology, sensor technology, network topologies, and different hardware are responsible for the interoperability in IoT systems. To minimize this problem one possible proposal is adopting a service-oriented architecture (SOA) that incorporates the business software of an IoT system (Spiess et al., 2009). Hence it is better to deploy different advanced new methods and technologies for IoT applications which will adopt these services that bridge each of the technologies needed for the development of the application.

13.2.1 Classification of IoT Services

IoT applications are enormous, from smart homes, retail applications, and smart agriculture, to industrial and smart health, to smart cities. A specific IoT service may be applied for each of these different applications to enhance the speed and quality of that application's development process.

Hence services are classified, based on the application, as identity-related services, information aggregation services, collaborative-aware services, and ubiquitous services (Xiaojiang et al., 2020).

13.2.1.1 Identity-related Services

The simplest and most IoT-based application service is an identity-related service. This means the designer will obtain a virtual detail (data/information) of each thing that is used in that application. This service is separated into two categories. One is an active identity-related service and the second one is a passive service (Gigli & Koo, 2011). These can be served to either an enterprise or individual that depends on this type of application. Every identity-related service has two important elements: "the things" and "read devices." The best example is the radio-frequency identification (RFID) tag, which is passive, whereas the read device will read the identity of that RFID tag, which is active (Gao et al., 2007).

13.2.1.2 Information Aggregation Services

As the name indicates, this service role is to collect the information from different things existing in an entire IoT network and send that information through the network to the IoT application for further processing. This service may utilize different kinds of things (sensors/actuators) and network devices to send their information through a common gateway service to the application (Gigli & Koo, 2011). The best example of this service is the IoT application developed by using multiple WSNs. In any WSN the information is collected and communicated via the terminals and host application platform to the other WSNs in that application for further processing. This is achieved by establishing a link between these WSNs by using an access gateway.

13.2.1.3 Collaborative-aware Services

This is the service which performs some action based on the decision retrieved from the collected information from the things in the IoT system. Such a type of service is called a collaborative-aware service. This is the basic distinction between the information aggregation service and the collaborative-aware service. This service in IoT applications requires information communication in the form of "terminal to terminal" or "terminal to person." The main characteristics behind the creation of collaborative-aware services are network security, terminal processing power, and the speed of information retrieval (Shen et al., 2010). The sensor terminals must be interleaved with the specific embedded devices so they may process the data within the network.

13.2.1.4 Ubiquitous Services

The ubiquitous service is the key essential service of the IoT. It is a collaborative-aware service for everyone and for everything. Since there is no specific architecture for the IOT, services like application programming interfaces (APIs), representational state transfer (REST), and Web services will couple "the things" of a specific application via the Internet, which can then be reutilized for any other application (Guinard, 2010). Therefore, ubiquitous services are the ultimate services which researchers are focusing on today for their IoT applications. For example, the IPv6 protocol will become one of the ultimate protocols to be utilized for the majority of IoT applications in the coming years and can be considered as one of the best ubiquitous services.

13.2.2 Prominent Applications of the IoT

13.2.2.1 The IIoT

The Industrial Internet of Things (IIoT) is the fourth industrial revolution and is termed "Industry 4.0." The IIoT applies IoT concepts and principles to industry, resulting in smart industry. In the IIoT smart machines run with the constant collecting of data from the smart sensors and actuators in that industry (Khan et al., 2020). This improves quality in industrial manufacturing and processing. Due to this, smart machines perform in a better way than the human beings working in that industry. Smart machines also deliver better analytics of the real time data that are accumulated in industrial applications like manufacturing, design automation, energy transformation, agriculture, and transportation. The IIoT has diverse technological features that give it an enhanced level of complexity, interoperability, and security. Figure 13.1 presents current IIoT applications.

13.2.2.2 The Internet of Medical Things (IoMT)

The "IoMT" refers to the integration of medical things, health apps, and the processing of medical data through intelligent systems. In IoMT systems, the data are captured from these things and used to monitor the health condition of patients with the help of artificial intelligence (AI) techniques (Rahmani et al., 2018).

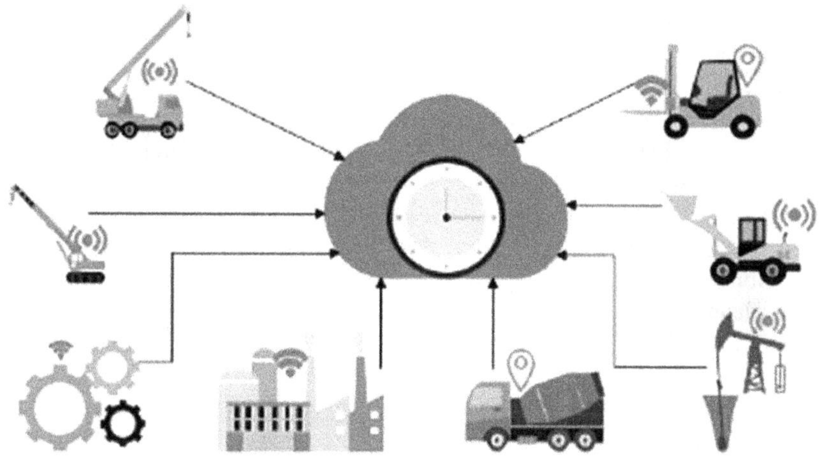

FIGURE 13.1 Industrial IoT applications.

FIGURE 13.2 Sensor arrangement in the IoMT.

The medical things in the IoMT are smart health sensors, as shown in Figure 13.2. These sensors are connected to a patient via IoT communication and connectivity protocols and measure body temperature, motion, the electrical activity of the heart, muscles, and the height and weight of the patient continuously. The data from the sensors are communicated to a carer or a doctor through IoT technology systems and other computer devices. Finally, the data aggregated from these medical things proceed through AI algorithms, which suggest decisions based on priorities. Hence with this IoMT, the heart rate, invasive blood pressure, and other health information of the patient could be measured automatically; the same can be used for preventing patient health problems and reducing patients visits to doctors.

13.2.2.3 Smart Cities

In the current situation, urbanization is required by society as it has several advantages. But due to this urbanization there are a lot of challenges to the urban development authorities. As the population increases, this leads to the increase in the demand for usage of energy resources, domestic water, maintaining sanitation, quality education, cost effective health services, and other public services like transportation and parking. The best solution for these challenges is "smart cities" (Hammi et al., 2017). In smart cities, the systems are designed using IoT principles; but these IoT applications require adequate network connectivity with a wide range of coverage. Also, it needs efficient resource allocation and utilization, specifically in terms of power, since almost all devices in IoT applications run on battery power. Hence researchers who want to develop smart city IoT applications need to focus on these parameters and find the best solution for covering a long range of connectivity with less power consumption and secure systems (Anagnostopoulos et al., 2017).

13.3 SMART ENVIRONMENT SYSTEMS

"Smart environment" means connecting sensors, actuators, and computational devices in a network through IoT technologies to enable the constant monitoring and understanding of variations in the natural environment which comprises non-human-made things like air, water, radiation, land, and trees (Gomez et al., 2019). Smart environment monitoring is the study and analysis of variations in environmental parameters like air quality, land erosion, water quality, wild life, and meteorological conditions (Elmustafa & Mujtaba, 2019). This is done with the help of IoT devices like smart environment sensors and terminals connected through embedded devices to cloud and running data analytics algorithms for the understanding of the variations in these parameters (Ullo & Sinha, 2020). There is a lot of scope for researchers to apply the smart environment framework, which they can use to develop effective smart environment monitoring systems and promote research at these frontiers. A sample smart environment system with WSN and IoT devices is shown in Figure 13.3.

13.3.1 Water Contamination Monitoring System

One of the smart environment examples is the smart water contamination monitoring and control (SWCMC) system. A detailed presentation of SWCMC is given in Figure 13.4. This system is designed with the help of IoT smart sensors, WSN principles, and cloud computing techniques (Jamil et al., 2015). The water contamination is sensed with the help of smart aqua-sensors placed in different water resources and these sensors are connected through a WSN framework (Ullo & Sinha, 2020). The collected data from the aqua-sensors are communicated to the cloud via IoT connectivity and communication protocols. The cloud-based system is embedded with data analytics and machine learning techniques which analyze the aggregated data in the cloud; based on the contamination level found, the respective departments will take the necessary action. After the quality check is done, the respective department may initiate water contamination control devices via the IoT-based control systems and the actuators present in the control system will minimize the causes of the contamination.

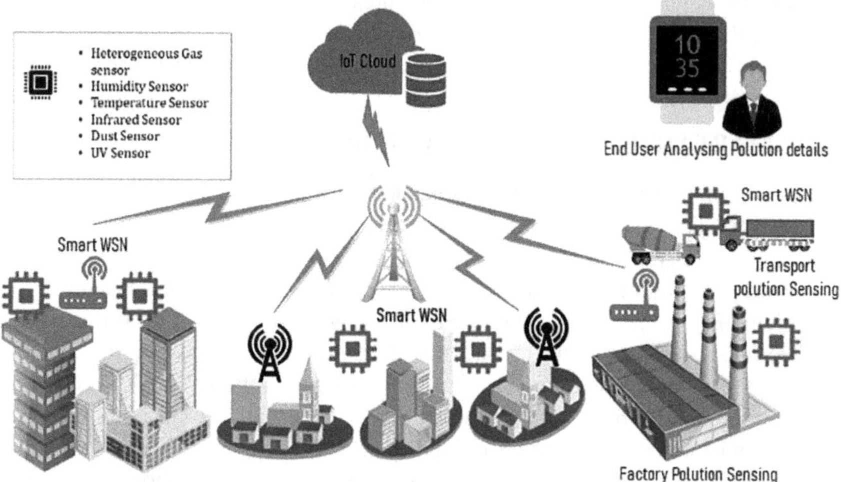

FIGURE 13.3 A smart environment system with WSN and IoT devices.

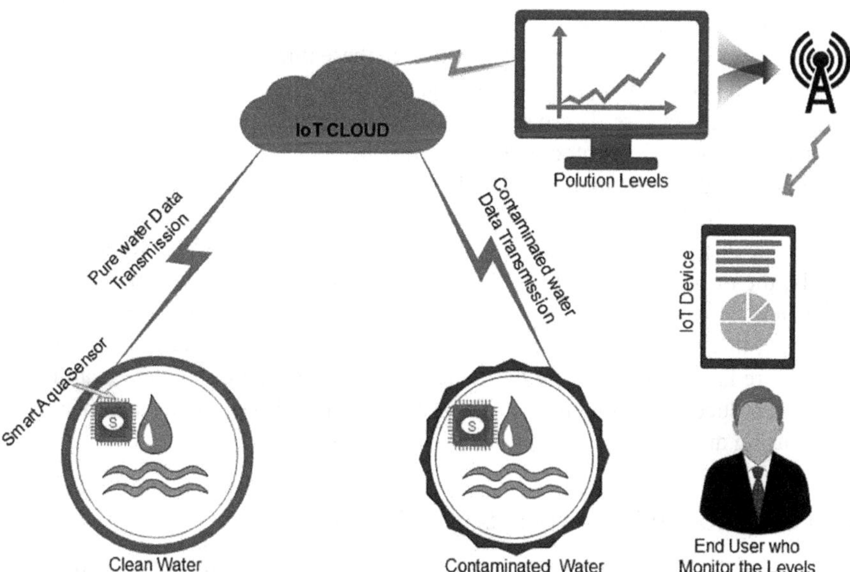

FIGURE 13.4 Smart water pollution monitoring systems with smart aqua-sensors and IoT devices.

13.3.2 Air Pollution Monitoring System

A smart air pollution monitoring (SAPM) system is another good example of a smart environment. The air quality is estimated by using heterogeneous smart sensors, IoT devices, and machine learning techniques (Amado & Dela Cruz, 2018). The air pollution levels can be predicted using either fixed or mobile WSNs so that the air quality can be determined in both situations present in the environment. Data collected from the smart edge nodes are processed and analyzed with the help of data analytics and machine learning algorithms executed in the respective cloud.

13.4 SMART INTELLIGENT HEALTHCARE SYSTEMS

Health is the fundamental thing which is needed for a good life style. But due to different reasons, like inadequate health services, especially in rural areas, and the unavailability of physicians as well as paramedical personnel during emergencies in both rural as well as urban areas, people face trouble during illness (Islam et al., 2020). Previously information and communication technology (ICT) was much used in the healthcare sector for storing patients' details in the form of electronic health record (EHR) systems, but the continuous monitoring of patients' health is not possible when they are at home or somewhere remote (Zeadally et al., 2019). In such situations smart health services are needed in which the system accesses patients' data with smart medical sensors which are connected with smart phones as well as IoT devices. Then the healthcare sector will be improved by the introduction of the IoT, which we may then call smart healthcare and which will definitely provide better health services to unreachable patients. The integration of AI in smart healthcare systems is the technology known as the IoMT (Rahaman et al., 2019).

In IoMT systems, the data are captured from these things which monitor the health conditions of patients with the help of AI techniques. The medical things in the IoMT are smart health sensors. These sensors are connected to a patient via IoT communication and connectivity protocols and measure the body temperature, motion, electrical activity of the heart, muscles, and height and weight of the patient continuously (Zeadally & Bello, 2019). The data from the sensors are communicated to a care taker or a doctor through IoT technology systems and other computer devices. Finally, the data aggregated from these medical things is proceed through AI algorithms and suggested decisions based on the priorities. Hence with this IoMT, the heart rate, invasive blood pressure and other health information of the patient could be measured automatically and the same is used for preventing patient health problems and reduced the patients visits to doctors. Figure 13.5 shows a smart healthcare model using an IOT-based human-body sensor network.

13.4.1 Research Frontiers in Smart Healthcare

Research in the smart healthcare sector involves the amalgamation of different technologies, like the IoT, mobile technology, smart sensors, WSNs, wearable computing, cloud computing, big data analytics, machine learning (ML) and AI (Islam et al., 2015). There is wide scope for current researchers to provide smart solutions for problems that are identified in this smart health domain, by applying IoT connectivity and communication principles along with wearable devices, smart sensors, WSNs, and mobile terminals for retrieving and transmitting medical information for further processing. Implementing these technologies in smart hospital systems (SHSs) such that the SHS can collect real-time environment conditions and patients' physiological parameters (Chaudhury et al., 2017). Researchers can address these applications

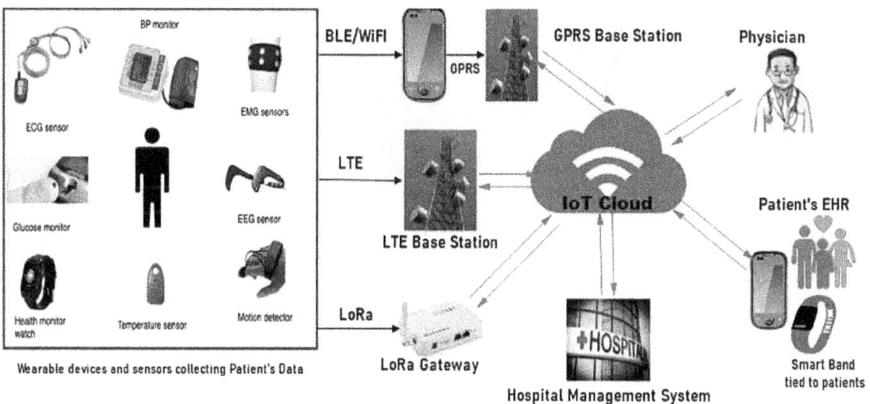

FIGURE 13.5 Smart healthcare model.

and adopt advanced technologies like cloud computing, big data analytics, ML, and AI for obtaining better solutions.

13.4.2 The Role of Big Data Analytics in Smart Healthcare

The data processing in any healthcare system is a challenging task since the data generated from IoT devices are accumulated in exponential orders. Therefore, researchers need to find suitable big data analytic approaches, such as Hadoop map reduction, Hive, or Apache Spark, to perform analysis of this abundant data (Olaronke & Oluwaseun, 2016). Since the datasets gathered are patients' data, evidence-based care has to be taken such that we can find anomalies in the datasets and provide cleaned data for better treatment. This process, involving new machine intelligence methods, leads to the saving of both time and financial burdens.

13.4.3 The Role of Security and Privacy in Smart Healthcare Systems

Data gathering from different smart sensors fixed to the patient or in surroundings using an IoT-based system is performed through the Internet. Thus, the accessed patient data during gathering or in transmission may involve unauthorized usage or tracking. This may damage the patient's safety and privacy (Zeadally et al., 2019). Therefore, data needs to be protected by designing systems with high end autonomous intelligent IoT devices. Also, the applications designed for use with IoT devices are very vulnerable because of wireless communication technologies. Intensive research is essential to ensure privacy, trust, and security throughout the healthcare environment.

13.5 IoT COMMUNICATION AND NETWORK PROTOCOLS

IoT protocol standards that function at different layers of the networking stack incorporate layers like a medium access control (MAC) layer, a network layer, and a session layer. The details of protocols used in these layers are introduced in this section.

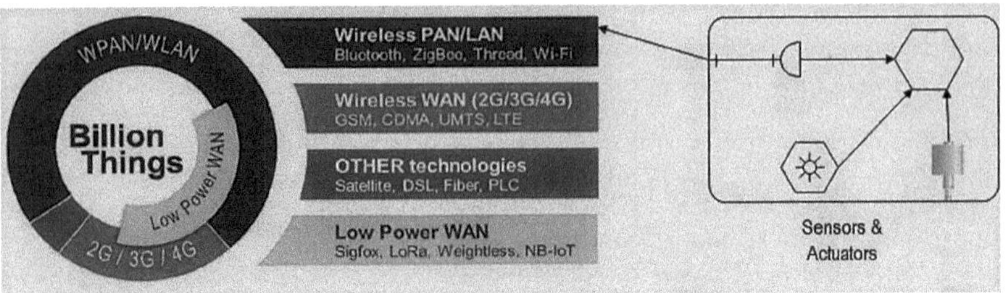

FIGURE 13.6 IoT protocols and standards.

TABLE 13.1
IoT Protocol Stack

Session		Type of Protocol	Security	Management
Network	Encapsulation	6LowPAN, 6Lo, Thread	TCG	IEEE1905
	Routing	RPL, CARP	SMARK	IEEE1451
Data link		WiFi, BLE, Z-Wave, ZigBee Smart	ISASecure	
		3G/LTE/5G.NFC, RFID, LoRaWAN	DTLS	

The protocol standards are offered by different professional organizations like the Internet Engineering Task Force (IETF), the Institute of Electrical and Electronics Engineers (IEEE), and the International Telecommunication Union (ITU) (Salman & Jain, 2015). The different IoT protocols and standards are depicted in Figure 13.6.

The protocols in IoT are basically classified as network connectivity protocols and communication protocols. The network connectivity protocols are further classified into data link protocols and network layer protocols. The protocol stack at different layers is depicted in Table 13.1.

Further, the communication protocols at the session layer are responsible for the interchange of messages between the clients or between a client and a sensor. In this section elaborated guidelines and direction are given to researchers who intend to work in this domain. The explanations may enable the choice of the right category of either connectivity protocol or communication protocol for research problems. The protocols are differentiated based on some bench-mark criterion like the protocol standard, its network topology, power utilization, connectivity range, type of modulation used, or security aspect.

13.5.1 Network Connectivity Protocols

The datalink and network layers play a major role in data transfer between the sensors and edge gateway. The data link layer connects two IoT sensors or it may be a sensor with a gateway edge that connects to different sensors via the Internet.

Specific protocols are designed for establishing routing amongst things (sensors) which come under the routing layer. The IoT paradigm depends on existing network connectivity technologies for both short range as well as long range. The short-range connectivity protocols currently used are Bluetooth, Zigbee, Z-waves, Wi-Fi (wireless fidelity), NFC (near field communication), RFID, and Thread. The long-range connectivity protocols, like LoRa, Sigfox, and cellular long-term evolution-advanced (LTE-A), have been used for different applications where connectivity is established between sensors which are located remotely (Al-Fuqaha et al., 2015). The details of some protocols mentioned in this section are presented below.

13.5.1.1 Bluetooth

Bluetooth classic is one of the mostly popular wireless connectivity standards in IoT applications and is one of short-range communication with a reduced power consumption protocol. The protocol is operated at a short-range frequency band of 2.4 GHz. Practically it is operated in the range of 2400.0–2483.5 MHz frequency bands, which include the guard band (Aguilar et al., 2017). The IEEE standard of the Bluetooth protocol is 802.15.1 and is now maintained by special interest groups (SIGs). For the majority of IoT applications, the Bluetooth Low Energy (BLE) standard is used by current researchers for their application development: it is the new version of Bluetooth technology and is operated under very low power and is almost 10 to 15 times more efficient than Bluetooth Classic. BLE is also known as Bluetooth Smart. The latest releases of this technology are Bluetooth 5.2, which was released on 7 January 2020, and Bluetooth Mesh 1.9, which was released on 20 September 2019 (Yin et al., 2019). The new features of the latest version have been enhanced to longer ranges and speeds, with a relaxation in transmitter power constraints; a channel coding feature is also incorporated. Both these versions accept Mesh topology. Those researchers who are intending to do their work in IoT wireless applications with reduced power usage are directed to utilize this technology for their research work.

13.5.1.2 Wi-Fi

Wi-Fi means wireless fidelity and is a radio technology that is used in local area network applications and devices. The IEEE standard for Wi-Fi is 802.11. This standard is maintained by the Wi-Fi Alliance and is operated in the frequency range of 2.4 to 5.0 GHz. This protocol supports a high speed data rate in the range of megabytes to gigabytes. The security methods incorporated in this standard are Wi-Fi Protected Access 2 (WPA2), Wi-Fi Protected Setup (WPS), and WPA Enterprise. This protocol operates at low power ratings in the range of 5 to 200 mA. This protocol is mostly used in digital communication devices like mobile phones, tablets, laptops, and smart televisions (Salman & Jain, 2015). It is well suited for some IoT applications where high band width as well as low latency are required. But the power ratings mentioned above are relatively high when comparing with the power ratings of other standards; therefore for those IoT applications where power consumptions and frame overheads are important then this standard is not suitable.

13.5.1.3 Zigbee

Zigbee is a global standard for control and sensor networks designed for low data rate control networks. It got its name from honeybees! This is in addition to the IEEE 802.15.4 specification layer and offers a standard procedure for functions, including network formation, messaging, and device discovery. This protocol is used in applications of wireless personal area networks (WPANs) and device-to-device networks (Samie et al., 2016). It is a packet based radio protocol of low cost and operates on 2.40 to 2.48 GHz frequencies. There are three different Zigbee devices or modules present in any Zigbee network. They are Zigbee Coordinator (ZC), Zigbee Router (ZR), and Zigbee End Device (ZED). ZC is responsible for starting a centralized security network and managing node joining and key distribution for the network. This coordinator is a ZigBee logical device type that includes the functionality of a trust center. ZR is responsible for managing node joining. Its main feature is extending the network. It joins two Zigbee networks. ZED is a Zigbee logical device that can only join an existing network. Zigbee Smart is for protocols used in IoT applications like home automation (smart homes), smart healthcare systems, and any remote-controlled systems. The star topology, mesh topology, and cluster network topology are suitable for this protocol standard.

13.5.1.4 Z-Waves

The Z-Wave protocol was developed by Zensys. This protocol comes under the category of a low power media access protocol (MAC), which is specifically designed for wireless smart home automation applications. Nearly 50 nodes can be connected with this protocol for any IoT communication application, like smart home or commercial applications (Ab Rahman and Jain, 2015). The data packets transmitted in this technology are very small in size and travel with a rate of 100 kbps, which is a slow speed across a 30 meter point-to-point communication distance. Hence it is relatively appropriate for IoT applications where small message transmissions are involved. Such applications are healthcare, energy control, domestic light control, and so on. Z-Waves depends on either controlling the device or the slave device. The slave devices are nodes which can't initiate the messages (Salman & Jain, 2015). Hence, they are called low cost devices, whereas the controlling commands are executed by the control devices of that network. In general, the Z-Wave protocol supports the mesh network topology.

13.5.1.5 Near Field Communication (NFC)

NFC is a short range wireless radio communication protocol. It is an inbuilt communication technology in the majority of mobile phones as well as in some portable gadgets (Cerruela García et al., 2016). When two NFC-enabled gadgets are brought together to a distance of less than 10 cm, the data will be transmitted between them. This protocol is operated at frequencies of 13.56 MHz and the allowable data rate is in the order of 100 to 420 Kbps. Hence it comes under very short-range wireless communication technology. The principle of this protocol is similar to RFID, and each NFC gadget has a tag which consists of a small quantity of data. As in RFID this tag is a read-only action, or it can be used as a rewritable one that depends on the device that it is using (Want, 2006). Therefore, NFC technology operates in three modes, one is the passive mode, also called the emulation mode; the second one is the active mode, also called the read/write mode; and the third category is the peer-to-peer mode. The major application of this technology today is contactless payment at the point of sale (POS).

13.5.1.6 LoRa

LoRa means long range wireless technology and has been introduced recently. It is operated as a LoRaWAN network, built for the majority of current IoT and M2M applications. This technology standard is designed for long range radio communication with low power utilization. A LoRaWAN network has the capacity to handle millions of end nodes using the LoRa gateway (Alliance, 2015). Generally, in LoRaWAN networks the nodes are not linked to a specific gateway. That means the data transmitted by the end nodes are received by multiple LoRa gateways. These data packets are then forwarded by each gateway to the cloud-based network server via either the cellular network or the ethernet; therefore the signals can be sent over long distances in the range of 10 to 15 km (Miorandi et al., 2012). To build a LoRaWAN one needs a LoRa gateway, a LoRa shield, and a LoRa GPS shield. It is clear there is much scope for research to develop applications like smart cities and industrial and commercial solutions with this new technology. Hence, it is clear that LoRa wireless technology will definitely play a vital role in IoT research.

13.5.1.7 6LowPAN

This protocol has special significance and comes under the network layer encapsulation protocol category (Le et al., 2012). The name 6LowPAN is derived from it being an "IPv6 over Low Power Wireless Personal Area Network." It follows the IEEE 802.15.4 standard that is specifically for devices which are compatible with this IEEE standard. It perfectly encapsulates the IPv6 long headers in the small frames of IEEE 802.15.4. Since it is an IP-based internetworking protocol that supports large number of IP addresses (2^{128} IP addresses), it is more suitable for the short range networks of IoT applications. The relative positioning of selected IoT-focused wireless technologies by capability are represented in Table 13.2.

13.5.2 IoT Communication Protocols

The data received from the network layer are send to the transport/session layer, where the session layer protocols enable these messages to further communicate between different IoT things or elements of that communication subsystem (Al-Sarawi et al., 2017). Hence these are also called data transfer messaging protocols. In general, the HTTP protocol standard is used for Internet-based communication, but for all IoT systems this protocol is not technically suitable. Due to this problem different communication protocols are designed for meeting the message transmission issues in IoT systems. Such IoT communication protocols are message queuing telemetry transport (MQTT), secure message queuing telemetry transport (SMQTT), advanced message queuing protocol (AMQP), and constrained application protocol (CoAP). Further security and management standards and protocols are being

TABLE 13.2
Different Parameters of Connectivity Protocols

Power	Range	Data Rate	Latency	Spectrum	Licensed/Unlicensed
Wi-Fi	Moderate	Moderate to Long	High	Low	Unlicensed
LTE Cat-M	Low	Moderate to Long	Moderate	Low	Licensed
LTE Cat-IoT	Very Low	Long	Low	Very Low	Licensed
LoRa	Very Low	Long	Low	Low	Unlicensed
SigFox	Very Low	Long	Very Low	Very Low	Unlicensed
BLE	Very Low	Short	Low	N/A	Unlicensed
Zigbee	Very Low	Short	Low	N/A	Unlicensed
6LoWPAN	Very Low	Short	Low	N/A	Unlicensed

Source: Maravedis and Rethink research (https://www.maravedis-bwa.com/2020/04/22/the-role-of-wifi-in-iot/)

developed for IoT systems (Triantafyllou et al., 2018). The details of some protocols mentioned in this section are presented below.

13.5.2.1 MQTT

MQTT is a machine-to-machine (M2M)/IoT connectivity protocol. It is a standardized network protocol by which short messages/commands can be transmitted. Hence it is considered a light weight protocol (Tukade & Banakar, 2018). This protocol was designed and proposed by IBM and Eurotech. It is suitable for connections with remote locations where a small coded footprint is required and/or network bandwidth is at a high demand. It can support thousands of remote clients and is capable of being supported by a single server. It is ideal for use in constrained environments where network bandwidth is low or where there is high latency and with remote devices that might have limited processing capabilities and memory. To keep things simple and allow bi-directional message passing, MQTT uses a publish and subscribe model, where clients subscribe to topics and have a connection to a broker server. As new messages are sent to the broker, they include the topic with the message, allowing the broker to determine which clients should receive the message (i.e., those subscribed to the same topic will receive the published messages). Messages are pushed to the clients through the always-on TCP connection. This mechanism is depicted in Figure 13.7.

13.5.2.2 SMQTT

SMQTT is a secure MQTT, which means the security feature in MQTT is enhanced and released as the protocol SMQTT (Singh et al., 2015). Since MQTT is a light weight protocol, the SMQTT is adopted with encryption which is considered to be a lightweight attribute-based encryption. Here, a single message is encrypted and broadcast to multiple other nodes. This is a very much required feature for almost all IoT applications. This principle is called broadcast encryption. It is a protocol with four modules which are termed: setup module, encrypt module, publish module, and finally decryption module. During the setup module, both subscribers and publishers

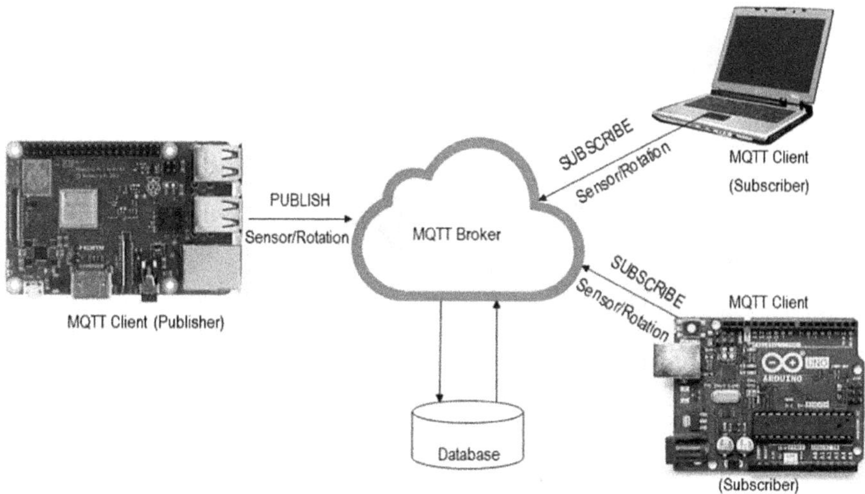

FIGURE 13.7 MQTT protocol.

need to register with the broker and they have to receive the secret master key, which is generated by a secret algorithm decided upon by the designer. Hence when the data is published to the broker it is first encrypted and then transferred to the subscribers. At the end, the decryption takes place and the same secret master key is made available to the subscriber.

13.5.2.3 AMQP

AMQP is a protocol that also comes under the category of session layer protocol (Oasis, 2012). The main aim of designing this protocol is for financial applications. It supports a publish/subscribe model and is almost equivalent to the MQTT protocol. Compared to MQTT it is different in broker. In this protocol broker has two different components: queues and an exchange. The messages from the publisher are received by an exchange and are based on some conditions they allocate to queues. The topics are represented in queues and are subscribed to as shown in Figure 13.8. All the sensory data will be obtained by subscribers whenever they are in queues.

13.5.2.4 CoAP

CoAP is a protocol that also comes under the category of session-layer protocol. The first release of CoAP was from the Internet Engineering Task Force (IETF) (Shelby et al., 2014). This protocol was designed for offering a lightweight RESTful interface, like HTTP. REST means "representational state transfer." The interface between an HTTP client and a server is provided by REST. But for lightweight IoT applications, REST could bring about critical overheads and power utilization. Hence one can understand that this CoAP is designed for sensors of a low power type, such that the RESTful services can be used.

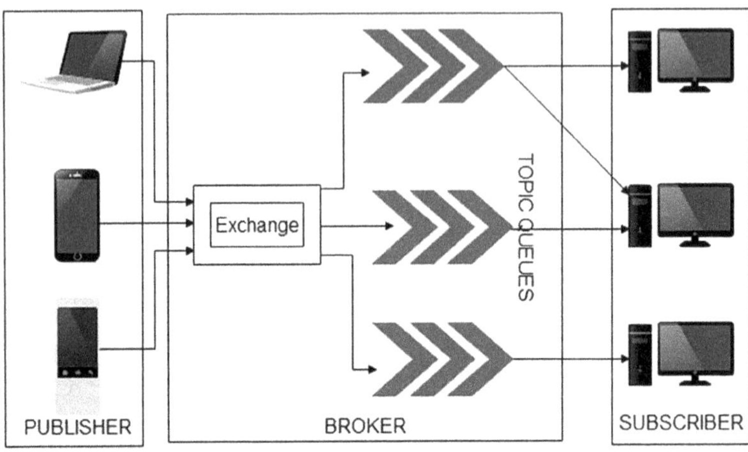

FIGURE 13.8 AMQP protocol.

13.6 IoT INDUSTRIAL APPLICATIONS AND NETWORK EDGES

The IIoT is the utilization of smart sensor edges for enhancing industrial design and processes in the manufacturing of industrial equipment or systems, which is also coined today as Industry 4.0. The IIoT leverages the utilization of real-time analytics and smart machines such that it takes advantage of the data that machines have generated (Khanna & Kaur, 2020). The fascinating viewpoint in this IIoT is that smart machines are better at analyzing and capturing data in real time compared to human beings; also they are good at collecting significant information that can be employed to make business decisions more accurately and faster. This fourth industrial revolution is the most troublemaking factor in industrial automation and affects the industries of healthcare, energy, transportation, and manufacturing. The components in IIOT ecosystems are able to sense the information generated from public as well as private infrastructure and transfer that information for storage and for further analytics. Based on the communication infrastructure and application, the analytics will be performed on information collected from the raw data and people. The IIoT is a division of IoT and spotlights especially on industrial applications such as manufacturing, energy, transportation, and agriculture. Remarkably, the IIoT has diverse technological requirements that give it its increased level of complexity, interoperability, complexity, and security wants. Both the IoT and IIoT have industrial problems, but complexities and threats to self-regulation and independent industrial application are intrinsically worse.

Use Case #1: Smart Metering:

Smart meters have gained a tremendous attractiveness in recent years. Industries ought to recognize and take advantage of smart meters (Suresh et al., 2020). Organizations have already invested around USD62 billion towards the infrastructure of these smart meters. A smart meter is more an Internet proficient tool as compared to traditional meters; they are even capable of reporting how a number of vital sources, such as energy, irrigation, power, and natural gases, are consumed. Power consumption industries use smart meters for reducing their functioning prices immensely. Smart meters help in alleviating the difficulty of quarterly and monthly meter calculations. Obviously, these smart meters may be located anywhere in the home. They are equipped with dashboards in real time, so that owners as well as tenants may know the status of their respective energy use. And most importantly, the insight provided by the smart meters in the IoT infrastructure can be employed dynamically in customer-centric structures.

Use Case #2: Machine Control and Predictive Maintenance:

Predictive maintenance analyzes the prior condition of the equipment and has the ability to lessen operational industrial expenditure significantly, which results in considerable monitoring benefits for manufacturers (Suresh et al., 2020). Industrial manufacturers implement predictive management, maintenance, and analysis of systems to reduce the likelihood of wear and tear of the assets and their downtime by using sensor devices to watch cameras, operational circumstances, storage of significant data in the cloud, and information analysis functions. These aspects make it possible to conclude when a piece of equipment will fail in the future. IIoT-enabled systems provide warnings and signs before problems arise. They are a blend of long-lasting equipment, a more successful exercising of field techniques, that avoids highly priced lay-offs. During 2016–2022, the yearly growth of the predictive maintenance rate is supposed to be 39%. Where technology adopts the IIoT it may need to spend nearly USD10.96 billion by 2022. This estimation is taken collectively by the predictive maintenance blended revenue of the foremost industries in this field: seven technologies and 13 industries. The information data that stream from the sensors and other devices assist in evaluating present circumstances to give alerts, acknowledge warning signs, and activate the proper predictive maintenance process without human intervention. The IoT is changing maintenance into a dynamic action that is done automatically and also yields improved equipment lifecycles by minimizing accidents.

Use Case #3: Real-time Asset Tracking:

According to the latest study, it is known that estimated financial values of USD1.9 trillion could be produced by IIoT machinery and equipment monitoring systems in worldwide supply chains (Suresh et al., 2020). In addition, a study by the Indian Institute of Management Ahmedabad (IIMA) and Infosys, states that 85% of manufacturing industries and companies are very conscious about analyzing asset tracking efficiency; however, 15% of these had already executed it at a more systematic level. In the services enabled by the IIoT, the assets have a crucial role in the contemporary economy as never before. Many companies have distributed their assets over large geographical sectors and they are having to face a host of problems that have an effect on their staffing, productivity, and operational costs. These issues could have happened due to the poor health of industrial assets, content leakage, poor quality of products, excessive pollution, and damage caused by a natural crisis. These tracking assets are industrial tools that allow business and industrial processes to offer information concerning the working status of machinery and other equipment, the security of physical devices, efficiency, and performance. In industrial transportation, IoT sensors serve to provide the position, placement, and status of ships at sea, on a large as well as small scale; they can present physical conditions such as the temperature of cargo containers, and this helps to preserve them at stable temperatures and consequently perishable goods stay fresh. Refrigerated containers require being well furnished with mobile transmitters, temperature sensing devices, along with a processing element. If the measure of the temperature goes beyond the desired limit, the crew will be informed of immediate corrective action that needs to be done.

Use Case #4: Fleet Management:

Fleet management is crucial in industries that depend on transportation. When based on IIoT, it assists manufacturers to reduce and overcome the risks associated with productivity. A study by the Swedish M2M/IoT research firm (Suresh et al., 2020) states that North America is supposed to attain around 13 million vigorous fleet management IoT systems, realized in industrial vehicle fleets, by 2020. This study also reveals that in Latin America, the number of dynamic fleet management systems is deemed to be reaching around 4.0 million by 2020. The connected technology of the IIoT has given power to fleet businesses with better efficiencies in managing the fleet, and decreasing staff and transportation costs significantly.

Use Case #5: Connected Vehicles:

The interconnected vehicles in the IIoT are furnished by the Web for automating normal driving tasks. They employ different network technologies to speak to the driver (Suresh et al., 2020). Accidents caused by vehicles on the highways of the United States amounted to 37,461 people in 2016. Department of Transportation research in the United States stated that 94% of severe accidents are due to human mistakes. A few advantages of automated vehicles are:

1. Lessening the necessity of constructing new infrastructure and equipment, reducing maintenance expenditures, and preventing accidents.
2. Reliable driving and augmented motor vehicle safety, by means of analyzing the surroundings continuously, preventing errors in the driver's attention.
3. Self-driving motor vehicles reformulate car ownership and create new prospects for vehicle sharing.

Use Case #6: Optimizing Industrial Robotic Systems in Manufacturing:

Possibly the difficulty in implementing AI-use cases concerns automated manufacturing business process optimization (Suresh et al., 2020). One important concern of this optimization is through robots which are autonomous machines. The most important thought behind these independent assets is that they imitate repetitive human functionalities in the product manufacturing processes, bringing cost benefits. Before employing robots in production, these autonomous machines carry out the same task repeatedly until they achieve sufficient accuracy. The underpinning training and learning technology are frequently used to teach robots and independent machinery. Using these techniques, a robot can moderately or quickly be trained to perform a task under the direction of a human being. The "brains" of such robots are usually neural networks.

13.7 SUMMARY

In this chapter the details of different services offered by the IoT have been presented with explanations of the role of that service in a real time application. Also provided some directions for IoT application development in terms of the role of Web services and gateways while they interface with network terminals. An attempt has been made to explain the importance of IPv6 and how it is responsible for connecting more devices in an IoT application. We have also presented different prominent IoT applications which are currently being developed.

Smart environment case studies have been presented which pointed out that the common issues that researchers face in both SWCMC and SAPM is interoperability. The reason for this interoperability in these two case studies is due to the usage of heterogeneous sensors in the IoT environment. Hence researchers need to look into this aspect. Also, there is good research scope for this domain because air pollution is increasing day by day due to huge emissions from vehicles and industries, which indicates that there is the possibility for effective research in proposing systems which will minimize these issues. Therefore, those who want to develop such systems can take this up as their research problem and contribute better solutions for reducing this issue and helping society as well as mankind.

Similarly, designing and implementation of healthcare systems with the IoT definitely have significant returns in terms of providing cost reduced services, and efficient, quality, and sophisticated services, to under-privileged rural and urban people and other healthcare stakeholders. While the healthcare domain is gradually leveraging IoT and other thrust technologies like big data and ML to become more effective, there are other challenges that need to be addressed by researchers before smart healthcare becomes a prevalent reality.

The different IoT protocols used at the connectivity level as well as the communication level and their comprehensive details have been presented. The majority of these protocols have been developed and released by technical professional organizations like the IEEE, IETF, and ITU. The explanation of each protocol has focused on how the developers and researchers have applied these in their respective applications as well as the details of each protocol characteristic. Finally, we have tried to present some challenges that still exist in IoT systems and have suggested that current researchers solve them in their future work in this domain.

An overview of IIoT implementation challenges, as well as sample use cases, have been lucidly explained. Moving in the direction of Industry 4.0 fetches different industrial, technological, and scientific challenges, together with the higher impacts on today's manufacturing industry, especially in the security domain.

13.8 CONCLUSIONS AND FUTURE SCOPE

The developing concepts of the IoT is creating a versatile path in our daily life, leading to the rapid improvement of our quality of living through the interconnectivity of people, mobile devices, smart systems, and their applications. Ultimately the IoT has become part and parcel of human life. It is making everything autonomous in all the activities of this modern environment. This chapter has presented brief detail of the evidence of these IoT concepts in different applications. These helpful technologies, protocols, methods, and their contribution to recent research developments address different IoT ideas and applications, which will in turn give some proper direction and foundation for those researchers who are interested in this IoT. They can understand the different frontiers of IoT technologies, protocols, and applications.

A detailed description has been presented of the areas of smart environments, IoT services, intelligent systems, smart health and smart homes, smart agriculture, networking protocols, communication protocols, and autonomous industrial applications. Additionally, some of the difficulties and challenging issues that affect IoT system design and development have been discussed. Moreover we have tried to throw some light on the role of big data and security in IoT applications. Finally, different use cases of the IIoT and smart health and environments have been presented. Hence, we conclude that extensive research in this IoT will create a colossal leap in human lifestyles without their involvement in the systems that they are using.

REFERENCES

Ab Rahman, A. B., & Jain, R. (2015). Comparison of Internet of Things (IoT) data link protocols. Technical report, Washington University in St Louis, 1–21.

Agrawal, S., & Vieira, D. (2013). A survey on Internet of Things. *Abakós*, *1*(2), 78–95.

Aguilar, S., Vidal, R., & Gomez, C. (2017). Opportunistic sensor data collection with Bluetooth low energy. *Sensors*, *17*(1), 159.

Al-Fuqaha, A., Guizani, M., Mohammadi, M., Aledhari, M., & Ayyash, M. (2015). Internet of things: A survey on enabling technologies, protocols, and applications. *IEEE Communications Surveys & Tutorials*, 17(4), 2347–2376.

Alliance, L. (2015). A technical overview of LoRa and LoRaWAN. White Paper, November 2015.

Al-Sarawi, S., Anbar, M., Alieyan, K., & Alzubaidi, M. (2017, May). Internet of Things (IoT) communication protocols. In *2017 8th International Conference on Information Technology (ICIT)* (pp. 685–690). IEEE.

Amado, T.M.; Dela Cruz, J.C., (2018). Development of Machine Learning-based Predictive Models for Air Quality Monitoring and Characterization. In *Proceedings of the TENCON 2018*, 2018 IEEE Reg. Jeju, Korea, pp. 668–672.

Anagnostopoulos, T., Zaslavsky, A., Kolomvatsos, K., Medvedev, A., Amirian, P., Morley, J., & Hadjieftymiades, S. (2017). Challenges and opportunities of waste management in IoT-enabled smart cities: A survey. *IEEE Transactions on Sustainable Computing*, *2*(3), 275–289.

Atzori, L., Iera, A., & Morabito, G. (2010). The internet of things: A survey. *Computer Networks*, *54*(15), 2787–2805.

Cerruela García, G., Luque Ruiz, I., & Gómez-Nieto, M. Á. (2016). State of the art, trends and future of bluetooth low energy, near field communication and visible light communication in the development of smart cities. *Sensors*, *16*(11), 1968.

Chaudhury, S., Paul, D., Mukherjee, R., & Haldar, S. (2017, August). Internet of Thing based healthcare monitoring system. In *2017 8th Annual Industrial Automation and Electromechanical Engineering Conference (IEMECON)* (pp. 346–349). IEEE.

Elmustafa, S. A. A., & Mujtaba, E. Y. (2019). Internet of things in smart environment: Concept, applications, challenges, and future directions. *World Scientific News*, *134*(1), 1–51.

Gao, J., Liu, F., Ning, H., & Wang, B. (2007, December). RFID coding, name and information service for internet of things. In *2007 Proceedings of the IET Conference on Wireless, Mobile and Sensor Networks (CCWMSN '07)*, Shanghai, China (pp. 36–39).

Gigli, M., & Koo, S. G. (2011). Internet of things: services and applications categorization. *Advances in Internet of Things*, *1*(2), 27–31.

Gomez, C., Chessa, S., Fleury, A., Roussos, G., & Preuveneers, D. (2019). Internet of Things for enabling smart environments: A technology-centric perspective. *Journal of Ambient Intelligence and Smart Environments*, *11*(1), 23–43.

Guinard, D. (2010, March). Towards opportunistic applications in a Web of Things. In *2010 8th IEEE International Conference on Pervasive Computing and Communications Workshops (PERCOM Workshops)* (pp. 863–864). IEEE.

Hammi, B., Khatoun, R., Zeadally, S., Fayad, A., & Khoukhi, L. (2017). IoT technologies for smart cities. *IET Networks*, *7*(1), 1–13.

Islam, S. R., Kwak, D., Kabir, M. H., Hossain, M., & Kwak, K. S. (2015). The internet of things for health care: a comprehensive survey. *IEEE access, 3*, 678–708.

Islam, M. M., Rahaman, A., & Islam, M. R. (2020). Development of smart healthcare monitoring system in IoT environment. *SN Computer Science*, *1*, 1–11.

ISO/IEC JTC1 information technology document. (2015), "Internet of Things (IoT)-Preliminary Report 2014", © ISO 2015, Published in Switzerland, pp-7.

Jamil, M. S., Jamil, M. A., Mazhar, A., Ikram, A., Ahmed, A., & Munawar, U. (2015). Smart environment monitoring system by employing wireless sensor networks on vehicles for pollution free smart cities. *Procedia Engineering*, *107*, 480–484.

Khan, W. Z., Rehman, M. H., Zangoti, H. M., Afzal, M. K., Armi, N., & Salah, K. (2020). Industrial internet of things: Recent advances, enabling technologies and open challenges. *Computers & Electrical Engineering*, *81*, 106522.

Khanna, A., & Kaur, S. (2020). Internet of Things (IoT), applications and challenges: A comprehensive review. *Wireless Personal Communications*, *114*, 1687–1762.

Le, A., Loo, J., Lasebae, A., Aiash, M., & Luo, Y. (2012). 6LoWPAN: a study on QoS security threats and countermeasures using intrusion detection system approach. *International Journal of Communication Systems*, *25*(9), 1189–1212.

Madakam, S., Lake, V., Lake, V., & Lake, V. (2015). Internet of Things (IoT): A literature review. *Journal of Computer and Communications*, *3*(05), 164.

Miorandi, D., Sicari, S., De Pellegrini, F., & Chlamtac, I. (2012). Internet of things: Vision, applications and research challenges. *Ad hoc networks*, *10*(7), 1497–1516.

Oasis, O. S. (2012). Oasis advanced message queuing protocol (amqp) version 1.0. *International Journal of Aerospace Engineering*, Hindawi, www.hindawi.com, *2018*.

Olaronke, I., & Oluwaseun, O. (2016, December). Big data in healthcare: Prospects, challenges and resolutions. In *2016 Future Technologies Conference (FTC)* (pp. 1152–1157). IEEE.

Rahaman, A., Islam, M. M., Islam, M. R., Sadi, M. S., & Nooruddin, S. (2019). Developing IoT based smart health monitoring systems: A review. *Revue d'Intelligence Artificielle*, *33*(6), 435–440.

Rahmani, A. M., Gia, T. N., Negash, B., Anzanpour, A., Azimi, I., Jiang, M., & Liljeberg, P. (2018). Exploiting smart e-Health gateways at the edge of healthcare Internet-of-Things: A fog computing approach. *Future Generation Computer Systems*, *78*, 641–658.

Salman, T., & Jain, R. (2015). Networking protocols and standards for internet of things. *Internet of Things and Data Analytics Handbook*, *7*, 14–18.

Samie, F., Bauer, L., & Henkel, J. (2016, October). IoT technologies for embedded computing: A survey. In *2016 International Conference on Hardware/Software Codesign and System Synthesis (CODES+ ISSS)* (pp. 1–10). IEEE.

Shelby, Z., Hartke, K., & Bormann, C. (2014). The constrained application protocol (CoAP). *Internet Engineering Task Force (IETF)*, 1–112.

Shen, J., Lu, X., Li, H., & Xu, F. (2010, August). Heterogeneous multi-layer access and RRM for the internet of things. In *2010 5th International ICST Conference on Communications and Networking in China.*

Singh, M., Rajan, M. A., Shivraj, V. L., & Balamuralidhar, P. (2015, April). Secure mqtt for internet of things (iot). In *2015 Fifth International Conference on Communication Systems and Network Technologies* (pp. 746–751). IEEE.

Spiess, P., Karnouskos, S., Guinard, D., Savio, D., Baecker, O., De Souza, L. M. S., & Trifa, V. (2009, July). SOA-based integration of the internet of things in enterprise services. In *2009 IEEE International Conference on Web Services* (pp. 968–975). IEEE.

Suresh, A., Nandagopal, M., Raj, P., Neeba, E. A., & Lin, J. W. (2020). *Industrial IoT Application Architectures and Use Cases.* CRC Press.

Triantafyllou, A., Sarigiannidis, P., & Lagkas, T. D. (2018). Network protocols, schemes, and mechanisms for internet of things (iot): Features, open challenges, and trends. *Hindawi Journal of Wireless Communications and Mobile Computing*, 2018, 1–24.

Tukade, T. M., & Banakar, R. (2018). Data transfer protocols in IoT—An overview. *International Journal of Pure and Applied Mathematics*, *118*, 121–138.

Ullo, S. L., & Sinha, G. R. (2020). Advances in smart environment monitoring systems using iot and sensors. *Sensors*, *20*(11), 3113.

Want, R. (2006). An introduction to RFID technology. *IEEE Pervasive Computing*, *5*(1), 25–33.

Whitmore, A., Agarwal, A., & Da Xu, L. (2015). The Internet of Things—A survey of topics and trends. *Information Systems Frontiers*, *17*(2), 261–274.

Xiaojiang, X., Jianli, W., & Mingdong, L. (2020). Services and key technologies of the Internet of Things. *ZTE Communications*, *8*(2), 26–29.

Yin, J., Yang, Z., Cao, H., Liu, T., Zhou, Z., & Wu, C. (2019). A survey on Bluetooth 5.0 and mesh: New milestones of IoT. *ACM Transactions on Sensor Networks (TOSN)*, *15*(3), 1–29.

Zeadally, S., & Bello, O. (2019). Harnessing the power of Internet of Things based connectivity to improve healthcare. *Internet of Things*, 100074. https://doi.org/10.1016/j.iot.2019.100074

Zeadally, S., Siddiqui, F., Baig, Z. and Ibrahim, A. (2019). Smart healthcare: Challenges and potential solutions using internet of things (IoT) and big data analytics. *PSU Research Review*, Vol. 4 No. 2, pp. 149–168. https://doi.org/10.1108/PRR-08-2019-0027

14 A Spatio-temporal Model for the Analysis and Classification of Soil Using the IoT

M. Umme Salma and Subbaiah Rachana
CHRIST (Deemed to be University)

Narasegouda Srinivas
Jyoti Nivas College

CONTENTS

14.1 Introduction ... 221
14.2 Literature Survey ... 222
14.3 Proposed Methodology ... 225
14.4 Experimental Work ... 226
    14.4.1 Experimental Setup ... 226
    14.4.2 Procedure ... 227
14.5 Comparative Study and Results ... 227
14.6 Data Analysis ... 228
14.7 Discussion ... 232
14.8 Conclusion and Future Work ... 232
References ... 233

14.1 INTRODUCTION

The Internet of Things (IoT) is a collection of sensing, operating, and networking devices embedded within physical entities and which are connected through guided and unguided networks, with the help of the same protocol that is connected to the Internet.

There exists no general definition for the IoT; the basic notion is that it is a collection of everyday objects connected by sensors, networking devices, and computing devices via the Internet to accomplish the desired objective. The basic concepts of the IoT have existed for years. The IoT has been used in the manufacturing and industry sectors where items such as huge cranes and large items carrying livestock were tracked using radio-frequency identification (RFID) tracking devices.

DOI: 10.1201/9781003146711-14

There exist myriad applications of the IoT including but not restricted to agriculture, automation, energy, healthcare, habitat management, and disaster management. Among a large group of applications supported by the IoT, agricultural IoT is of high relevance especially in countries like India where the majority of the rural population relies on it. The IoT provides the connected physical entities with the ability to see, hear, think, and accomplish a desired task by enabling them to communicate with each other, allowing the system to perform efficient decision making. In the near future it is believed that all the activities of individuals and organizations will be connected and executed through the IoT, making the world not only digital but smart.

The IoT has reshaped the agricultural sector with the help of precision farming, telemetry, and smart cultivation, resulting in a remarkable shift in agriculture-based economies. Farmers have begun to upgrade themselves from the traditional approach to the smart approach guided by technology. Smart agriculture, enabled with ubiquitous devices, helps farmers to protect their crops from weeds, invaders, and natural and manmade disasters. The IoT in agriculture also helps farmers to check the quality of seeds, yield, and soil. Soil quality checking is of high importance as quality soil provides the main food source for plants to grow. With the help of the IoT one can determine the quality of the soil by analyzing its nutrient composition. One can also classify the soil based upon parameters such as pH, temperature, and moisture content.

This chapter mainly address a spatio-temporal model for the analysis and classification of soil using the IoT; the following sections cover: the state of the art, a proposed methodology, experimental work, a comparative study with results, data analysis, discussion, and conclusion.

14.2 LITERATURE SURVEY

Liqiang et al. (2011) proposed the implementation of two types of hubs and built a sensor network model. The programming framework involves TinyOS, an embedded operating system, which involves gadget drivers, framework portion, and other applications embedded together. The dissemination protocol is the correlative activity to gathering. The objective of a dissemination protocol is to directly transfer the information regarding synchronization directions and bit control mechanisms to each node in the network.

Karim and Karim (2017) present work related to irrigation management. The main idea is to facilitate the farmer by providing timely updates about irrigation in the form of a dashboard showing the required statistics and graphs. This software monitors real-time variations of the soil. The farmer is also notified about any changes related to the water level. For instance if the level approaches a critical point, the farmer is notified immediately so that the crop doesn't face water stress.

Satyanarayana and Mazaruddin (2013) discuss how data is transmitted from the field to the central monitoring system (CMS) using either a global positioning system (GPS) module or a general packet radio service (GPRS) module with Zigbee technology. Shi et al. (2015) propose the application of the IoT in creepy crawly bug control and horticultural malady. The proposed framework is intended to monitor farming infection and bug problems. An infection and bug control framework in view of the IOT is presented, which comprises three levels and three frameworks. Sasikumar and Priya (2017) propose another sensor-data accessing mechanism where a framework is developed using the IoT, distributed computing, and the cloud to send and receive the sensor data. This type of framework is much discussed now in the field of the Indian-based cultivation and cordiality industry. In India around 70% of the population relies on cultivation and 33% of the country's capital originates from this. Issues concerning horticulture have continually hampered the progress of the country.

Channe et al. (2015) highlight that, in spite of the fact that precision farming has been implemented in a couple of nations, the horticultural field in India demands modernization. This could be achieved using big data analysis on an agro-cloud. The analysis can then be used for providing information to the farmer related to manure prerequisites, crop examination, making best yield arrangements, being up to date regarding the livestock and crop market, and so on. Baruah and Dutta (2009) note that, with the increase in world population, interest in an expanded supply of nourishment has propelled researchers and specialists into planning new strategies to support rural generation. In this regard the IoT can suggest to farmers how to select the proper amount of pesticides and fertilizers so as to increase the agricultural yield. The authors also provide information related to nanotechnology and its usefulness in converting persistent chemicals into useful compounds which are harmless to the crops.

Duhan et al. (2017) state that nanotechnology is an interdisciplinary area of research and can be combined with the IoT to provide impactful solutions in agriculture. Lobsey et al. (2010) show how the IoT can be used for soil analysis. It is a known fact that the traditional way of examining soil is a tedious, time consuming, and costly task. With the help of soil sensors one can investigate the quality of the soil; the results obtained are of high value especially in precision agriculture. Mo (2011) proposes a system of cutting-edge agrarian technology, which highlights the current circumstances regarding the coordinating of farming. This article brings up the fundamental issues existing in conventional horticultural items, and exhibits the SCOR method of advanced rural items in view of the IoT. To enhance the aggressive capacity of the agrarian items in India, we can construct renowned brands, screen their quality and security, increment ranchers' wages for farming items ahead of time, and build a cutting-edge inventory administered by the IOT.

Holland et al. (2012) discuss active optical sensors (AOSs) that have become an effective tool at contributing to agribusiness, not only for determining the required agrochemicals but also for mapping the soil with the corresponding crop. These sensors control the individual wavelength reaction information that can provide helpful data with respect to the separation of varieties between the sensor and the objective. Nakutis et al. (2015) propose a remote farming mechanism where the distribution exhibits the farming procedure involving the usage of machines, actuators, and sensors associated with the IoT running on a server. Sensors and actuators are exceptionally broad and don't have to take account of any knowledge identified with the procedure under control.

Chen and Jin (2012) propose a structure for computerized horticultural application in view of the IoT. The authors claim that the IoT will without doubt be a data revolution involving electronic devices and the Internet. Jaiganesh et al. (2017) survey the applying of a cloud-based IoT application within the scope of the agricultural sector. This paper provides information related to the agro-cloud that updates horticultural data on the server, making storage and computation efficient and less time consuming. Khan et al. (2019) propose an IoT-based model applying a continuous stream of data for examining the cultivation of cotton yield. This will provide the ongoing outcomes which will help the cultivator to recognize events on the farm. Salam and Shah (2019) note that the developments and new learning in the field of the IoT are crucial to empower a thorough and well-coordinated innovation in precision horticulture. In order to achieve precision farming, mastery in numerous areas, including agronomy, ecology, soil ripeness, entomology, information mining, and artificial intelligence (AI), is required. All endeavors need to interface ably in these areas and to be able to discern significant innovation, improvement, and appropriation challenges in precision horticulture. The use of the most recent advancements and innovation in the agricultural IoT helps in detecting anomalies and leads to upgraded farming. Araby et al. (2019) propose a framework dependent on the joining of the IoT and AI to anticipate late scourge in potatoes and tomatoes, which decreases expenses by providing the farmer with a warning so as to apply defensive pesticides which help to spare yield generation and decrease the use of superfluous pesticides. It is to be noted that there exists myriad IoT models in agriculture, each having their own advantages and limitations. The choice of the technology, sensors, and protocols for setting up the application depends on the use (Glaroudis et al., 2020).

Agricultural IoT involves many sensing devices. Various sensors used in agriculture which are available in the literature are given in Table 14.1.

TABLE 14.1
Different Agricultural Sensors

Agricultural Sensor Type	Sensor Details
Location	The location sensors, with the help of satellites, are able to provide details related to latitude, altitude, and longitude of any entity in the given area.
Optical	The optical sensors, also called light sensors, utilize light to measure the properties of the soil. They are used to determine the organic and inorganic components in the soil. These sensors can be easily deployed in smart devices, drones, and robots to study the soil.
Electrical and chemical	This type of sensor helps in examining the electrical and chemical properties of the soil. They provide information related to the nutrient value of the soil and its pH level by examining the various ions available.
Mechanical sensors	This type of sensor is used to study the mechanical properties of the soil. With the help of a mechanical sensor one can determine the mechanical resistance and/or compactness of the soil.
Soil	This type of sensor is also called a dielectric soil sensor and records the moisture values of the soil.
Air flow	This type of sensor is used to measure the permeability of air.

14.3 PROPOSED METHODOLOGY

Sample Collection: In order to study the quality of the various soils, eight different soil samples were collected from different geographical parts of India: Bangalore, Kodagu, Puducherry Beach, Mangalore Beach, Shivamogga, Andhra Pradesh, Tamil Nadu, and Assam. The samples collected were exposed to temperature, humidity, and pH sensors to generate the respective readings. Based upon the temperature, humidity, and pH value, the samples were classified as agricultural and non-agricultural soil. The details of the soils collected are provided in Table 14.2.

Sensors Used: The study included three sensors: temperature, humidity, and pH.

Data Acquisition: A DHT11 sensor was used on the top level of the soil sample where the humidity was measured. If the soil is moist the sensor will be damaged, so that is why the sensor is placed on top of the sample.

A pH sensor is used to obtain the acidity or alkalinity quality of the soil sample by getting a water measurement. The test is done on the water mixture of the soil.

There are three sensors used for temperature, humidity, and pH.

The humidity and temperature (i.e., DHT11) sensor identifies the humidity and temperature reading of the different soil samples. If the range of temperature is less than 36 degrees it is considered to be agricultural soil; otherwise it is considered to be non-agricultural soil. A humidity range less than 1.00039 is considered to be agricultural soil; otherwise it is treated as non-agricultural soil. The relative humidity is calculated using the formula given in Equation 14.1.

$$H_r = \exp^{\left(7.3\times10^{-6}\times H\right)} \tag{14.1}$$

where Hr denotes relative humidity, exp is an exponential function, and H is the humidity reading obtained from the humidity sensor.

The pH sensor is used to measure the alkaline and acidic quality of the soil which is collected from different parts of India. A pH value of less than 3.96 is identified as agricultural soil; otherwise it is considered as non-agricultural soil.

TABLE 14.2
Details of Soil Samples

Serial No.	Location	Categories	Color	Type
1	Bangalore	Loamy and clayey loamy	Dark red	Non-agricultural
2	Shivamogga	Deep clayey and loamy soil	Black	Agricultural
3	Kodagu	Deep clayey and loamy soil	Black	Agricultural
4	Tamil Nadu	Loamy and clayey loamy	Black	Agricultural
5	Mangalore Beach	Light sandy yellow soil	White sand	Non-agricultural
6	Puducherry Beach	Light sandy black soil	Black sand	Non-agricultural
7	Assam	Deep rich loamy soil	Brown	Non-agricultural
8	Andhra Pradesh	Loamy and clayey loamy	Black	Agricultural

All the samples collected have undergone preprocessing of data by removing the repeated values; only unique values in each row and column are given for prediction using a feedforward neural network(FNN) combined with an AND logical operation.

Acquiring Temperature Values: The temperature reading of the given soil samples is acquired using a DHT11 sensor, which has three pins: a voltage common collector (VCC), data, and ground. A VCC pin is the first pin used for the power supply of about 3.5 to 5.5 V. Pin 2 is a data pin which provides both humidity and temperature information through serial data. Pin 3 is the ground pin which is connected to the ground pin of the circuit. The sensor is factory calibrated and is easy to interface with the microcontroller of the circuit. The outputs obtained are all in serial data.

Acquiring Humidity Values: The same DHT11 sensor is used to record the humidity values of the soil samples. As the readings obtained are in a higher range, in order to bring them to the actual level, a relative correction is done using the formula given in Equation 14.2 (see Section 14.6).

Acquiring pH Values: A pH sensor is commonly used for measuring the acidity and alkalinity of the given solution. Each change in the unit of the data is represented as a tenfold change in acidity or alkalinity. The pH value is the negative logarithm of the hydrogen-ion concentration activity.

Data Storage: Data is stored on the cloud system using the Blynk app for further usage, which provides the real-time analysis of the soil sample.

Data Preprocessing: After collection from the sensors, the data undergoes preprocessing by removing the repeated entry of data in the same column in the Excel sheet.

Computational Model: The computational model used for this research is a simple feedforward neural network; the details of the model are provided in Section 14.4.

14.4 EXPERIMENTAL WORK

The experiment was conducted on a dataset of 26,698 rows and 4 columns generated collectively from the eight different soil samples using temperature, humidity, and pH sensors.

The FNN model with three input neurons, three hidden neurons, and one output neuron was used to classify the data. A sigmoid function was used as a training function in order to get the classification result 0 or 1. Based on the output generated, the soil is classified as agricultural or non-agricultural. The value 1 indicates agricultural soil and 0 indicates non-agricultural soil.

14.4.1 Experimental Setup

For conducting the experiment, a Jupyter Notebook with Python code is used along with three variable AND logic gates for analyzing the three values (temperature, humidity, and moisture) of the data acquired from the sensors, which was given as input.

TABLE 14.3
FNN Accuracy for Various Data Splits (%)

Data Splitting	Accuracy of FNN
50:50	92.30
70:30	97.00
80:20	96.75

14.4.2 Procedure

The data collected from all eight samples was divided into different percentages of training and testing samples, as given in Table 14.3. Finally, 70% of the training data (18,200 × 4) and 30% of the testing data (7800 × 3) was subjected to an FNN classifier which implements three input-based AND gates; the highest accuracy of the algorithm was found to be 97%. A feedforward algorithm was combined with the AND logical operation; based on the conditions provided for temperature, humidity, and pH value the soil samples were differentiated as agricultural or non-agricultural soil.

The FNN model was also compared with a back-propagation and multi-layered perceptron. The results of the comparative study are discussed in Section 14.5. The classified data was later subjected to a spatio-temporal model for soil analysis and is discussed in Section 14.6.

14.5 COMPARATIVE STUDY AND RESULTS

The FNN algorithm was compared with two other neural network techniques, namely back propagation (BP) and multilayer perceptron (MLP), using the same parameters: input neurons set to 3, hidden neurons set to 3, and one output neurons. All the models were fed with the same number of input, hidden, and output neurons; a sigmoid function was used as the training function; A 70:30% data split was used. It was observed that FNN provided an accuracy of 97%, MLP provided an accuracy of 93.88%, and BP an accuracy of 92%. By observing the comparative results shown in Figure 14.1 we can say that FNN provides more accurate results compared to its counterparts.

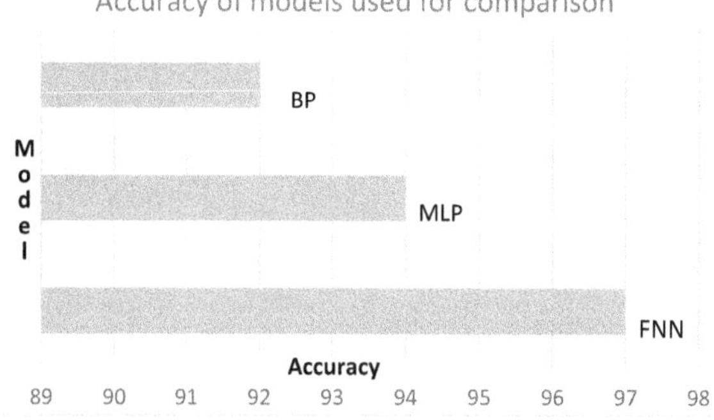

FIGURE 14.1 Comparative study of different neural networks used.

Based on the output received through FNN, the soil sample is classified as agricultural soil or non-agricultural soil with an accuracy of 97%.

14.6 DATA ANALYSIS

Primary Analysis – Spatio-temporal analysis: For the analytical modeling of agricultural data, a well-defined spatio-temporal model was used. This is a data model that includes both the spatial and temporal details of the data. The mathematical form of a general spatio-temporal model is shown in Equation 14.2.

$$y(s,t) = \mu(s,t) + e(s,t) \tag{14.2}$$

In Equation 14.2, s denotes the spatial domain, t denotes the temporal domain for the model y, μ indicates the mean structure, and e indicates the error structure.

Spatio-temporal analysis was done for two domains, namely space fixed with time varying (e.g., keeping one soil constant and varying the time stamps) and time fixed with space varying (e.g., with the same time stamp observing different soil types).

In this work the spatial characteristics of agricultural data with different time spans for a given season was analyzed. Here the change in pH, temperature, and humidity of eight different soil samples with respect to various time frames in a day was monitored and analyzed. The analysis of the patterns observed for agricultural and non-agricultural samples is given in Figures 14.2 to 14.7. The dark lines indicate temperature readings; the lighter ones indicate humidity readings.

The graph in Figure 14.2 shows the non-agricultural pattern for temperatures from 2 p.m. to 10 p.m., where the maximum temperature is 47.26 degrees Celsius and the minimum is 26.99 degrees Celsius. Similarly the maximum humidity reading is found to be 1.00057 and the minimum is 1.00043.

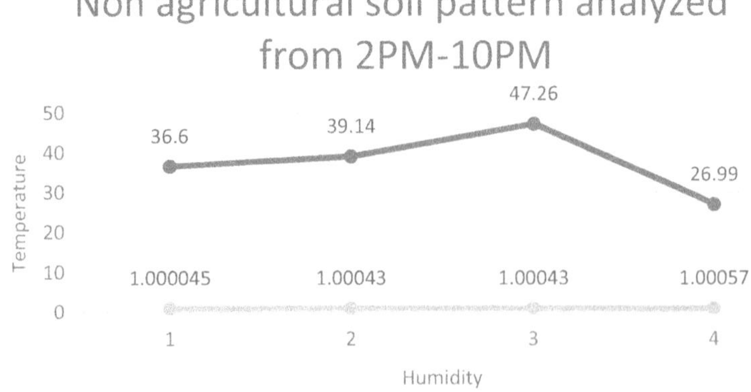

FIGURE 14.2 Non-agricultural soil pattern analyzed from 2 p.m. to 10 p.m.

The graph in Figure 14.3 shows the agricultural pattern for temperatures from 2 p.m. to 10 p.m., where the maximum temperature is 34.1 degrees Celsius and the minimum is 25.11 degrees Celsius. Similarly the maximum humidity reading is found to be 1.00032 and the minimum is 1.00025.

The graph in Figure 14.4 shows the non-agricultural pattern for the temperature from 10 p.m. to 6 a.m. where the maximum temperature is 39.24 degrees Celsius and the minimum is 26.32 degrees Celsius. Similarly the maximum humidity reading is found to be 1.00048 and the minimum is 1.00031.

The graph in Figure 14.5 shows the agricultural pattern for the temperature from 10 p.m. to 6 a.m. where the maximum temperature is 30.76 degrees Celsius and the

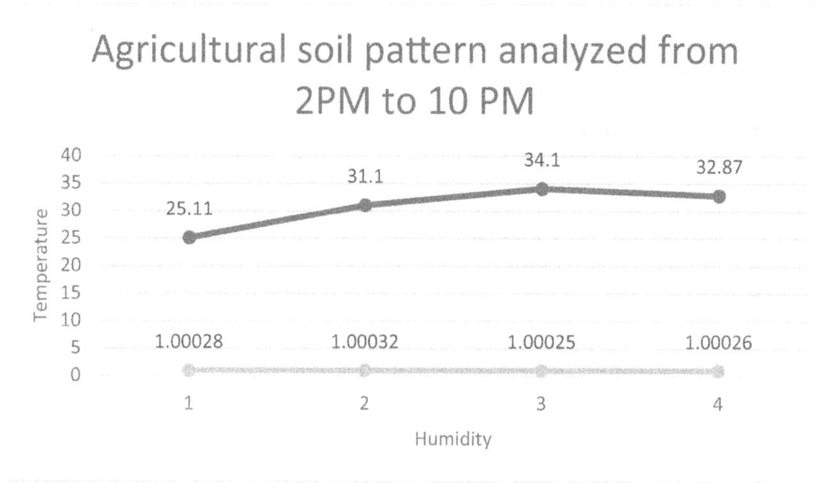

FIGURE 14.3 Agricultural soil pattern analyzed from 2 p.m. to 10 p.m.

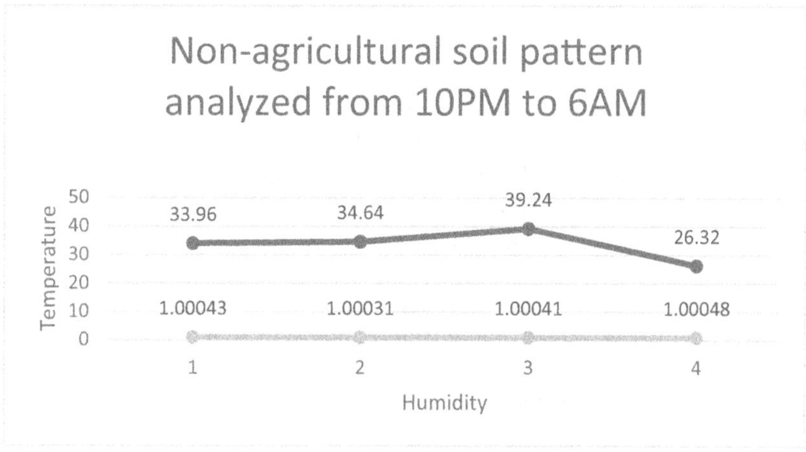

FIGURE 14.4 Non-agricultural soil pattern analyzed from 10 p.m. to 6 a.m.

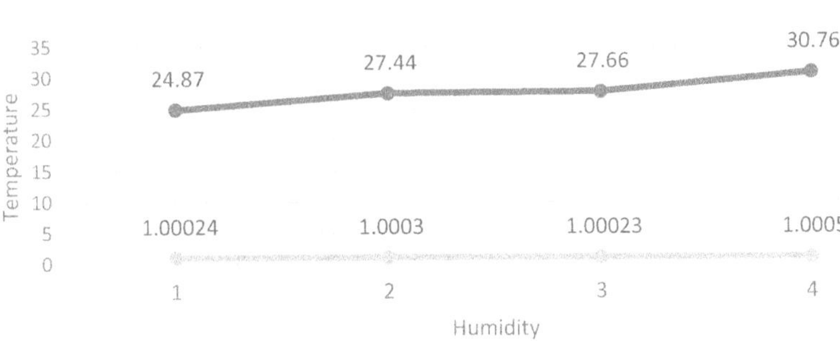

FIGURE 14.5 Agricultural soil pattern analyzed from 10 p.m. to 6 a.m.

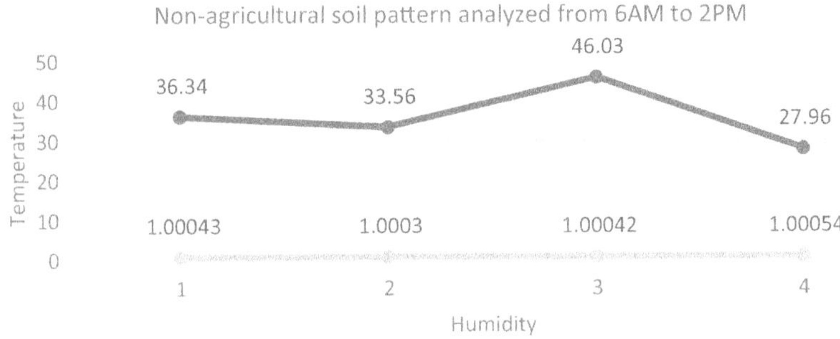

FIGURE 14.6 Non-agricultural soil pattern analyzed from 6 a.m. to 2 p.m.

minimum is 24.87 degrees Celsius. Similarly the maximum humidity reading is found to be 1.0005 and the minimum is 1.00023.

The graph in Figure 14.6 shows the non-agricultural pattern for the temperature from 6 a.m. to 2 p.m. where the maximum temperature is 46.03 degrees Celsius and the minimum is 27.96 degrees Celsius. Similarly the maximum humidity reading is found to be 1.00054 and the minimum is 1.0003.

The graph in Figure 14.7 shows the agricultural pattern for the temperature from 6 a.m. to 2 p.m. where the maximum temperature is 33.86 degrees Celsius and the minimum is 30.02 degrees Celsius. Similarly the maximum humidity reading is found to be 1.00032 and the minimum is 1.00024.

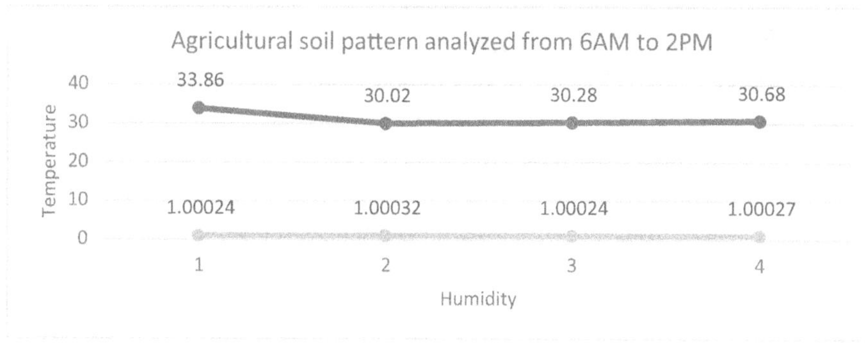

FIGURE 14.7 Agricultural soil pattern analyzed from 6 a.m. to 2 p.m.

TABLE 14.4
Range of Soil Value for the Time-Fixed and Space-Varying Criteria

Place	Temp (°C)	Humidity	pH	Soil Type
Bangalore	30–33	1.00043–1.00044	4.36	Non-agricultural
Tamil-Nadu	30–32	1.00026–1.00029	3.78	Agricultural
Andhra	29–30	1.00031–1.00032	3.19	Agricultural
Coorg	31–32	1.00024–1.0025	3.86	Agricultural
Mangalore Beach	46–49	1.00033–1.00035	4.24	Non-agricultural
Puducherry Beach	45–46	1.00041–1.00043	4.41	Non-agricultural
Shivamogga	30–31	1.00028–1.00032	3.96	Agricultural
Assam	17–19	1.00056–1.00059	4.02	Non-agricultural

Similarly, the details for the average readings of temperature, humidity, and pH collected over 24 hours to satisfy the time-fixed and space-varying criteria is given in Table 14.4, separately for all eight soil samples.

With space fixed and time varying, the range of the agricultural soil temperature is 25–34 degrees Celsius, the average humidity is 1.00024–1.00032, and the range of pH is 3.19–3.96. With time fixed and space varying, the range of temperature was found to be around 29–32, the humidity 1.00026–1.00059, and the pH is 3.19–4.36. We can observe that there is no change in humidity and pH.

Secondary Analysis: This is done by collecting the information related to the temperature, humidity, and pH required to grow various crops. The details of the temperature specific to different crops which are collected from various Web sources (Jagranjosh, 2021) are summarized in Table 14.5. Based on the results in Tables 14.4 and 14.5 we can see that rice and millet crops require a high temperature and that coffee and wheat crops require a low temperature for cultivation.

TABLE 14.5
List of Various Indian Crops with their Temperature Range

Crop Type	Temperature (°C)
Rice	22–32
Wheat	10–15
Millets	27–32
Grams	20–25
Sugar cane	21–27
Cotton	21–30
Tea	20–30
Coffee	15–28
Oil seeds	20–30
Pepper	18–26

14.7 DISCUSSION

The main aim of this chapter was to provide an analysis of soil using a spatio-temporal model applied on the different samples collected from various parts of Karnataka and other states of India. The analysis was done by using temperature, humidity, and pH sensors in order to find whether soil quality is best for agriculture or non-agriculture. Classification was done using FNN by considering the temperature, humidity, and pH features of the soil and by taking the time and space domains into consideration. The spatiotemporal analysis with different time slots results in observing the variation in temperature and humidity of the soil samples. It is to be noted that the pH remains constant.

Therefore, the soil with more fertility will be used as agricultural soil. In this research we have collected eight different samples and, based on the features, four samples are categorized as agricultural soil. As a result the soil sample in the above-mentioned range will be significant for growing different crops.

14.8 CONCLUSION AND FUTURE WORK

This chapter has presented a detailed spatio-temporal analysis of the characteristics of various soil samples. The soil samples were classified as agricultural or non-agricultural based on the spatial parameters at various time frames. Three different IoT sensors were used (temperature, humidity (DHT11), and pH) for identifying the soil quality. This experimental result shows that this model works correctly and reliably. The design objective is basically to provide farmers with knowledge as to which soil is suitable for growing crops. Further, the research can be done by considering different seasons and varieties in different climates to identify which is the suitable crop and which provides maximum yield during which season. In terms of analysis we can extend the spatio-temporal model into two more domains (space and time varying and space and time fixed) to understand seasonal and regional variation. The same analysis can be done for rainfall.

Based upon spatio-temporal analysis we can confirm that Shivamogga soil with a temperature of 31 degrees Celsius, a humidity of 1.00032, and a pH of 3.96 is more suitable for areca cultivation, whereas Kodagu soil with a temperature of 32 degrees Celsius, a humidity of 1.00024, and a pH of 3.86 is more suitable for coffee and pepper cultivation. Based on the results obtained from primary data and secondary data analysis we can confirm that Kodagu soil is suitable for coffee and pepper crops, Shivamogga soil is suitable for areca and sugar cane crops, Tamil-Nadu soil is suitable for sorghum and millet crops.

REFERENCES

Araby, A. A., Elhameed, M. M. A., Magdy, N. M., Abdelaal, N., Allah, Y. T. A., Darweesh, M. S., & Mostafa, H. (2019). Smart IoT monitoring system for agriculture with predictive analysis. In *2019 8th International Conference on Modern Circuits and Systems Technologies (MOCAST)*, 1–4. IEEE.

Baruah, S., & Dutta, J. (2009). Nanotechnology applications in pollution sensing and degradation in agriculture: a review. *Environmental Chemistry Letters, 7*(3), 191–204.

Channe, H., Kothari, S., & Kadam, D. (2015). Multidisciplinary model for smart agriculture using internet-of-things (IoT), sensors, cloud-computing, mobile-computing & big-data analysis. *International Journal of Computer Technology & Applications, 6*(3), 374–382.

Chen, X.-Y. & Jin, Z.-G. (2012). Research on key technology and applications for internet of things. *Physics Procedia,* 33, 561–566.

Duhan, J. S., Kumar, R., Kumar, N., Kaur, P., Nehra, K., & Duhan, S. (2017). Nanotechnology: The new perspective in precision agriculture. *Biotechnology Reports, 15*, 11–23.

Glaroudis, D., Iossifides, A., & Chatzimisios, P. (2020). Survey, comparison and research challenges of IoT application protocols for smart farming. *Computer Networks, 168*, 107037.

Holland, K. H., Lamb, D. W., & Schepers, J. S. (2012). Radiometry of proximal active optical sensors (AOS) for agricultural sensing. *IEEE Journal of Selected Topics in Applied Earth Observations and Remote Sensing, 5*(6), 1793–1802.

Jagranjosh. (2021). https://www.jagranjosh.com/general-knowledge/list-of-major-crops-of-india-temperature-rainfall-soil-1473918924-1. Accessed on 14-02-2021.

Jaiganesh, S., Gunaseelan, K., & Ellappan, V. (2017). IOT agriculture to improve food and farming technology. In *2017 Conference on Emerging Devices and Smart Systems (ICEDSS)*, 260–266. IEEE.

Karim, F., & Karim, F. (2017). Monitoring system using web of things in precision agriculture. *Procedia Computer Science, 110*, 402–409.

Khan, F. A., Abubakar, A., Mahmoud, M., Al-Khasawneh, M. A., & Alarood, A. A. (2019). Cotton crop cultivation oriented semantic framework based on IoT smart farming application. *International Journal of Engineering and Advanced Technology, 8*(3), 480–484.

Liqiang, Z., Shouyi, Y., Leibo, L., Zhen, Z., & Shaojun, W. (2011). A crop monitoring system based on wireless sensor network. *Procedia Environmental Sciences, 11*, 558–565. Elsevier.

Lobsey, C. R., Rossel, R. V., & McBratney, A. B. (2010). Proximal soil nutrient sensing using electrochemical sensors. *Proximal Soil Sensing,* 77–88.

Mo, L. (2011, August). A study on modern agricultural products logistics supply chain management mode based on IOT. In *2011 Second International Conference on Digital Manufacturing & Automation,* 117–120. IEEE.

Nakutis, Z., Deksnys, V., Jaruevicius, I., Marcinkevicius, E., Ronkainen, A., Soumi, P., & Andersen, B. (2015). Remote agriculture automation using wireless link and IoT gateway infrastructure. In *2015 26th International Workshop on Database and Expert Systems Applications (DEXA)*, 99–103. IEEE.

Salam, A, & Shah, S. (2019, April). Internet of things in smart agriculture: Enabling technologies. In *2019 IEEE 5th World Forum on Internet of Things (WF-IoT)*, 692–695. IEEE.

Sasikumar, V. & Priya, S. (2017). IOT applications for Indian based farming and hospitality industry. *International Journal of Advance Research, Ideas and Innovations in Technology, 3*(4), 739–745.

Satyanarayana, G. V. & Mazaruddin, S. D. (2013). Wireless sensor based remote monitoring system for agriculture using Zigbee and GPS. In *Proceedings of the Conference on Advances in Communication and Control Systems-2013,* 110–114. Atlantis Press.

Shi, Y., Wang, Z., Wang, X., & Zhang, S. (2015, May). Internet of things application to monitoring plant disease and insect pests. In *2015 International Conference on Applied Science and Engineering Innovation,* 31–34. Atlantis Press.

15 A Critical Survey of Autonomous Vehicles

Nipun R. Navadia and Gurleen Kaur

Department of Computer Science and Engineering, Dronacharya Group of Institutions, Greater Noida, Uttar Pradesh, India

Harshit Bhardwaj and Aditi Sakalle

USICT, Gautam Buddha University, Greater Noida, Uttar Pradesh, India

Yashpal Singh

Department of Computer Science and Engineering, MIET, Greater Noida, Uttar Pradesh, India

Taranjeet Singh

Department of Computer Science and Engineering, GL Bajaj, Greater Noida, Uttar Pradesh, India

Arpit Bhardwaj and Divya Acharya

Bennett University, Greater Noida, Uttar Pradesh, India

CONTENTS

15.1 Introduction....................236
15.2 History of Autonomous Vehicles....................236
15.3 Technologies Used in Autonomous Vehicles....................239
15.3.1 Sensors in Autonomous Vehicles....................239
15.3.2 Forms of Camera....................240
15.3.2.1 Omnidirectional Cameras....................240
15.3.2.2 Cameras for Accidental Situations....................241
15.3.3 Vehicular ad hoc Networks (VANET) in Autonomous Vehicles....................242
15.4 Datasets Used for Autonomous Driving....................243
15.5 Limitations and Challenges to Implementation....................243
15.5.1 Cost of Vehicles....................243
15.5.2 Approvals....................245
15.5.3 Litigation, Liability, and General Opinion....................246
15.5.4 Security....................247

DOI: 10.1201/9781003146711-15

15.5.5 Privacy 248
15.6 Research Gaps 249
15.7 Conclusion 250
Bibliography 250

15.1 INTRODUCTION

The engineering and automobile companies have made significant strides over the last few years in introducing computerization into what has been solely a biological process for over a couple of centuries. Autonomous means the agent can perform tasks independently without any or minimum human intervention (Singh and Kumar, 2016). Technology like automatic ships or cruise management and parking assistance programs that enable cars to drive or be shepherded automatically into parking bays are increasingly featured in modern car models. By developing autonomous vehicles (AVs, usually named automatic or self-driving automobiles) that can handle themselves independently on highways and on certain kinds of roads and environmental events with virtually no direct human feedback, some businesses have stretched the boundaries further. By avoiding deadly accidents, by ensuring vital mobility for the aged and disabled, by dealing with a growing road capacity, reducing gasoline, and reducing pollution, AVs could radically alter transportation networks. Complementary developments in sharing drives and vehicles will take us from cars to an on-demand service as a proprietary commodity. Investment in facilities and technological upgrades, transport options and demands for parking, type of land use, and driving trucks and other operations may be impacted. Besides, it is possible to convert the passenger cabin: former drivers can work comfortably on their computers, enjoy a meal, read novels, view television, and call their families (Howard and Dai, 2014).

A huge advancement has been achieved in developing autonomous vehicles (AVs) since before the first successful demonstrations in the 1980s. Despite these advancements and ambitious business targets, truly autonomous transportation has not been realized to date in general settings. There are two explanations for this. Initially, AVs operating in complex, diverse environments require models that can promptly generalize unexpected circumstances and can reason. Second, rational judgments require detailed interpretation, but most of the latest image processing versions are comparable to physical reality and reasoning (Abou-Zeid et al., 2010).

15.2 HISTORY OF AUTONOMOUS VEHICLES

Analogous to Carl Benz's development of the car in 1886, autonomous technology in automobiles ensures that we have a substantial effect on our potential. We will briefly study, from 1925 to 2019, the history of AVs and self-driving cars.

In 1925, Houdina Radio Control revealed the program-controlled car that moved down Broadway in New York, but which was hindered by a driver in another automobile. The first demonstration of a driverless vehicle was recorded in 1925 (Time USA, 1925). General Motors contacted Norman Bel Geddes many years later for him to outline his concept of maneuverability in the future, which resulted in Futurama, known as one of the most promising exhibitions in 1939, the famous New York World Fair. This vision sketched out wireless or radio-controlled electric automobiles that cruised on a road by electromagnetic circuits. This vision led to many versions, such as the GM-Firebird-II in the year 1956 and the wire-controlled car from RCA Labs in the year 1960, as well as the presentation of Citroen with its DS-19 and Demag/Cabinentaxi1 MBB in 1970. However, owing to its limited scalability and high prices, the principle of framework-based automated navigation was generally limited to a particular scenario, like ground transportation at airports, controlled services, or park shuttles.

The first designs for self-driving cars that did not depend on dedicated networks reached the ground in 1986. The Navlab team at Carnegie Mellon University in the USA directed this groundbreaking initiative. In 1995, by traveling from Washington to San Diego, with 98% automatic and manual longitudinal power, the Americas tour was completed (Pomerleau and Jochem, 1996, 2015). The Navlab team (Thorpe et al., 1988) accomplished another significant milestone. With ALVIN (Pomerleau, 1988), an imitation learning technique was demonstrated. A reasonably narrow-scaled neural network had an optimized edge-to-edge algorithm to hold the automobiles on the path depending on users' demos. On the other hand, a commutable solution was proposed by Dickmanns, in which a vehicle and a model of the road were used to continually estimate the condition of and monitor vehicles (Dickmanns 1995). The first automated ride from Munich to Odense was demonstrated by the PROMETHEUS team in 1995, at speeds of up to 180 km/h with around 94% automated driving (Dickmanns et al., 1994, 1990; Franke et al., 1994).

Furukawa and Ponce (2010) define a real-time or actual-time vision framework for automated driving in dynamic metropolitan traffic conditions, inspired by PROMETHEUS, to drive autonomously on specific highways. Although highway situations have been extensively examined, metropolitan areas have not been historically discussed. Their device provided stereo detection and identification of deep-based obstacles and a monocular recognition and detection mechanism for pertinent artifacts, like traffic signals. Several methods established through these initiatives to address the complex challenges of autonomous driving are addressed and discussed by Bertozzi et al. (2000). They concluded that there is progressively ample processing capacity, but that challenges such as reflections, wet paths, direct sunlight, tunnels, and shadows, still render it challenging to read results. They also proposed the improvement of sensor capabilities. They also found that there is a need to closely examine the regulatory issues around autonomous driving and its effect on human passengers. In summary, automation is expected to be confined to unique infrastructures but will eventually be expanded.

Although complete self-driving has remained unanswered to date, market progress has been accomplished by driver-assistance technologies, rich driving comfort, and protection. Mitsubishi developed the first LiDAR-based distance control in 1995 (Mitsubishi Corporation 1998), and Mercedes-Benz introduced radar-based flexible driver control in 1999. Navigation systems and automated roadmaps became usable in 2000. Today, differential GPS in conjunction with inertial measurement units makes it possible to locate in good conditions to a precision of 5 cm, allowing the usage of accurate road-level lane maps (HD maps) and ensuring redundancy for blurry position-based vision.

The US Department of Military's Defense Advanced Research Projects Organization (DARPA) began planning and funding a sequence of three races in

2004 to support self-driving technology (DARPA, 2014). At the Darpa Grand Challenge held in 2004, the first competition was restricted to US participants. For the first squad to automatically complete a 240-km-long path from California to Nevada, driven using GPS coordinates, DARPA proposed prize money of $1 million. The final race in this season, the Darpa Urban Challenge (Buehler et al., 2009), was hosted in 2007 by DARPA, where foreign competitors were also permitted.

Compared to previous competitions, this race allowed vehicles to travel a 96-km path across a George Air Force Base mock-up town while complying with traffic rules, overcoming barriers, communicating with many vehicles, and ending up in traffic. The CMU team was placed first this time, followed by a Stanford team, which was ranked second. In particular, many active teams depended on the multibeam LiDAR established by Velodyne in a groundbreaking attempt. This rotating LiDAR sensor allowed accurate depth calculations to be obtained around a 360-degree field around the car, which was critical for urban navigation.

Google assumed the initiative in 2009 and recruited a host of star scientists participating in the Darpa Challenges. They designed their self-driving car, which involved a new driving area and a specialized, inexpensive, LiDAR multibeam scanner being created. According to crash reports, Google's autonomous systems were involved in 14 accidents (Department of Motor Vehicles CA, 2019), whereas 13 were triggered by others up to 2016.

In 2010, at the University of Parma, the VisLab team headed by Alberto Broggi organized the VisLab Intercontinental Autonomous Challenge (VIAC) (Broggi et al., 2010). VIAC (Bertozzi et al., 2011) was an attempt to travel semi-autonomously from Parma to Shanghai, based on experience with different concept vehicles (Braid et al., 2006; Broggi et al., 1999; Grisleri and Fedriga, 2010). In this presentation, a second vehicle autonomously followed a visually-based path on some GPS coordinates transmitted by the primary vehicle by a manually operated lead vehicle. The on-board device enabled barriers, lane markers, ditches, and berms to be detected and the previous vehicle's location and direction to be recognized.

In the same year, at 4300 meters above sea level, Audi displayed an auto-driven car trip to the summit of Pikes Peak, and the Technological University of Braunschweig showed their Stadtpilot (Braunschweig, 2010) which was able to maneuver around a tiny geo-fenced inner-city region focused on LiDAR, cameras, and HD charts. The PROUD project (Broggi et al., 2015), a presentation of intercity and highway driving in Parma, was undertaken by the VisLab team in 2015.

The Grand Cooperative Driving Challenge (Lauer, 2011), a competition for autonomous cooperative driving actions, was founded in 2011 by TNO. It was organized for the first time in Helmond, the Netherlands, in 2011, and for the second time in 2016. The auto-driven vehicles had to navigate convoys, guide convoys, and join convoys throughout the competition. Although longitudinal control was separate, a human safety driver had lateral control.

The KITTI Vision Benchmark3 (Geiger et al., 2012, 2013) was written in 2012. For the first time, scientists and innovators worldwide were able to test their success rationally and reasonably on various tasks relating to self-driving vision (including reconstruction, motion prediction, and object recognition). Simultaneously, several areas, including computer vision and robotics, began to be revolutionized by deep learning, which laid the groundwork for significant advances, particularly in precision, robustness, and runtime of AVs' perception peripherals.

V-Charge, an EU-funded joint project led by Volkswagen, Bosch, and many academic collaborators (ETHZ, Oxford, Parma, Braunschweig), aimed at wholly automated charging and parking of electric vehicles. An total operating framework was demonstrated for this project, including vision-only localization, routing, navigation, and power. The project championed many fascinating topics, including calibration, stereo, restoration, site logging and monitoring (SLAM), and free space detection (Furgale et al., 2013).

In 2014, the Society of Automotive Engineers published its self-driving device classification into six Society of Autonomous Engineers (SAE) autonomy tiers, varying from level 0 to level 5, as seen when Mercedes unveiled its S Class in the same year. Tesla introduced its Autopilot (Tesla, 2014) with level 2 (the driver has to control the machine at all times), providing automatic driving, lane maintenance, acceleration, and highway braking. Ride-hailing firm Uber unveiled its self-driving project one year later (Uber, 2015), attracting many Carnegie Mellon University (CMU) robotics experts. All automobiles manufactured by Tesla have been fitted with eight cameras, 12 ultrasonic sensors, and a forward-facing radar since October 2016 to allow complete self-driving in the future. Uber and Tesla, though, also encountered tragic collisions in which neither the pilot was vigilant nor the self-driving device operated correctly.

In 2016, the self-driving efforts of Google became Waymo, a stand-alone division of Alphabet Inc. after completing over 1.5 million miles. Today, Waymo provides admission to its early rider program (Waymo, 2019) to 400 Phoenix residents, including complete self-driving in many geo-fenced Phoenix regions with a rear-seat safety pilot.

In the same year, utilizing a single convolutionary neural network, NVIDIA (Bojarski et al., 2016) demonstrated a 98% auto-driven trip from Holmdel to Atlantic Highlands. Through imitation learning, the framework learned to predict automobile control directly from input pictures. Several last-mile distribution ventures were unveiled in 2018, including Nuro (Metz, 2018), a project created by two former self-driving car developers from Google and Scout (Scott, 2019), an entirely electric distribution system built to distribute goods utilizing automated delivery systems to Amazon customers securely. Bosch and Daimler unveiled a fleet of auto-driven cars in 2019, offering passengers on designated roads in California a taxi service of AVs (Daimler, 2019).

15.3 TECHNOLOGIES USED IN AUTONOMOUS VEHICLES

The following are the technologies used for the enhancement of AVs.

15.3.1 Sensors in Autonomous Vehicles

A sensor suite containing sensing devices, cameras, tire odometry, and spectrum sensors typically addresses AV transportation (LiDAR, RADAR, and SONAR). Tesla, for example, utilizes multiple sensors, RADAR, and ultrasonic testing for its sophisticated autopilot driver-assistance system. The fusion of information from several sensors makes it possible to leverage their complementary features and resolve the shortcomings of distributed sources, such as the failure of camera iterative reconstruction or the lack of color information in date information. Wheel odometry tests a wheel's movement, which may be used to compute the location the automated vehicle covers. However, wheel odometry does not include the car's complete orientation (i.e., all degrees of freedom) and is usually paired with optical odometry or SLAM strategies. Additional knowledge regarding the design of real-time surroundings and configuration is given by range sensors (i.e., SONAR, RADAR, LiDAR). Ultrasonic (SONAR) devices produce high-frequency sound waves and monitor the distance traveled to neighboring objects by electromagnetic radiation. Because the speed of seismic waves is identified, the distance to artifacts is measured using the driving time. The same concept runs LiDAR and RADAR, but instead of sound waves, they utilize electromagnetic waves and laser light beams. RADAR sensors profit from a greater working distance than LiDAR and SONAR because of the larger wavelength, but at a lower precision price.

These are an appealing sensor option for self-driving vehicles, as sensors are inexpensive, passive, and simple to install, and many current automobiles depend on lane-keeping or motion detection cameras. Other very influential camera styles are now briefly addressed, and a quick summary of standard measurement pipelines for the estimation of positive and negative Internet signals is provided.

15.3.2 Forms of Camera

Many traditional cameras have an aperture and one or two lenses and like the pinhole camera model. By exploiting mirrors or special lenses, omnidirectional cameras enable the field of view to be significantly increased. Event cameras allow for amplitude adjustments at very high temporal resolutions to be obtained. We will provide a short description of the omnidirectional and case cameras below. We refer to a detailed discussion of the pinhole camera model and projective geometry (Andrew, 2001; Szeliski, 2011).

15.3.2.1 Omnidirectional Cameras

In autonomous driving, a panoramic field of view is ideal for obtaining optimum knowledge about the local environment for secure navigation. Omnidirectional cameras with a 360-degree field of view have increased coverage by minimizing the need for separate cameras or electronically convertible cameras. The styles of omnidirectional cameras are distinct. A regular camera with a shaped mirror, such as a parabolic, hyperbolic, or elliptical mirror, is paired with a catadioptric camera. In contrast, a dioptric camera utilizes solely fish-eye lenses. To have a total spherical angle of vision, polydioptric cameras use several cameras with conflicting fields of view.

All central catadioptric structures have a unifying principle in the form of a coherent projection model which is commonly utilized by numerous calibration toolboxes (Geyer and Daniilidis, 2000). Scaramuzza and Martinelli (2006) consider using the Taylor series extension to model the imaging function. Mei and Rives (2007) strengthen the unified projection process (Geyer and Daniilidis, 2000) by predicting distortions to account for real-world mistakes. Schönbein et al. (2014) suggest a quick approximation to non-central camera models that are computationally costly.

In automated driving science, omnidirectional cameras are gaining prominence. A wide field of view allows for the extraction and matching of interest points from all over the vehicle for feature-based applications such as navigation, motion prediction, and visualization. Omnidirectional cameras have thus been used effectively to enhance vehicle ego-motion estimation (Scaramuzza and Siegwart, 2008) and 3D static scene reconstruction (Häne et al., 2014; Schönbein et al., 2014).

15.3.2.2 Cameras for Accidental Situations

Unlike standard frame-based cameras, event cameras create a stream of simultaneous events of brightness adjustments exceeding a predefined microsecond resolution threshold. The venue, symbol, and time stamp of the shift involve a case. This representation can decrease transmission and retrieval demands when activities are scarce in both space and time. The more excellent spatial resolution helps highly reactive structures to evolve.

Dynamic and active-pixel vision sensors (DAVIS) display all complementary metal oxide semiconductor (CMOS) pictures and simultaneous events at a set bit rate, integrating both sensors' advantages. Mueggler et al. (2017) use a range of DAVIS-captured actual and synthetic datasets to advance study into event-based approaches. The DAVIS driving dataset is presented by Binas et al. (2017) and illustrates end-to-end camshaft position understanding. For function tracking (Gehrig et al., 2018) and SLAM (Vidal et al., 2018), recent work exploits DAVIS, enhancing precision and robustness over the usage of a single modality.

Many strategies have been established that take advantage of the high temporal resolution and the event sensor's asynchronous design for various problems. Most of these strategies are used in uncrewed aerial vehicles (UAVs), as very efficient methods are needed to maneuver these structures. Event-based cameras are often used in this sense to estimate ego-motion (Mueggler et al., 2015), SLAM (Rebecq et al., 2016), as well as to locate corresponding features (Gallego et al., 227). More recently, the advantages of event-based detectors have also been explored for driverless cars by an earlier partial understanding of steering angles (Maqueda et al., 2018).

The issue of measuring intrinsic and extrinsic parameters of one or more sensors to link 3D environment points to 2D observations correctly is geometric calibration. To promote parameter evaluation, fiduciary labels and checker boards are also used.

Since the early 1970s, it has been easy to find different methods for camera tuning. The first to recognize the whole calibration pipeline, including control point extraction, model layout, and picture correction, were Heikkila and Silven (1997). To acquire the specifications of a real camera and fix the issue of compensating for image distortions, they suggested a four-step method.

To improve robustness and coverage, new cars are usually fitted with several different sensors. To meet the needs of such broad sensor suites, multiple calibration procedures have been suggested. Kassir and Peynot (2010) suggest the first complete automated camera-to-range calibration systems, whereas early approaches focused on the bottom of the body of feature vectors in laser beams. It was also shown that a rig utilizing two cameras and a standard range transmitter such as a Kinect or

Velodyne light sensor could be automatically adjusted. The issue of measuring the intrinsic and extrinsic specifications of a multi-camera rig, despite compromising the angle of vision, is discussed by Heng et al. (2013). Heng et al. (2015) expand this work by eliminating the need to use a chart and geological areas rather than fiducial marks to change the setting.

15.3.3 Vehicular ad hoc Networks (VANET) in Autonomous Vehicles

One of the most widely employed methods for AV identification has been the study of camera frames utilizing multiple techniques. Semantic segmentation, which marks each pixel in a picture with the corresponding artifact category, is one suggested process. To decide if a particular pixel belongs to a car, an individual, or some other class of items, a sufficiently sizeable contextual window is established to display the pixel's surroundings and, accordingly, make an educated decision on the entity class of the pixel. Techniques focused on Markov random fields (MRFs) and conditional random fields (CRFs), and several graphical models are provided in Munoz et al. (2010) to ensure that pixel marking is compatible in the overall picture sense. Furthermore, Kumar and Koller (2010) have developed various methods for pre-segmenting images into superpixels, which are used to derive categories and characteristics from both individual segments and neighboring segment combinations.

Alternately, by obtaining a long-range spatio-temporal regularization of semantic video segmentation, Kundu et al. (2016) tried to establish a 3D reconstruction of complex scenes since both the object and the image are in motion. The principle established is to combine deep convolutionary neural networks (CNNs) and CRF to execute sharp object boundaries at the pixel level. The suggested solution reduced the distances between the characteristics correlated with the respective points in the scene and improved the dense CRF function space used. Deep learning has shown the highest success in inferring artifacts from previously untrained scenes for this purpose. The segmentation of the input images was accomplished in Vogel et al. (2015) by viewing the dynamic scene as a series of rigidly rotating planes. The geometry/3D motion was restored jointly when another scene was over-segmented. Instead of pixel-based depictions, as partially used, the built piece-wise rigid scene depicts real-world scenes with individual entity motions.

A general-purpose object detection device defined by a resolution classifier and the use of two wholly linked networks has been developed. On top of a 24-layer convolutional network, these two networks are created, followed by two wholly linked layers. In addition, with several sub-network detectors and several output layers for multiple entity type identification, a single multi-scale deep CNN for real-time object detection is built.

LiDAR scanners are another commonly employed sensing tool for the 3D recreation of ecosystems. In distance ranging, obstacle detection and avoidance, and route planning, 3D LiDAR-generated point clouds have already been used and have thus been imported into autonomous driving systems. For the processing and identification of 3D LiDAR point cloud objects, 2D CNNs have been developed. However, since they need a model to restore the initial geometric relationships, this approach is not ideal. Vote3Deep is built for quick point cloud entity detection utilizing 3D CNN to retain LiDAR's core strength as distance and 3D shapes of artifacts and depth detection. The KITTI Vision Benchmark Suite provided point-cloud raw LiDAR and labeled artifacts.

To achieve object identification on LiDAR, stereo cameras or radar sensors, and scene flow estimation of artifacts on paths are required. They can also detect hidden elements, such as private cars or pedestrians, considering the significant improvement in both technologies. Owing to limited vision circumstances, such as bad weather conditions (e.g., very bright light, fog, heavy rain/snow) and similar colors in the surrounding nature, camera-dependent systems can even struggle to identify geometrical line-of-sight entities (e.g., the cause of the prototype accident). Finally, the camera and LiDAR methods struggle to identify road/traffic hazards (e.g., red traffic signals, speed limit changes) that may cause traffic violations and even fatalities. As described above, if these sensing technologies are complemented by accurate in-flow information from both nearby vehicles and traffic infrastructure through VANETs, all such problems can be solved. To build efficient chips responsible for collecting and fusing data retrieved from radars, cameras, and LiDARs, the Google self-driving car project called WAYMO operates with Intel. Researchers proposed an approach to geometrically aligning LiDAR scan points to points taken from a 360-degree camera.

In addition, on seven control channels running over a dedicated 75 MHz spectrum band of about 5.9 GHz, VANETs have various forms of vehicle-to-vehicle (V2V) and vehicle-to-infrastructure (V2I) protection messages. In this chapter, our goal is to present an augmented understanding of scene flow and entity mapping by considering LiDAR and cameras and V2V beacons shared between vehicles based on dedicated short-range communications (DSRC).

15.4 DATASETS USED FOR AUTONOMOUS DRIVING

By presenting problem-specific examples of ground reality, datasets have played a vital role in developing many research areas. Quantitative analyses of various approaches provide critical observations about their capacities and weaknesses. Table 15.1 shows several datasets which were used by researchers in the domain.

15.5 LIMITATIONS AND CHALLENGES TO IMPLEMENTATION

The latest automated cars could provide significant benefits for the community. They could result in improvements in riders' actions. However, should any country create AV schemes, they would come with similar obstacles, such as buying costs. There are security risks associated with AVs, especially from the point of view of safety and privacy. It needs special analysis with a seamless and reasonably fast rollout to tackle protection and privacy issues. This topic sheds light on many issues that AVs are faced with.

15.5.1 Cost of Vehicles

Another large-scale factor adoption that hasn't taken place is due to the costs involved with AV networks. AV technologies require modern sensors, connectivity and guidance technology, and applications for each car. According to Shchetko and Nick

TABLE 15.1
Different Datasets Analyzed According to the Required Processing

Dataset	Auto-driving	Traffic Signal Detection	Road Detection	Lane Detection
HCI Benchmark	Yes	—	—	—
Cityscapes	Yes	—	—	—
Euro City Persons Dataset	Yes	—	—	—
Mapillary	Yes	—	—	—
Apollo Scape	Yes	—	—	Yes
NuScenes	Yes	—	—	—
Berkeley Deep Drive	Yes	—	Yes	Yes
German Traffic Sign Recognition Benchmark	Yes	Yes	Yes	Yes
German Traffic Sign Detection Benchmark	Yes	Yes	Yes	Yes
Tsinghua-Tencent 100K	Yes	Yes	Yes	Yes
SYNTHIA	Yes	—	—	—
Playing for Data	Yes	—	—	—
Playing for Benchmarks	Yes	—	—	—
Caltech Lanes Dataset	Yes	—	—	Yes
VPGNet Dataset	Yes	—	—	Yes
MOTChallenge		—	—	—
Caltech Pedestrian Detection	Yes	—	—	—
Argoverse	Yes	—	—	—
Waymo Open Dataset	Yes	—	—	Yes
KITTI	Yes	—	Yes	—
VirtualKITTI	Yes	—	—	—

(2014), LiDAR's expense on top of AV is $30,000 to $85,000 per unit. The goods will become even more competitive as the LiDAR sensor's price becomes lowered (Dellenback and Steven, 2013). There are reports that most existing military and civilian AV programs cost over $100,000 (Boesler and Matthew, 2012). Most citizens cannot afford the costly, albeit high-end, cars. While more costly methods are feasible, it was reported by Chengalva et al. (2009) that the hardware cost is expected to be below $20,000.

This is why it is predicted that this development of electric vehicles (Evs) will become well known in the future. Shchetko and Nick (2014) predict that the average expense of an AV is between $25,000 and $50,000, although it definitely would not decline below $10,000 for at least ten years. To compensate for the investment, savings on insurance, petrol, and car park prices can be made. The average maintenance and running costs for 2012 are between $6000 to $13,000, depending on the car; insurance and gasoline expenses are over $900 to $1100 (American Automobile Association, 2012). Both expenses could be decreased by 50% for premiums and parking, and maybe 13% for gasoline costs, resulting in significant further savings.

If rates for AVs are similar to classic cars, there would be consumers willing to purchase AVs. After expecting an extra expense of $3000, respondents to a study would still buy an automated car (J.D. Power and Associates, 2012). This is the final expense that Volvo senior engineer Erik Coelingh thinks of to purchase a self-driving vehicle. Hensley et al. (2009) suggested that electric car prices have been decreasing at 6–8% annually, indicating that in five to seven years, per-person electric vehicle cost could vary from $10,000 to $3000. On top of other amenities, it also raises the vehicle's selling price by as much as $12,500, from a base price of $47,800 (BMW of North America, 2013). These features will help to perform some tasks, but the human operator will still be in charge.

As AVs move from custom retrofits to mass-manufactured ones, expenses could come somewhere near Coelingh and J.D. Power and Associates' goal price range of $3,000 to $4,000 for each car (KPMG, 2012). This is a big obstacle in the adoption of renewable energy since it is unaffordable for many people.

15.5.2 Approvals

As of July 2014, California and Nevada passed legislation to establish AV testing, while Florida has passed legislation that facilitates it. In Michigan, SB 0169 requires AV testing. And in Washington, D.C., a bill has been proposed which will mandate it. There are also proposals making their way through state legislatures in California, Indiana, New Jersey, Washington, Massachusetts, South Dakota, Minnesota, South Carolina, Maryland, Hawaii, Louisiana, Georgia, and New York. However, at this point, the states have left much of the enforcement issues, such as needing equipment to be licensed before the sale, to the state departments of motor vehicles, or health. Particular laws may have a range of goals, distinct from state to state. For example, Nevada's initial legislation included only 23 lines of meanings and limited directions to its Department of Motor Vehicles (DMV), whereas California's is a total of six pages and is a similar guide to its DMV. Without a straightforward approval process and uniform collection of protection for recognition, AV producers may be confronted with regulatory confusion and needless duplication, among other concerns.

California needs even more protection and wellness interventions for non-sharing cars. Senate Bill no. 1298 specifies that all AVs should be checked on actual roads and built with fail-safe mechanisms. In addition, all AV details need to be preserved and processed beforehand so that any injuries may be conveniently analyzed. This piece of legislation calls for the California DMV to consider potential rules on a wide variety of concerns, including the overall number of AVs utilizing public highways, the number of AVs licensed in the state, licensing standards for the drivers of AVs, the feasibility of revoking the operators' licenses, and the possibility of refusing licenses. California state's law includes a subsection that allows driverless vehicles to be checked and renders their driverless cars susceptible to stringent supervision by the state's DMV. California has recently issued licenses to Mercedes-Benz, Audi, and Google to test AVs on public roads. In around four years, AV rule making will be required by California to create a level of treatment for AV sales to the general public.

In Nevada, AV research licenses have been given to Google, Parks Car, and Audi, all three for public roads. Nevada's e-pass software enables the car to engage full-time in autonomous driving and high-level road activity that does not include humans.

By analyzing traffic control systems and recognizing the nature of people and livestock, going into a building site and through a school zone, drivers are using multiple traffic control devices and the involvement of pedestrians, bikes, and other artifacts. They realize that the speed limit is different when the path is closed for development, where there is a school zone, or a Christmas Parade. In addition to these regional and environmental restrictions, the state can also issue testing licenses and approval under certain environmental and geographic limitations (e.g., autonomous service only on the state's interstate highways, daytime travel clear of snow and ice). Through addressing what AVs will someday be able to do, this chapter illustrates the importance of state entities ensuring that AVs can securely function in any scenario. The gradual existence of these developments is becoming a fact. Although several of these states are finding constructive ways to cope with e-cigarette legislation, if all these states have disparate legislative policies, e-cigarette makers would be required to waste time and resources designing and evaluating four distinct collections of e-cigarette items.

Many states adhere to a mutual arrangement with drivers' licenses accepted by other states as a licensing model authorized by the same jurisdiction. Thus, technically, a driver in one state may be permitted to drive in another state, even though that state may not grant driver licenses. Smith (2012) indicated that as of 2012, existing laws probably do not preclude AVs from being used in states without specific AV licenses, even though failure to specify regulations could hurt these vehicles' usage.

15.5.3 Litigation, Liability, and General Opinion

Once AVs are fully licensed for safe operation by a state DMV or other regulatory body, several new insurance and liability problems will emerge, including getting insurance companies to embrace the technology as a safe way to continue to drive. While autonomous driving is still in its growing stages, there will still be occasions where incidents are prevented. For example, when a deer runs into the path of one's vehicle, does one miss the deer, swerve out of the way, or strike the deer? Can the climate all around and the pavement figure in your choice? What if I use the slip road to avoid an impending crash?

In comparison, AVs have sensors, visual analysis tools, and algorithms that allow them to make more educated decisions (in contrast to self-driving cars). And if the AV wasn't really "at fault," taking responsibility for a faulty piece of equipment can be very complicated. Many other ethical issues often emerge with too many citizens dying in automobile accidents, including whether or not AVs can prefer mitigating injury to their passengers over other crash-involved groups. And should owners be permitted to alter the temperature of such a smoke sensor?

Regardless of the reality that these cars ought to go through a safety check to travel on public highways, they will be deemed possibly hazardous if they do not have a driver. Perception problems may have also been at stake in such a strategy. Thus, adoption may be postponed. If AVs are kept at a higher quality, it is possible that they would cost more because of awareness of their problems. Those who cannot afford an AV will still not be willing to buy one. With relation to responsibility, much effort has been taken to stop it. Before an accident, a vehicle is expected to store 30 s of sensor data, assuming that the device and the sensor have been adequately and accurately calibrated to ensure that the car is steering along a straight line and would not crash into something. As businesses begin to establish autonomous cars, semi-autonomous systems such as parking assist and adaptive cruise control will undoubtedly have initial test cases that could contribute to how completely AVs will be kept accountable.

15.5.4 Security

Electronic protection is also the concern of transport regulators, vehicle designers, and potential AV drivers. More commonly, disgruntled workers, terrorist groups, computer hackers, and aggressive countries can target AVs and intelligent transport systems, triggering collisions and traffic disturbances. A two-stage computer virus may be configured as a worst-case situation to first disperse dormant software through automobiles over a week-long duration, virtually infecting the entire US fleet then forcing all in-use AVs to accelerate up to 70 mph concurrently and veer left. As each AV in the fleet represents an entry point to specific networks, it might not be possible to build an entirely safe device.

It is essential to approach the issue from an effort-and-impact perspective and consider mitigating approaches widely employed in comparable vital infrastructure programs of national significance to understand the scale of this hazard. Present cyber attacks are most generally acts of espionage (obtaining unauthorized entry to a device for the intent of knowledge collection) rather than sabotage (actively undermining the regular activity of a system), according to Hickey and Jason (2012), vice president of the software protection company Vínsula. For example, disrupting a vehicle's contact or sensors will take a more complicated and nuanced assault than one intended to merely capture details. It would also be more challenging to interrupt the control commands of the vehicle. It is possible that designing an assault to concurrently compromise a vehicle fleet, either from a point source (e.g., compromising all vehicles around an infected AV) or from a system-wide transmission over infected networks, will face far more significant challenges for a future intruder. Nevertheless, the danger is possible, and a security compromise may have lasting implications.

Fortunately, strong protection can make it much tougher to stage assaults. The USA has shown that massive, vital, national infrastructure networks, including electricity grids and air traffic control systems, can be managed and protected. A system for upgrading essential data protection systems is currently being established by the National Institute of Standards and Technology (NIST). The guidelines arising from this framework can be integrated into autonomous and connected automated vehicles. Although protection controls were primarily introduced as an after thought and in an ad hoc fashion for personal computers and Internet contact (Hickey and Jason, 2012), V2V and V2I protocols were built with security implemented in the initial development process (National Highway Traffic Safety Administration, 2011). This and other protection mechanisms (such as the isolation of mission-critical and communication systems) could make it especially challenging for large-scale attacks on AVs and related networks. While both Grau and Alan (2012) and Hickey and Jason (2012) agree that there is no "silver bullet," these steps make it far more challenging to pull off attacks while limiting the harm that can be done.

15.5.5 Privacy

Consumer Watchdog, a consumer protection and advocacy group headquartered in California, posed questions regarding privacy during a new phase of AV-enabling regulations (Brandon and John, 2012). If AVs and non-autonomous connected vehicles become more popular and data sharing becomes widespread, such issues are likely to expand. This gives rise to five issues relevant to data: Who can hold or monitor the vehicle's data? What are the kinds of data stored? Who can these collections of data be exchanged with? In what forms are they going to make those data available? And, what ends are they going to be used for?

Crash records would undoubtedly be owned or rendered accessible to manufacturers of AV technology. They are likely to be liable for harm in the case of a crash, given that the auto-drivers were at fault. However, privacy issues exist if a person is operating a car with automated capability while the collision happens. Nobody wants the location tracking of his or her car to be included in court against themselves, but this is only an expansion of an established problem: about 96% of new passenger cars traded in the USA have similar data recorders that provide information of vehicle behavior recorded in the seconds before and after a collision, and the National Highway Traffic Signal Administration (NHTSA) is proposing mandating the performance of data recorders in late 2014. Although certain states limit access (and include a subpoena for access) to specific data from insurance providers, data rights and control remain undefined in most US states (Kaste and Martin, 2013).

It is possibly more controversial to provide AV travel details, like paths, locations, and periods of the day, to centralized and government-operated structures, notably if the information is registered and processed. Although there is still some degree of individuals' activity detected by roadside Bluetooth sensors and triangulation of mobile phone towers, continuous surveillance could carry this situation somehow to a completely whole new dimension. Without sufficient protection, this knowledge may be embezzled for tracking people by government officials or be accessible to law enforcement agents for rampaging surveillance. Automobile movement information has broad-spectrum monetary implications, such as personalized advertisements, that could be problematic for individuals.

Simultaneously, responsible distribution and utilization of AV data would benefit administrators and designers of transportation networks. This knowledge could be used to promote a change from a petrol tax to a vehicle miles traveled (VMT) charge or, by place and time of day, theoretically enforce traffic pricing schemes. For example, someone who programs traffic signal systems might use such data to enhance the system's reliability and the travel standard for travelers. Continuously linked AVs or connected traditional vehicles, on the other hand, could illuminate persistent vehicle paths and velocity changes and thus alert new strategies in signaling systems. Besides, such data may support infrastructure planning to determine potential changes, resulting in more effective project decisions and transport policies. Similar data may indeed help law enforcement, and commercial advertisement revenues could push down AV prices. There are tradeoffs in exchanging this information, and any selection to improve traveler protection must always be weighed against the advantages of standard mutual information.

15.6 RESEARCH GAPS

Although AVs could be profitably accessible for five to six years, in several respects the relevant research lags. Part of this is attributed to the ambiguity innate in new situations: AVs are not yet active in freight streams, except for a few research cars, and it is impossible to foresee the future after such a disruptive change. Besides, technological innovation and related policy decisions may affect performance, which may generate more significant confusion. With these caveats in mind, to properly plan for the introduction of AVs, it will help to recognize the crucial holes in current investigations.

A systematic consumer penetration appraisal is one of the most urgent requirements. As KPMG (2012), O'Brien (2012), the Nissan Motor Company (2013), Carter and Marc (2012), and others make plain, in the next decade AVs will possibly drive on roads, but it is unclear whether they would make up a large proportion of the U.S. fleet. Further consequential forecasts of market seepage can apply percentages and dates to intrusive, probable, and cautious AV endorsement scenarios. This would have a fair set of data for traffic planning and policy-makers to evaluate conflicting transit assets, strategies, and arrangements.

Alternate major study holes were established at the Autonomous Vehicles Symposium in 2014 (Transportation Research Board, 2014), with these large subject areas highlighted:

- Automated transport and collective auto-mobility.
- Provisional modeling and planning.
- Boulevard leadership and activities.
- Opportunities for vehicle robotics.
- Legal braking and accelerators.
- Presence of automated human influences in trucks.
- Near-term prospects for implementation.
- The commercialization of personal car robotics.
- Operating criteria for automation programs.
- Connected-automated vehicle road infrastructure specifications.

Many crucial, and sometimes cross-cutting, issues emerge from any of these subject domains. For starters, if auto-driven bright taxis were legally and financially and technically feasible, several trips currently operated by privately owned cars may be served. This would eliminate parking and ownership demands and have implications that reach the priority areas of transport and shared transportation, community development, highway maintenance, and promotion. The country would be limited in its preparations to incorporate auto-driven vehicles into the automobile structure effectively. Many of these and other vital issues remain unsolved.

15.7 CONCLUSION

AVs tend to provide the possibility of fundamentally changing transportation to another level. This technology may reduce pollution and energy consumption. Since many things are becoming autonomous, ranging from manufacturing to transportation, vehicles are most important. AVs at the start of this decade have highlighted several impressive mile stones that have been achieved. A combination of VANETS and AVs could provide us with specific solutions to a vehicle's automation and safety. AVs predict (see) the environment so as to know where they can drive and where it is not possible to do so, which comprises several sensors to perform various other tasks. Automotive sensors fall into two categories: active and passive sensors. Sensors like ultrasonic sensors, GNSS, and LIDAR are also included. For building a 360-degree view of the vehicle's environment, visible light cameras have been used that can also detect road markings. A lot more can be done in this particular domain of the automation of vehicles to converge various technologies and algorithms for vehicles and safety in communication among vehicles in vehicular ad hoc networks. Big data can be highlighted as the future of AVs.

BIBLIOGRAPHY

Abou-Zeid, M., Ben-Akiva, M., Bierlaire, M., Choudhury, C., Hess, S., 2010. Attitudes and value of time heterogeneity. In: De Voorde, E.V., Vanelslander, T. (Eds.), *Applied Transport Economics: A Management and Policy Perspective*. De Boeck, pp. 523–545.

American Automobile Association, 2012. *Your Driving Costs: How Much are you Paying to Drive?* Heathrow, FL.

Andrew, A. M. (2001). Multiple view geometry in computer vision. Kybernetes.

Bertozzi, M., A. Broggi, and A. Fascioli (2000). "Vision-based intelligent vehicles: State of the art and perspectives." *Robotics and Autonomous Systems (RAS)* 32(1): 1–16.

Bertozzi, M., L. Bombini, A. Broggi, M. Buzzoni, E. Cardarelli, S. Cattani, ... P. Versari (2011). "VIAC: An out of ordinary experiment". In: *Proc. IEEE Intelligent Vehicles Symposium (IV)*. 175–180.

Binas, J., D. Neil, S. Liu, and T. Delbrück (2017). "DDD17: End-To-end DAVIS driving dataset". *Proc. of the International Conf. on Machine Learning (ICML) Workshops*.

BMW of North America, 2013. *Build Your Own 2013* 528i Sedan. Woodcliff Lake, NJ.

Boesler, Matthew, 2012. The 27 Best Selling Vehicles in America, Business Insider. Bose, Arnab, Ioannou, Petros, 2003. Analysis of traffic flow with mixed manual and semiautomated vehicles. *IEEE Trans. Intel. Transport. Syst.* 4, 173–188.

Bojarski, M., D. D. Testa, D. Dworakowski, B. Firner, B. Flepp, P. Goyal, L. D. Jackel, M. Monfort, U. Muller, J. Zhang, X. Zhang, J. Zhao, and K. Zieba (2016). "End to end learning for self-driving cars". *arXiv: 1604.07316 [cs.CV]*.

Braid, D., A. Broggi, and G. Schmiedel (2006). "The TerraMax autonomous vehicle". *Journal of Field Robotics (JFR)*. 23(9): 693–708.

Brandon, John, 2012. Privacy Concerns Raised over California "Robot Car" Legislation, Fox News. September 14.

Braunschweig, T. U. (2010). "Project Stadtpilot". https://www.tu-braunschweig.de/stadtpilot. Online: accessed 18-October- 2019.

Broggi, A., M. Bertozzi, A. Fascioli, and G. Conte (1999). *Automatic Vehicle Guidance: The Experience of the Argo Vehicle*. Singapore: World Scientific.

Broggi, A., P. Cerri, S. Debattisti, M. C. Laghi, P. Medici, ..., A. Prioletti (2015). "PROUD – Public Road Urban Driverless-Car Test". IEEE Trans. on *Intelligent Transportation Systems (T-ITS)*. 16(6): 3508–3519.

Broggi, A., P. Medici, E. Cardarelli, P. Cerri, A. Giacomazzo, and N. Finardi (2010). "Development of the control system for the Vislab Intercontinental Autonomous Challenge". In: *Proc. IEEE Conf. on Intelligent Transportation Systems (ITSC).* 635–640.

Buehler, M., K. Iagnemma, and S. Singh (2007). *The 2005 Darpa Grand Challenge: The Great Robot Race.* Vol. 36. Springer.

Buehler, M., K. Iagnemma, and S. Singh (2009). "The DARPA urban challenge". DARPA Challenge. *Advanced Robotics* 56.

Carter, Marc, 2012. Volvo Developing Accident-Avoiding Self-Driving Cars for the Year 2020, Inhabitat. December 5 <http://inhabitat.com/volvo-developing-accident-avoiding-self-driving-cars-for-the-year-2020/>.

Center, G. H. (2017). "Self-driving cars, in 1956?" https://www.gmheritagecenter.com/featured/Autonomous_Vehicles.html. Online: accessed 18-October-2019.

Chengalva, Mahesh, Bletsis, Richard, Moss, Bernard, 2009. Low-cost autonomous vehicles for urban environments. *SAE Int. J. Commer. Veh.* 1 (1), 516–527.

Corporation, M. M. (1998). "Mitsubishi motors develops 'new driver support system'". https://www.mitsubishi-motors.com/en/corporate/pressrelease/corporate/detail429.html. Online: accessed 17-May-2019.

Daimler AG (2019). "Bosch and Daimler. Metropolis in California to become a pilot city for automated driving". https://www.daimler.com/innovation/case/autonomous/pilot-city-for-automated-driving.html. Online: accessed 18-October-2019.

DARPA (2014). "The DARPA grand challenge: Ten years later". https://www.darpa.mil/news-events/2014-03-13. Online: accessed 18-June-2019.

Dellenback, Steven, 2013. Director, Intelligent Systems Department, Automation and Data Systems Division, Southwest Research Institute. Communication by Email, May 26.

Department of Motor Vehicles CA (2019). "Report of traffic collision involving an autonomous vehicle". https://www.dmv.ca.gov/portal/dmv/detail/vr/autonomous/autonomousveh_ol316. Online: accessed 17-May-2019.

Dickmanns, E. D. (1995). "Dynamic machine vision". http://dyna-vision.de/. Online: accessed 18-June-2019.

Dickmanns, E. D., B. D. Mysliwetz, and T. Christians (1990). "An integrated spatio-temporal approach to automatic visual guidance of autonomous vehicles". *IEEE Trans. on Systems, Man and Cybernetics (TSMC).* 20(6): 1273–1284.

Dickmanns, E. D., R. Behringer, D. Dickmanns, T. Hildebrandt, M. Maurer, F. Thomanek, and J. Schiehlen (1994). "The seeing passenger car 'VaMoRs-P'". *In: Proc.* IEEE Intelligent Vehicles Symposium (IV).

Franke, U., S. Mehring, A. Suissa, and S. Hahn (1994). "The Daimler-Benz steering assistant: A spin-off from autonomous driving". *In: Proc.* IEEE Intelligent Vehicles Symposium (IV).

Furgale, P. T., U. Schwesinger, M. Rufli, W. Derendarz, H. Grimmett, P. Mühlfellner, S. Wonneberger, J. Timpner, S. Rottmann, B. Li, B. Schmidt, T. Nguyen, E. Cardarelli, S. Cattani, S. Bruning, S. Horstmann, M. Stellmacher, H. Mielenz, K. Köser, M. Beermann, C. Hane, L. Heng, G. H. Lee, F. Fraundorfer, R. Iser, …, R. Siegwart (2013). "*Toward automated driving in cities using close-to-market sensors: An overview of the V-Charge Project*". In: *Proc. IEEE Intelligent Vehicles Symposium (IV).*

Furukawa, Y. and J. Ponce (2010). "Accurate, dense, and robust multi-view stereopsis". *IEEE Trans. on Pattern Analysis and Machine Intelligence (PAMI).* 32(8): 1362–1376.

Gallego, G., H. Rebecq, and D. Scaramuzza (2018). "*A unifying contrast maximization framework for event cameras, with applications to motion, depth, and optical flow estimation*". In: *Proc. IEEE Conf. on Computer Vision and Pattern Recognition (CVPR).*

Gehrig, D., H. Rebecq, G. Gallego, and D. Scaramuzza (2018). "*Asynchronous, photometric feature tracking using events and frames*". In: *Proc. of the European Conf. on Computer Vision (ECCV).*

Geiger, A., P. Lenz, and R. Urtasun (2012). "*Are we ready for autonomous driving? The KITTI vision benchmark suite*". In: *Proc. IEEE Conf. on Computer Vision and Pattern Recognition (CVPR).*

Geiger, A., P. Lenz, C. Stiller, and R. Urtasun (2013). "Vision meets robotics: The KITTI dataset". *International Journal of Robotics Research (IJRR).* 32(11): 1231–1237.

Geyer, C. and K. Daniilidis (2000). "*A unifying theory for central panoramic systems and practical implications*". In: *Proc. of the European Conf. on Computer Vision (ECCV).*

Grau, Alan, 2012. President, *Icon Labs. Telephone Interview*, October 12.

Grisleri, P. and I. Fedriga (2010). "*The BRAiVE platform*". In: *Proc. of the IFAC Symposium on Intelligent Autonomous Vehicles (IFAC).*

Häne, C., L. Heng, G. H. Lee, A. Sizov, and M. Pollefeys (2014). "*Real-time direct dense matching on fisheye images using plane-sweeping stereo*". In: *Proc. of the International Conf. on 3D Vision (3DV).*

Heikkila, J. and O. Silven (1997). "*A four-step camera calibration procedure with implicit image correction*". In: *Proc. IEEE Conf. on Computer Vision and Pattern Recognition (CVPR).*

Heng, L., B. Li, and M. Pollefeys (2013). "*CamOdoCal: Automatic intrinsic and extrinsic calibration of a rig with multiple generic cameras and odometry*". In: *Proc. IEEE International Conf. on Intelligent Robots and Systems (IROS).*

Heng, L., P. T. Furgale, and M. Pollefeys (2015). "Leveraging image-based localization for infrastructure-based calibration of a multi-camera rig". *Journal of Field Robotics (JFR).* 32(5): 775–802.

Hensley, Russel, Knupfer, Stefan, Pinner, Dickon, 2009. Electrifying cars: How three industries will evolve. *McKinsey Quart.* 3, 87–96.

Hickey, Jason, 2012. Vice President, Vínsula. Telephone Interview, October 11.

Howard, D., Dai, D., 2014. *Public perceptions of self-driving cars: the case of Berkeley, California.* In: *93rd Annual Meeting of the Transportation Research Board*, Washington, D.C.

J.D. Power and Associates, 2012. 2012 U.S. Automotive Emerging Technology Study.

Kassir, A. and T. Peynot (2010). "*Reliable automatic camera-laser calibration*". In: *Proc. IEEE Australasian Conf. on Robotics and Automation (ACRA).*

Kaste, Martin, 2013. Yes, Your New Car has a 'Black Box'. Where's the Off Switch? National Public Radio. March 20.

KPMG, CAR, 2012. *Self-Driving Cars: The Next Revolution.* Ann Arbor, MI.

Kumar, M. P., & Koller, D. (2010, June). *Efficiently selecting regions for scene understanding.* In *2010 IEEE Computer Society Conference on Computer Vision and Pattern Recognition* (pp. 3217–3224). IEEE.

Kundu, A., Vineet, V., & Koltun, V. (2016). *Feature space optimization for semantic video segmentation.* In *Proceedings of the IEEE Conference on Computer Vision and Pattern Recognition* (pp. 3168–3175).

Lauer, M. (2011). "Grand cooperative driving challenge 2011". Proc. *IEEE Intelligent Transportation Systems Magazine (ITSM).* 3(3): 38–40.

Maqueda, A. I., A. Loquercio, G. Gallego, N. N. García, and D. Scaramuzza (2018). "Event-based vision meets deep learning on steering prediction for self-driving cars". *In: Proc.* IEEE Conf. on Computer Vision and Pattern Recognition (CVPR).

Mei, C. and P. Rives (2007). "*Single view point omnidirectional camera calibration from planar grids*". In: *Proc. IEEE International Conf. on Robotics and Automation (ICRA).*

Metz, C. (2018). "A Toaster on wheels to deliver groceries? Self-driving tech tests practical uses". https://www.nytimes.com/2018/12/18/technology/driverless-mini-car-deliver-groceries.html. Online: accessed October 18 2019.

Mueggler, E., G. Gallego, and D. Scaramuzza (2015). "*Continuous-time trajectory estimation for event-based vision sensors*". In: *Proc. Robotics: Science and Systems (RSS).*

Mueggler, E., H. Rebecq, G. Gallego, T. Delbrück, and D. Scaramuzza (2017). "The event-camera dataset and simulator: Event-based data for pose estimation, visual odometry, and SLAM". *International Journal of Robotics Research (IJRR)*.

Munoz, D., Bagnell, J. A., & Hebert, M. (2010, September). Stacked hierarchical labeling. In *European Conference on Computer Vision* (pp. 57–70). Springer, Berlin, Heidelberg.

National Highway Traffic Safety Administration, 2011. USDOT Connected Vehicle Research Program: Vehicle-to-Vehicle Safety Application Research Plan. DOT HS 811 373.

Nissan Motor Company, 2013. Nissan Announces Unprecedented Autonomous Drive Benchmarks (Press Release). http://nissannews.com/en-US/nissan/usa/releases/nissan-announces-unprecedented-autonomous-drive-benchmarks.

O'Brien, Chris, 2012. Sergey Brin Hopes People Will Be Driving Google Robot Cars in Several Years, Silicon Beat.

Pomerleau, D. (1988). "*ALVINN: An autonomous land vehicle in a neural network*". In: *Advances in Neural Information Processing Systems (NeurIPS)*. 305–313.

Pomerleau, D. and T. Jochem (1996). "Rapidly adapting machine vision for automated vehicle steering". *IEEE Expert*. 11(2): 19–27.

Pomerleau, D. and T. Jochem (2015). "Look, ma, no hands". https://www.cmu.edu/news/stories/archives/2015/july/look-ma-no-hands.html. Online: accessed June 18 2019.

Rebecq, H., T. Horstschaefer, G. Gallego, and D. Scaramuzza (2016). "*EVO: A geometric approach to event-based 6-DOF parallel tracking and mapping in real-time*". In: *IEEE Robotics and Automation Letters (RA-L)*.

Scaramuzza, D. and A. Martinelli (2006). "*A toolbox for easily calibrating omnidirectional cameras*". In: *Proc. IEEE International Conf. on Intelligent Robots and Systems (IROS)*.

Scaramuzza, D. and R. Siegwart (2008). "Appearance-guided monocular omnidirectional visual odometry for outdoor ground vehicles". *IEEE Trans. on Robotics*. 24(5): 1015–1026.

Schönbein, M. and A. Geiger (2014). "*Omnidirectional 3D reconstruction in augmented Manhattan worlds*". In: *Proc. IEEE International Conf. on Intelligent Robots and Systems (IROS)*.

Schönbein, M., T. Strauss, and A. Geiger (2014). "*Calibrating and centering quasi-central catadioptric cameras*". In: *Proc. IEEE International Conf. on Robotics and Automation (ICRA)*.

Scott, S. (2019). "Meet scout". https://blog.aboutamazon.com/transportation/meet-scout. Online: accessed October 18 2019.

Shchetko, Nick, 2014. Laser eyes pose price hurdle for driverless cars, *Wall Street Journal*. July 21.

Singh, T., & Kumar, A. (2016). Survey on characteristics of autonomous system. *International Journal of Computer Science & Information Technology (IJCSIT)* 8.

Smith, Bryant Walker, (2012). *Automated Vehicles Are Probably Legal in the United States*. Center for Internet and Society, Stanford, CA.

State of Montana (2011). Montana Code Annotated 2011: 61-5-401. Driver License Compact.

Szeliski, R. (2011). *Computer Vision – Algorithms and Applications*. Texts in Computer Science. Springer.

Tesla (2014). "Tesla autopilot". https://www.tesla.com/autopilot. Online: accessed October 18 2019.

Thorpe, C., M. H. Hebert, T. Kanade, and S. A. Shafer (1988). "Vision and navigation for the Carnegie-Mellon Navlab". *IEEE Trans. on Pattern Analysis and Machine Intelligence (PAMI)*. 10(3): 362–372.

TIME USA (1925). "Science: Radio auto". http://content.time.com/time/magazine/article/0,9171,720720,00.html. Online: accessed October 18 2019.

Uber (2015). "Advanced technologies group". https://www.uber.com/de/de/atg/. Online: accessed October 18 2019.

Vidal, A. R., Rebecq, H., Horstschaefer, T., & Scaramuzza, D. (2018). Ultimate SLAM? Combining events, images, and IMU for robust visual SLAM in HDR and high-speed scenarios. *IEEE Robotics and Automation Letters*, 3(2), 994–1001.
Vogel, C., Schindler, K., & Roth, S. (2015). 3d scene flow estimation with a piecewise rigid scene model. *International Journal of Computer Vision*, 115(1), 1–28.
Waymo (2019). “Be an early rider”. https://waymo.com/apply. Online: accessed October 18 2019.

16 A Meta-learning Approach for Algorithm Selection for Capacitated Vehicle Routing Problems

Neha Sehta and Urjita Thakar
Shri G.S. Institute of Technology and Science, Indore

CONTENTS

16.1 Introduction 255
16.2 The Capacitated Vehicle Routing Problem 257
16.3 Meta-learning and Algorithm Selection 259
16.4 Meta-learning Process 260
16.4.1 Meta-features of the CVRP 260
16.4.2 Meta-labels 262
16.4.3 Meta-learning Technique 263
16.5 Meta-learning Experiment 263
16.5.1 Meta-knowledge Acquisition 264
16.5.2 Learning Phase 265
16.6 Results and Discussion 265
16.7 Conclusion, Limitations, and Future Scope 266
References 267

16.1 INTRODUCTION

The vehicle routing problem (VRP) is a challenging discrete combinatorial optimization task with many relevant real-world implementations, such as waste disposal, school bus routing, supply chain routing, dial-a-ride systems, and the collection and dispatching of goods/services. Dantzig and Rasmer (1959) proposed the first VRP which was to find the routes for gasoline trucks. Since then, there have been many variants of VRP based on the constraints imposed and objectives defined, such as the capacitated vehicle routing problem (CVRP), the distance constrained VRP (DCVRP), the heterogeneous fleet VRP (HVRP), the multiple depots VRP (MDVRP), and the pickup-and-delivery VRP (PDVRP). In this chapter, CVRP is considered

DOI: 10.1201/9781003146711-16

where all the customers have deterministic demands and are known in advance. The capacity of all vehicles is the same. Each vehicle starts and finishes at the depot node. A vehicle serves until its capacity is exhausted or it cannot serve a customer fully. The solution is to determine a collection of routes that reduce the distance traveled when serving all customers.

The VRP is an NP-Hard problem. Two approaches are generally used to solve a hard problem: (i) the exact approach that solves the problem optimally and (ii) the heuristic approach which gives a faster and near-optimal solution. In the first approach, all possible solutions are generated and, out of them, the best is selected. Such an exhaustive search is not practical in cases with a vast number of customers due to computational time and memory constraints which are usually of exponential order. The second approach applies heuristics and stochastic search algorithms, commonly known as meta-heuristics, which is more viable as it is comparatively faster and provides a near-optimal solution. A significant number of heuristic algorithms have been developed to solve the VRP, such as the sweep algorithm (Gillett & Miller, 1974), the savings algorithm (Clarke & Wright, 1964), and the Christofides-Mingozzi-Toth two-phase algorithm. Heuristic algorithms are problem-specific while meta-heuristics are general strategies applied to solve the hard problems, which include simulated annealing, tabu search, genetic algorithms, and foraging algorithms. Christofides etal. (1981) and Laporte (1992) presented exact and heuristic solution approaches to the VRP.

It is often necessary to choose an algorithm from a given pool of algorithmic techniques to solve a CVRP instance in hand. No free lunch (NFL) theorems for optimization state that no technique can be said to perform better than the others for all problem instances (Wolpert & Macready, 1997). This is because the bias associated with a heuristic algorithm makes it perform better for a particular subset of problem instances (Kanda et al., 2010). One way to find the most appropriate heuristic to solve a problem instance in hand is to perform experiments with several heuristic or meta-heuristic algorithms and then select an algorithm which performs better than the others. But such an empirical evaluation of different algorithms is not preferred because of the high computational cost.

The issue of selection of the most appropriate algorithm for an instance is also known as the algorithm selection problem (Rice, 1976). The machine learning group accepted the algorithm selection challenge as a learning task and developed a meta-learning area based on "learning" the learning algorithm's performance. Using meta-learning the characteristics of problems are to be mapped to algorithm performance.

In this chapter, an approach based on meta-learning is proposed to select an algorithm out of two classical heuristic algorithms: the parallel Clarke and Wright (1964)saving algorithm (CW), and the Gillett and Miller (1974) sweep algorithm (SW) for CVRP instances. A binary classification model is trained on data consisting of problem specific features and algorithm performance. This model was then used to forecast one of the two heuristic algorithms that work better in a new case. Five classifiers namely –naive Bayes (NB), multi-layer perceptron (MLP), stochastic gradient descent (SGD), J48, and random forest (RF) – are learned and compared.

The remainder of this chapter is arranged as follows: CVRP and the key approaches to solve the problem are briefly defined in Section 16.2. In Section 16.3, existing work on meta-learning and algorithm selection for various optimization problems is addressed. The meta-learning process including meta-features studied for CVRP is detailed in Section 16.4. Section 16.5 outlines a meta-learning experiment on instances of CVRP, and the results are described in Section 16.6. Finally, in Section 16.7, inferences are outlined along with the limitations and future directions of this work.

16.2 THE CAPACITATED VEHICLE ROUTING PROBLEM

Transportation, scheduling, distribution, and logistics are some of the important areas where vehicle routing is applied. The goal is to determine routes for a set of vehicles that start and end at a depot(s) which serve a set of customers without violating the given constraints. The constraints are generally defined as (i) the capacity of the vehicle, (ii) the number of cities on a route, (iii) the travel distance/time of a route, (iv) the time window to visit a customer, and (v) the order of visiting customers. The aim is to minimize distance traveled, time, and so on. The CVRP is also defined using the graph G{V,E}, where V contains nodes and E is the edge set consisting of possible connections between the nodes. G is a simple undirected complete graph. There is one depotnode and the rest are customer nodes. Each edge (i, j) ε E has an associated cost c_{ij}. There are K vehicles, all of the same capacity C. For each customer there is a demand $d_i > 0$. One of the vehicles reaches each customer precisely once. A vehicle cannot be overloaded, therefore the total demand it serves must be less than or equal to its capacity. The demand of a customer cannot be served partially. The principal decision variable in the CVRP is x_{ijk}, where:

$$x_{ijk} = \begin{cases} 1 \text{ if vehicle k travels from customer i to customer j} \\ 0 \text{ otherwise} \end{cases} \tag{16.1}$$

The objective function of the mathematical model is:

$$\sum_{k \in K} \sum_{i \in V} \sum_{j \in V, i \neq j} c_{ij} \cdot x_{ijk}$$

subject to:

$$\sum_{k=1}^{K} \sum_{j=1}^{N} x_{1jk} \leq K \tag{16.2}$$

$$\sum_{k=1}^{K} \sum_{j=1}^{N} x_{j1k} \leq K \tag{16.3}$$

$$\sum_{i=1}^{N}\sum_{j=1}^{N} x_{ijk} \leq K \text{ for } k \in \{1,2,\ldots K\} \text{ and } i \neq j \tag{16.4}$$

$$\sum_{i=1}^{N}\sum_{j=1}^{N} d_j x_{ijk} \leq C \text{ for } k \in \{1,2,\ldots K\} \text{ and } i \neq j \tag{16.5}$$

Constraints (16.2, 16.3) define that there are at most K incoming and outgoing edges respectively at the depot (node 1). Constraint (16.4) ensures that an edge is traversed only once by one of the vehicles. Capacity constraint is represented by Equation (16.5). Figure 16.1 shows a schematic solution for a sample CVRP instance where 1 denotes the depot node and 2–9 are customer nodes. All the customers can be served by three vehicles.

The CVRP is a blend of two optimization problems: (i) the bin packing problem (BPP) and (ii) the traveling salesman problem (TSP). In BPP, products with varying sizes have to be packed into bins of a given volume such that the minimum number of bins is used. The TSP is defined so as to obtain the shortest tour for a traveling salesman who wants to visit a variety of cities precisely once, starting from his hometown, then returning there. The CVRP with one vehicle with unlimited capability is equal to the TSP or the TSP with a number of salesmen equivalent to the CVRP.

The CVRP can be solved using exact, heuristic, and meta-heuristic algorithms. Exact algorithms are those that generate an optimal solution. But the running time for exact algorithms is of exponential order. Branch and bound, branch and cut, column generation, and set partitioning are examples of exact algorithms. Exact approaches are inefficient for large CVRP instances. Laporte and Nobert (1987) presented a survey on the exact methods for the VRP. Heuristics are constructive methods of obtaining feasible solutions in polynomial time, but do not guarantee a solution's quality. A large number of heuristics are designed for the VRP. Two major groups are

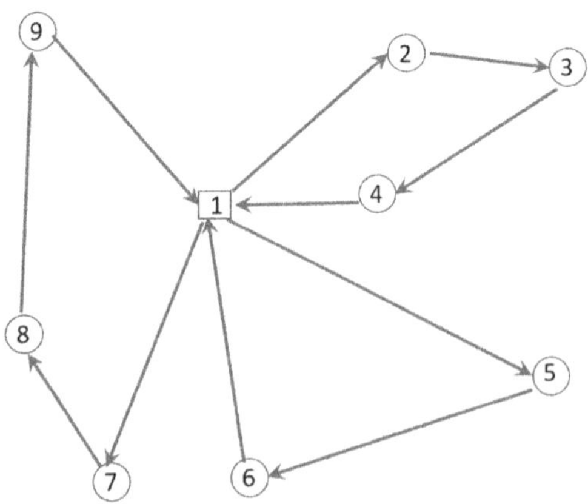

FIGURE 16.1 Schematic solution of a sample CVRP instance.

classical heuristics (1960–1990) and meta-heuristics (developed after 1990). While the heuristics work on the problem space, the meta-heuristics explore the solution space. Meta-heuristics are slower and solution quality depends on the initial solution. A good overview of all classical heuristics (Laporte & Semet, 2002) and meta-heuristics (Gendreau, Laporte, & Potvin, 2002) is in the book *The Vehicle Routing Problem* (Toth & Vigo, 2002).

In view of the available algorithms to solve the problem, selecting the best one is successfully dealt with by meta-learning which is a subfield of machine learning (Brazdil et al., 2008)and is discussed in the next section.

16.3 META-LEARNING AND ALGORITHM SELECTION

Meta-learning is to learn the performance of the learning algorithm (Brazdil et al., 2008). Initially meta-learning was applied to learning tasks such as classification and clustering. Later, the concepts of meta-learning were generalized to a broad range of non-learning problems, namely optimization problems, bioinformatics, cryptography, security, and sorting (Smith-Miles, 2009). The objective of meta-learning is to understand the relationship between problem features and algorithm performance. To apply meta-learning, meta-knowledge is acquired which consists of meta-features and meta-labels. Meta-features are the problem-specific characteristics that affect the performance of solution algorithms. Meta-features can be problem independent, like the characteristics of the solution space. The performance metrics of algorithms on problem instances are recorded as meta-labels. Meta-features are taken as input attributes while meta-labels are target or output attributes. A model is trained on acquired meta-knowledge. It can later be used to predict the performance of algorithms for a new case. Meta-learning models have significant uses for analyzing algorithm behavior, algorithm selection, and automated parameter configuration (Hutter et al., 2014). Such models are also referred to as empirical performance models (Leyton-Brown et al., 2009).

Meta-learning for algorithm selection has been applied to different optimization problems. Smith-Miles (2008) applied meta-learning to the quadratic assignment problem (QAP). She used 28 instances of the QAP which are represented by nine features. These nine features describe problem-specific features like the number of facilities, flow dominance, and the search space characteristics of the QAP. Instances are solved using three meta-heuristics. The difference between the solution cost obtained by meta-heuristics and the known optimal solution defined as the performance gap is used as an output feature. She framed the meta-learning task as a multiple regression task and used a neural network to train the model. Along with the QAP, Smith-Miles has also applied meta-learning to algorithm selection for the time tabling problem, traveling salesman problem, and graph coloring problem. Her work on the QAP is extended by Dantas and Pozo (2018a). They proposed a cascade classification scheme to train meta-learning models. In their next work multi-label classification is used (Dantas & Pozo, 2018b). They used the same nine features proposed by Smith-Miles (2008). Since the calculation of the search space characteristics is a computationally intensive task, Dantas and Pozo (2020) tried to reduce the computational cost of feature extraction. They also proposed a few new features.

Kanda et al. (2010) experimented with meta-learning on 535 instances of the TSP. The authors extracted 16 features from TSP instances and applied four meta-heuristic algorithms to find the solution cost. Each instance was then labeled with a best performing meta-heuristic. Since more than one meta-heuristic can give the best result, the instances were labeled with more than one meta-heuristic algorithm name. Thus, the meta-learning task was framed as a multi-label classification task. In their subsequent work, an MLP model was trained on the same meta-knowledge acquired, and predicted the meta-heuristics ranking for a new TSP instance (2011).

Musliu and Schwengerer (2013)applied meta-learning to select an algorithm for the graph coloring problem. The authors used 78 graph features. Six heuristic algorithms were applied to three publicly available sets of instances. The authors reported that the performance of all the algorithms was the same on average. They labeled each instance with the best performing algorithm name and proposed meta-learning as a classification problem. For the CVRP, Rasku et al. (2019)trained a classification model to select the best classical heuristic algorithm out of 15. Tuzun et al. (1997)trained a neural network model to select the best heuristic algorithm for a VRP instance based on its basic features. Gutierrez-Rodriguez et al. (2019) used basic and land-marking features and trained an MLP classifier for meta-heuristics selection to solve an instance of the CVRP with time window (CVRPTW).

Different meta-learning techniques have been used in the literature; however, the same meta-learning process is used in them all, which is described in the next section.

16.4 META-LEARNING PROCESS

Meta-learning is a subfield of machine learning (ML). Like any ML task, meta-learning is also performed in two phases: (i) the knowledge acquisition phase and (ii) the learning phase. The meta-learning process is illustrated in Figure 16.2.

In the first phase, knowledge is acquired. Each problem instance is represented as a meta-example. This comprises:

1. Meta-features: the features that describe a problem instance.
2. Meta-label: performance metric(s) obtained by executing candidate algorithms on that instance.

In the second phase, a meta-learner (a learning algorithm) is used on meta-examples to train a model. In the proposed work, meta-learning is experimented with on CVRP instances.

16.4.1 META-FEATURES OF THE CVRP

In this work, 23 meta-features identified by Meghan Steinhaus for the CVRP have been used (Steinhaus, 2015). Meta-features should be the problem features which affect the performance of different heuristics for the VRP. They should also be easily extractable from the problem data. Meta-features used in this work, basically describe the spatial distribution of customers, cluster formation within the instance, and

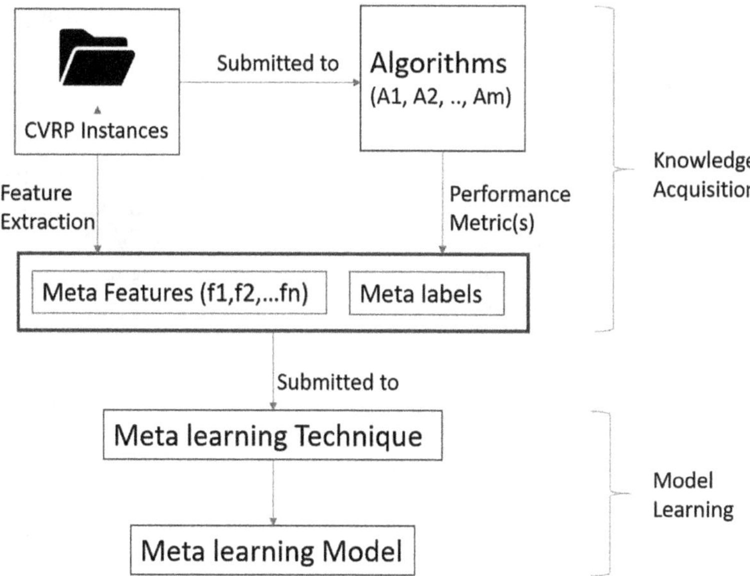

FIGURE 16.2 Meta-learning process.

demand distribution. The structure of routes is closely related to the distribution of customer demands. The performance of heuristic algorithms for the CVRP is affected by the spatial distribution of the customers. Nygard et al. (1990) reported that a sweep algorithm works better on uniformly distributed problems than on clustered ones. All instances are scaled to an area of (400*400) to make a fair comparison. Some of the problem features along with a brief description are:

1. Node count: This is the number of nodes in the problem instance. Node 1 is the depot node while the rest are the customer nodes.
2. Depot coordinates: the x and y coordinates of node 1.
3. Standard deviation of demand: The demand from every customer (node) is converted to a fraction of the vehicle capacity. For these normalized demands, the standard deviation is calculated.
4. Tightness: This describes the total demand in comparison to the total capacity.
5. Number of clusters: Clustering features of the instance are identified using a DBSCAN clustering algorithm with the following two parameters:

    a. epsilon(eps): the maximum distance between two nodes to be considered in the same cluster;
    b. minPoints: the minimum number of points that a cluster must contain.

    As all CVRP instances are scaled to 400*400 the values eps = 40 and minPoints = 4 are used (Steinhaus, 2015). A large number of clusters indicates that nodes are geographically dispersed.

6. Largest cluster demand: The demand distribution within clusters greatly affects the solution cost.
7. Cluster ratio: This describes the clusterability of customers and the depot and is determined by dividing the cluster count by the node count. This identifies the reach of the clusters.
8. Outlier ratio: Outliers are the points that do not belong to any cluster. The more the outlier, the more search effort is required to obtain a solution. The solution cost will also increase.
9. Outlier nodes demand: This feature is calculated by dividing total outlier node demands by total demand.
10. Mean radius of the clusters: This feature is calculated by averaging the distance from each customer to its cluster centroid.
11. Average number of customers per vehicle route: This characteristic reflects how long the routes would be for a given VRP instance. It depends on customer average demand with respect to vehicle capacity. The larger the demand, the fewer will be the customers per route.
12. Minimum number of trucks required: This is calculated by dividing total demand by vehicle capacity.
13. Area of the rectangle enclosing all nodes: This is equal to $(x_{max} - x_{min}) * (y_{max} - y_{min})$.
14. Distance matrix features: These contain the Euclidean distances between every pair of node coordinates. The standard deviation of the distances is taken as a feature.
15. Centroid features: The coordinates of the centroid is calculated by averaging the customers' coordinates.
16. Instance radius: This is calculated by averaging the distances from each customer to the centroid.
17. Nearest neighbor distance features: Here only the first nearest neighbor is considered.

The meta-labels are the performance metrics obtained on application of the candidate algorithm on the given problem instances.

16.4.2 Meta-labels

The meta-label in this algorithm selection task is the algorithm name that produces the solution with the lowest cost for a problem instance. Two heuristic algorithms chosen for this work are:

1. The parallel Clarke and Wright (1964)saving algorithm: This algorithm starts with all possible single customer routes which are formed from the depot to the customer and back to the depot. This is a constructive algorithm that builds a solution incrementally. Clarke and Wright (1964) defined a saving as that which occurs when two single customer routes are combined into one. The algorithm merges the routes which provide the largest saving. While merging restrictions are taken care of, if the merging of two routes with the largest saving violates the capacity constraint then it is dropped. The

remaining routes are merged in decreasing order of saving until there are no feasible merges left.

2. The Gillett and Miller (1974)sweep algorithm: In this algorithm the Cartesian coordinates (x, y) of customers with respect to the depot node are transformed into polar coordinates (r, θ) and then sorted by the θ value. A new route is initialized by taking an arbitrary customer. While moving in a clockwise (or anticlockwise) direction, adjacent non-routed customers are inserted until the capacity of the vehicle is exhausted or the customer has been served completely. This process continues until a feasible solution is obtained.

The classical heuristics have deterministic behavior; therefore, they are ideal for empirically verifying the suitability of the feature set (Rasku et al., 2019). The heuristic algorithm which produces the lower cost solution is selected as the meta-label for the instance. For each CVRP instance, the meta-features listed above are extracted and the meta-label identified, collectively called a meta-example. Figure 16.2 depicts the schema of the meta-example. The meta-knowledge acquired will be the set of meta-examples upon which a learning technique is applied.

16.4.3 Meta-learning Technique

A dataset containing instance features and the performance metric(s) of candidate algorithms can be prepared from previous experiments performed. A different learning mechanism can be applied to output predictive models. A learning algorithm can be applied to impose a hypothesis mapping characteristic onto a performance metric. A regression or classification task can be framed based on the nature of the performance metric collected. For example, if the algorithm performance is a continuous value, like a solution cost, runtime required, deviation from best solution, or error generated, then a regression task can be framed.

An alternative to this approach is to apply a lazy learning mechanism in which, instead of learning a predictive model, a meta-feature space is searched to find the feature vectors that satisfy some similarity criteria (called neighbors) to provide a new instance of a feature vector. The performance predicted for the new instance will be based on the performance of its neighbors.

In this work the meta-learning technique used is the binary classification problem. The dataset of 85 meta-examples is prepared for learning by the single label classification model. There are five classification algorithms: naive Bayes, stochastic gradient descent, multi-layer perceptron, J48, and random forest.

16.5 META-LEARNING EXPERIMENT

To perform the meta-learning experiment, CVRP benchmark problem sets are taken from a repository maintained by the Networking and Emerging Optimization (NEO) research group. The benchmark problem set consists of input and output files. An input file contains the coordinates of all nodes, the demand of customer nodes, vehicle capacity, and so on. A typical CVRP instance file is shown in Figure 16.3. An instance file name "A-n5-k2" has five customers including the depot and two vehicles. Customer coordinates and demands are listed in node_coord_section and

```
NAME : A-n5-k2
COMMENT : (Min no of trucks: 2)
TYPE : CVRP
DIMENSION : 5
EDGE_WEIGHT_TYPE : EUC_2D
CAPACITY : 100
NODE_COORD_SECTION
1 0 0
2 2 1
3 1 2
4 3 3
5 0 4
DEMAND_SECTION
1 0
2 20
3 35
4 30
5 45
DEPOT_SECTION
1
-1
EOF
```

FIGURE 16.3 A sample CVRP instance file.

demand_section respectively. The output file contains the best known solution and its cost. For this experiment, the problem sets used were: A, B, P (Augerat et al., 1995), F (Fisher, 1994), E, and M (Christofides & Eilon, 1969). For the instances of A, the node coordinates and demands are randomly generated, while nodes in the B instances are clustered. P instances are of size 12–101 and obtained by making changes in other instances where the number of vehicles is multiplied by 2 and vehicle capacities are adjusted accordingly. The customers in set E are from 22–101 and are distributed randomly with the depot located centrally. The set F consists of real life CVRP instances of size from 45–135 customers from National Grocers Limited.

Meta-learning is performed in two phases as discussed in Section 16.4. In the first phase meta-knowledge is acquired from CVRP instances and then in the second phase a model is trained using animal algorithm.

16.5.1 Meta-knowledge Acquisition

The meta-features described in Section 16.4.1were extracted using feature extractors that are implemented in Python (Rasku et al., 2016). All features extracted are numeric. Each instance class is labeled by the heuristic name which performed best. The solution cost for CVRP benchmark instances using the saving algorithm and sweep algorithm is given in Steinhaus (2015, Appendix A). Table 16.1 lists the resulting class distribution which shows that for 58 out of 85 instances, the Clarke and Wright saving algorithm outperforms the sweep algorithm.

TABLE 16.1
Instance Class Distribution

Class	Instances	Instances (%)
CW	58	68
SW	27	32

16.5.2 Learning Phase

Once we have our meta-data, the relationships among them are learned to produce rules or a model to assist automatic algorithm selection. For our first meta-learning attempt, this problem was framed as a binary classification problem. The software Weka 3.8.4 (Witten, 2016)was used. The five classifiers were naive Bayes (NB), multi-layer perceptron(MLP), stochastic gradient descent (SDG), J48, and random forest (RF); their default parameter settings were learned using a tenfold cross-validation (10 FCV) methodology.

16.6 RESULTS AND DISCUSSION

Meta-learning is a data intensive activity. Problem features were extracted from 85 CVRP problems and labeled with one of the two algorithm names which performed best for that instance. Binary classification was then applied to the dataset prepared. The classifiers learned were evaluated on three performance metrics: accuracy, area under the receiver operating characteristic curve (AUC), and the F-measure. Accuracy is defined as the fraction of correct predictions made out of the total predictions. The F-measure provides a way to combine both precision and recall into a single measure. It is the harmonic mean of the precision and recall. AUC reports the ability of the model to distinguish between the classes. The higher the AUC, the better the model is. The performance of the classification algorithms is shown in Table 16.2. For each classifier except MLP, accuracy is higher than the baseline accuracy (majority class distribution).

The classification tree generated by the J48 classifier is shown in Figure 16.4. A decision tree is used to explicitly represent decisions and decision making. The non-leaf nodes in the decision tree show the meta-features, based on its value branching.

TABLE 16.2
Classification Model Evaluation

Measures	NB	MLP	SDG	J48	RF
Accuracy	74.12%	67.06%	74.12%	74.12%	78.82%
F-measure	0.749	0.674	0.741	0.738	0.773
AUC	0.780	0.708	0.701	0.726	0.771

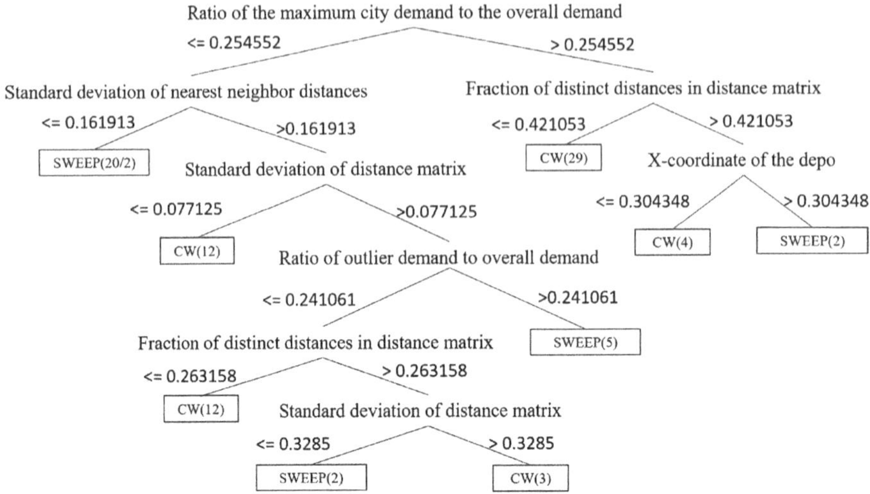

FIGURE 16.4 Decision tree generated by the J48 algorithm.

The leaf nodes show the number of instances assigned to a class along with the count of misclassified instances. Starting from the root node of the decision tree, meta-rules can be specified to find the class for a new instance. The feature importance is clear and relations can be viewed easily.

16.7 CONCLUSION, LIMITATIONS, AND FUTURE SCOPE

In this chapter, the issue of algorithm selection for a given instance of the optimization problem from an existing large pool of heuristics and meta-heuristic algorithms has been addressed. Binary classification was used as the meta-learning-based technique to select a heuristic algorithm to be applied to new CVRP instances. To be sure of selecting the best algorithm five classification techniques were applied to the dataset. The method used was found useful in selecting an appropriate algorithm. However, certain limitations were observed.

The meta-learning models constructed allow us to have a better understanding of the problem characteristics that affect the performance of optimization heuristics or meta-heuristics. Though a large number of ML tools are available, the performance of all learning algorithms is still influenced by the availability of training data and hyper-parameters. Thus, making larger amounts of data available to apply ML algorithms is the major requirement to build meta-learning models. The setting and fine-tuning of hyper-parameters is required to achieve better results. The selection of a learning algorithm out of many is also a challenging task.

Based on the experiment conducted, it has been found that meta-learning is a promising research area for algorithm selection. Results published on application of heuristic and meta-heuristic algorithms on CVRP by researchers can be utilized for training meta-learning models. Instead of designing new algorithms, emphasis should be given to algorithm performance analysis. However, the selection of algorithms depends on the selection of meta-features for developing the meta-learning model, hence the selection of the most appropriate meta-features from a large pool of problem features is a difficult task. Rigorous analysis of problem descriptors that affect the performance of solution methods is needed. Feature selection techniques and the design of algorithm portfolios with a mix of heuristic and meta-heuristic algorithms that are complementary to each other are required for the development of effective algorithm recommendation systems. The application of both supervised and unsupervised learning mechanisms to perform meta-learning seems a promising direction for future work.

REFERENCES

Augerat, P., Belenguer, J. M., Benavent, E., Corberan, A., Naddef, D., & Rinaldi, G. (1995). *Computational results with a branch and cut code for the capacitated vehicle routing problem* (Tech. Rep. No. R.495). Universite Joseph Fourier, Grenoble, France.

Brazdil, P., Carrier Christophe, G., Soares, C., & Vilalta, R. (2008). *Metalearning: Applications to data mining.*Springer Science & Business Media.

Christofides, N., & Eilon, S. (1969). An algorithm for the vehicle-dispatching problem. *Journal of the Operational Research Society, 20*(3), 309–318. doi: https://10.1057/jors.1969.75

Christofides, N., Mingozzi, A., & Toth, P. (1981). Exact algorithms for the vehicle routing problem based on spanning tree and shortest path relaxations. *Mathematical Programming, 20*(1), 255–282. https://doi.org/10.1007/BF01589353

Clarke, G., & Wright, J. W. (1964). Scheduling of vehicles from a central depot to a number of delivery points. *Operations Research*, 568–581.

Dantas, A. L., & Pozo, A. T. (2018a). Selecting algorithms for the quadratic assignment problem with a multi-label meta-learning approach. In *Proceedings of 2018 7th Brazilian Conference on Intelligent Systems.* 175–180. https://doi.org/10.1109/BRACIS.2018.00038

Dantas, A. L., & Pozo, A. T. (2018b). A meta-learning algorithm selection approach for the quadratic assignment problem. In *Proceedings of 2018 IEEE Congress on Evolutionary Computation (CEC)*1–8. https://doi.org/10.1109/CEC.2018.8477989

Dantas, A., & Pozo, A. (2020). On the use of fitness landscape features in meta-learning based algorithm selection for the quadratic assignment problem. *Theoretical Computer Science, 805*, 62–75.

Dantzig, G. B., & Rasmer, J. H. (1959). The truck dispatching problem. *Management Science*, 6(1), 80–91. https://doi.org/10.1287/mnsc.6.1.80

Fisher, M. L. (1994). Optimal solution of vehicle routing problems using minimum k-trees. *Operations Research, 42*(4), 626–642. https://doi.org/10.1287/opre.42.4.626

Gendreau, M., Laporte, G., & Potvin, J.-Y. (2002). Metaheuristics for the capacitated VRP. In *The vehicle routing problem,* 129–154, SIAM, https://doi.org/10.1137/1.9780898718515.ch6

Gillett, B. E., & Miller, L. R. (1974). A heuristic algorithm for the vehicle-dispatch problem. *Operations Research, 22*(2), 340–349.

Gutierrez-Rodriguez, A. E.-P., Ortiz-Bayliss, J. C., & Terashima-Marin, H. (2019). Selecting meta-heuristics for solving vehicle routing problems with time windows via meta-learning. *Expert Systems with Applications, 118*, 470–481. https://doi.org/10.1016/j.eswa.2018.10.036

Hutter, F., Xu, L., Hoos, H. H., & Leyton-Brown, K. (2014). Algorithm runtime prediction: Methods & evaluation. *Artificial Intelligence, 206*, 79–111. https://doi.org/10.1016/j.artint.2013.10.003

Kanda, J., Carvalho, A., Hruschka, E., & Soares, C. (2010). Using meta-learning to classify traveling salesman problems. In *Proceedings of Brazilian Symposium on Neural Networks*.73–78. https://doi.org/10.1109/SBRN.2010.21

Kanda, J. Y., Carvalho, A. C., Hruschka, E. R., & Soares, C. (2011). Using meta-learning to recommend meta-heuristics for the traveling salesman problem. In *Proceedings of 10th International Conference on Machine Learning and Applications and Workshops*.1, 346–351. https://doi.org/10.1109/ICMLA.2011.153

Laporte, G. (1992). The vehicle routing problem: An overview of exact and approximate algorithms. *European Journal of Operational Research, 59*(3), 345–358. https://doi.org/10.1016/0377-2217(92)90192-C

Laporte, G., & Nobert, Y. (1987). Exact algorithms for the vehicle routing problem. *North-Holland Mathematics Studies*, 132, 147–184, Elsevier.

Laporte, G. & Semet, F. (2002). Classical heuristics for the capacitated VRP. In Toth Paolo and Daniele Vigo (eds.), *The vehicle routing problem,*109–128, SIAM.

Leyton-Brown, K., Nudelman, E., & Shoham, Y. (2009). Empirical hardness models: Methodology and a case study on combinatorial auctions. *Journal of the ACM (JACM), 56*(4), 1–52, https://doi.org/10.1145/1538902.1538906.

Musliu, N., & Schwengerer, M. (2013). Algorithm selection for the graph coloring problem. In *Proceedings of International Conference on Learning and Intelligent Optimization*, 389–403, Springer. https://doi.org/10.1007/978-3-642-44973-442

Nygard, K. E., Juell, P., & Kadaba, N. (1990). Neural networks for selective vehicle routing heuristics. *ORSA Journal on Computing, 2*(4), 353–364.

Rasku, J., Karkkainen, T., & Musliu, N. (2016). *Feature extractors for describing vehicle routing problem instances.* In *Proceedings of 5th Student Conference on Operational Research.*

Rasku, J., Musliu, N., & Karkkainen, T. (2019). Feature and algorithm selection for capacitated vehicle routing problems. In *Proceedings of 27th European Symposium of Artificial Neural Networks, ESANN 2019.*

Rice, J. R. (1976). The algorithm selection problem. *Advances in Computers, 15*, 65–118.

Smith-Miles, K. A. (2008). Towards insightful algorithm selection for optimisation using meta-learning concepts. In *Proceedings of 2008 IEEE International Joint Conference on Neural Networks (IEEE World Congress on Computational Intelligence),*4118–4124.

Smith-Miles, K. A. (2009). Cross-disciplinary perspectives on meta-learning for algorithm selection. *ACM Computing Surveys (CSUR), 41*(1), 1–25. https://doi.org/10.1145/1456650.1456656.

Steinhaus, M. (2015). *The application of the self organizing map to the vehicle routing problem.* (Doctoral dissertation, University of Rhode Island). Retrieved from https://digitalcommons.uri.edu/oadiss/383.

Toth, P., & Vigo, D. (Eds.). (2002). *The Vehicle Routing Problem,* 9, Society for Industrial and Applied Mathematics. https://doi.org/10.1137/1.9780898718515.

Tuzun, D., Magent, M. A., & Burke, L. I. (1997). Selection of vehicle routing heuristic using neural networks. *International Transactions in Operational Research, 4*(3), 211–221.

Witten, I. H. (2016). *Data mining, fourth edition: Practical machine learning tools and techniques* (4th ed.). Morgan Kaufmann Publishers Inc.

Wolpert, D. H., & Macready, W. G. (1997). No free lunch theorems for optimization. *IEEE transactions on evolutionary computation, 1*(1), 67–82. doi:10.1109/4235.585893

17 Early Detection of Autism Disorder Using Predictive Analysis

T. Harshvardhan, G. Preethi, S. Aishwarya, T. Vinod, and K. S. Meghana
Jyothy Institute of Technology, Bangalore

CONTENTS

17.1 Introduction 269
17.2 Literature Review 270
17.3 Dataset Description 271
17.4 Algorithm Description 273
17.4.1 Support Vector Machine 273
17.4.2 Logistic Regression 273
17.4.3 K-Nearest Neighbor (k-NN) 275
17.4.4 Naive Bayes 276
17.4.5 Correlation Matrix 276
17.5 Results and Discussion 277
17.6 Conclusion 280
17.7 Future Scope 280
References 280

17.1 INTRODUCTION

Autism spectrum disorder (ASD) is a neuro-developmental disorder which refers to the lack of ability in socio-behavioral and communicative conduct. This is a reaction of infants to cerebral functions, where we usually see a deterioration of non-verbal and verbal communication and the recurrence of common behaviors. We can recognize that there isn't just one type of autism, but many subtypes, most stimulated via the aggregation of genetic and environmental factors. A spectrum disorder is a mental ailment of linked conditions where the elements of a spectrum either have a similar appearance or are caused by the same elementary mechanism. Because autism is a spectrum disorder, anybody with autism has their own set of strengths and challenges. People with ASD are born with a distinct brain development which is usually seen in the first couple of years of their life. The methods where humans with autism learn, identify, and resolve problems differ from the distinctly professional to the significantly challenged. Some people with ASD require significant assistance in

DOI: 10.1201/9781003146711-17

their everyday lives, while others require much less assistance and, in a few cases, stay totally independent.

Different people are affected in different ways by ASD. Certain researchers and doctors say that autism might be due to a genetic or non-genetic cause, or might be due to the influence of various environmental factors on the life of a child. Early symptoms may be not responding or reacting to parents, friends, or other children's activity.

With the aid of machine learning (ML) and statistical analysis we can study a dataset to build our required predictive model. The algorithms in ML, with the help of meaningful information, help us to process and predict the outcome of various diseases. Since ML and artificial intelligence have begun intervening in the health industry, we have seen numerous prediction models come into the picture and help doctors. If we do not detect ASD in babies that range from 20–60 months it is difficult to treat them. Early detection facilitates improvement and delivers advantages through the lifespan which in turn helps parents to understand and help their children better.

Speaking of prevention, having heaps of data at their disposal allows companies and institutes to identify whether behavioral patterns are linked with the occurrence of certain ailments.

17.2 LITERATURE REVIEW

In a paper written by Raj and Masood (2020), ASD was predicted by implementing various algorithms in ML, such as support vector machine (SVM), naive Bayes, convolutional neural network (CNN), logistic regression, k-nearest neighbor, or University of California at Irvine (UCI) ML repository datasets. Based on 20 common attributes between the three datasets of children, adolescents, and adults they achieved 99.53% accuracy using CNN on the ASD adult dataset, 98.30% using logistic regression on the ASD adolescents dataset, and 96.88% using CNN on the ASD children dataset. After the analysis of various performance metrics, they arrived at the conclusion that CNN and SVM have the same accuracy for the children dataset whereas it was the highest for the other two datasets. Hence, they concluded CNN to be the better algorithm for ASD prediction.

Heinsfeld et al. (2018) have proposed a method to predict ASD using deep learning techniques on the autism brain imaging data exchange (ABIDE) dataset. The dataset contains phenotypical information of 505 ASD individuals. The authors used SVM algorithms, a deep neural network, and random forest in their predictive model. After analysis and comparison, they arrived at the conclusion that DNN achieves 70% accuracy, 65% accuracy using SVM, and 63% accuracy using random forest.

Omar et al. (2019) have proposed a predictive-model-based mobile application which predicts ASD in people at any age. This model was built by combining three algorithms, namely decision tree-CART, random forest-CART, and random forest-ID3. This model was trained and tested with the AQ-10 dataset and 250 items of real-time data collected from people with and without autistic traits. The AQ-10 dataset consists of three groups: children, adolescents, and adults. This model has an accuracy of 92.26%, 93.78%, and 97.10% respectively, and the real-dataset-proposed model has an accuracy of 77.26%, 79.78%, and 85.10% respectively. The results showed marginal performance in terms of accuracy (77% to 85%) for the real dataset.

Kumar and Stella (2020) from the Department of CSE Cambridge Institute of Technology published a paper on "Prediction and Comparison Using AdaBoost and ML Algorithms with Autistic Children Dataset". They utilize three ML techniques (AdaBoost, random forest, and SVM) to compare and project the highest accuracy and provide a result for a given input. In predicting autism for children, they used the Autism Screening Data for Toddlers dataset, which comprises a total of 1054 records of infants from 12 to 36 months. The accuracy of random forest was 96.20%, of SVM was 96.68%, and of AdaBoost was 100%. The AdaBoost algorithm works efficiently with a large dataset and multiple feature selection, hence the AdaBoost classifier gains 100% accuracy.

Alteneiji et al. (2020), in "Autism Spectrum Disorder Diagnosis Using Optimal Machine Learning Methods", used AQ-10 ASD data. Using the ASD AQ-10 technique, data was gathered using a mobile application. This paper highlights using ML techniques to find individuals with certain ASD symptoms and to identify the best suited technique. The ML models are SVM, naive Bayes, neural networks, random forest, GBM, XgBoost, AdaBoost, and CV boosting and are used on the three different categories of datasets. Among all algorithms, the neural networks model had a higher sensitivity and specificity (99.03%) rate than the other remaining algorithms in all the available datasets.

17.3 DATASET DESCRIPTION

The project was performed in the following four phases.

First phase. We used three datasets from the UCI Machine Learning Repository on children, adolescents, and adults. The creators of these datasets used the Autism Spectrum Quotient Tool to predict whether the person is affected by autism or not. The AQ-10 screening test questions are a mixture of different domains, such as paying attention to detail, communication with people, switching attention, imagination, and social interaction. Only one point is awarded per question; based on the response, a score of 0 or 1 basically corresponds to yes or no for all ten questions. The datasets have 292, 104, and 704 instances for children, adolescents, and adults respectively. The datasets can be partitioned into ten different behavioral queries for all age groups, and for different factors that have an impact on the concluding assessment of the conditions that are employed in the database. The variables that have such an impact include age, gender, ethnicity, jaundice, and family history (see Figure 17.1).

Second phase. The collected data is then cleaned to make sure there are no null values and to remove any data that are irrelevant. In cases where there were missing values, we replaced them with the median of that feature. Results showed that by dropping null values and replacing irrelevant values with an appropriate one, we could achieve more accuracy.

id	A1_Score	A2_Score	A3_Score	A4_Score	A5_Score	A6_Score	A7_Score	A8_Score	A9_Score	A10_Scor	age	gender	ethnicity	jundice	austim	contry_of	used_app	result	age_desc	relation	Class/ASD
1	1	1	0	0	1	1	0	1	0	0	6	m	Others	no	no	Jordan	no	5	4-11 year	Parent	NO
2	1	1	0	0	1	1	0	1	0	0	6	m	Middle Ea	no	no	Jordan	no	5	4-11 year	Parent	NO
3	1	1	0	0	0	1	1	1	0	0	6	m	?	no	no	Jordan	yes	5	4-11 year	?	NO
4	0	1	0	0	1	1	0	0	0	1	5	f	?	yes	no	Jordan	no	4	4-11 year	?	NO
5	1	1	1	1	1	1	1	1	1	1	5	m	Others	yes	no	United Sta	no	10	4-11 year	Parent	YES
6	0	0	1	0	1	1	0	1	0	1	4	m	?	no	yes	Egypt	no	5	4-11 year	?	NO
7	1	0	1	1	1	1	0	1	0	1	5	m	White-Eur	no	no	United Kin	no	7	4-11 year	Parent	YES
8	1	1	1	1	1	1	1	1	0	0	5	f	Middle Ea	no	no	Bahrain	no	8	4-11 year	Parent	YES
9	1	1	1	1	1	1	1	0	0	0	11	f	Middle Ea	no	no	Bahrain	no	7	4-11 year	Parent	YES
10	0	0	1	1	1	0	1	1	0	0	11	f	?	no	yes	Austria	no	5	4-11 year	?	NO

FIGURE 17.1 Dataset description.

Third phase. To build the prediction model for autism detection we made use of the various algorithms in ML like logistic regression (LR), decision tree, SVM, k-nearest neighbor, support vector regression, and random forest to check which yielded the highest accuracy.

Fourth phase. The model built was tested with the collection of real-world data in terms of specificity, accuracy, and the rate of positivity.

1. Out of the 16 questions mentioned below any 10 questions are picked for the AQ-10 questionnaire for the three categories accordingly.How well or often do they notice and recognize the small sounds around them while others do not pay attention to these?
2. When a picture is shown to them, do they concentrate on the whole picture or pay attention to the small details in the picture?
3. Do they find it easy to accomplish more than one task at a time?
4. If there is an interference, can they switch back instantly to what they were doing previously?
5. Can they figure out that the person they are talking to is getting bored listening to them?
6. Do they find it effortless to "read between the lines" when a person is conversing with them?
7. Can they figure out the character's intentions easily while they are reading a story?
8. Do they like to gather data about different groups of things like car types, plant types, etc.?
9. Does s/he notice patterns in things around the clock?
10. Can s/he keep track of different conversations with different people in a large social gathering?
11. Can s/he switch back to the task they were on without a pause if there is any disturbance or interruption?
12. Does s/he find it challenging to keep the exchange of dialogues alive?
13. Is s/he good at group chit-chat?
14. Would s/he enjoy engaging in games that involved pretending with other children in their childhood?
15. Do they find it challenging to picture themselves in someone else's shoes?
16. Does s/he have trouble figuring out the character's intentions or feelings easily while they are reading a story?

17.4 ALGORITHM DESCRIPTION

17.4.1 Support Vector Machine

The SVM algorithm comes under the category of supervised learning algorithms which are most commonly used for regression as well as classification in ML.

The objective of the algorithm is to build the best decision boundary or the line that can separate an n-dimensional space into classes so that we can correctly categorize and add the new sample into the right category in the upcoming findings; this decision boundary is known as a hyperplane. We selected only the farthermost vectors or points while making the hyperplane. The vectors are termed "support vectors," hence the name "support vector machine." Figures 17.2 and 17.3) show how the data are divided using a hyperplane.

To evaluate the maximum marginal hyperplane(MMH), which is the important goal of the SVM algorithm, we further divided the datasets into distinct classes, which can be evaluated using these steps:

- Generate different hyperplanes in an iterative process which differentiates the classes in the best way.
- Select a hyperplane that actually differentiates the classes correctly.

17.4.2 Logistic Regression

LR is a classification ML algorithm that uses statistical methods to analyze and classify. It is the best go-to method for binary classification problems. It uses the logistic

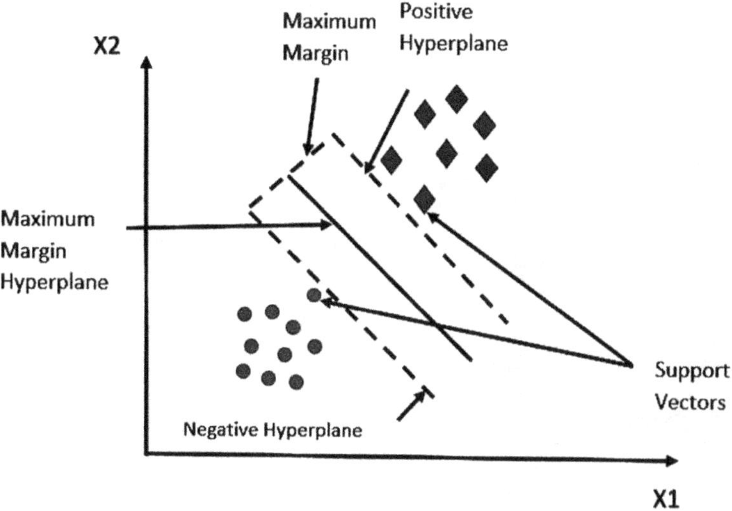

FIGURE 17.2 Support vector machine hyperplane.

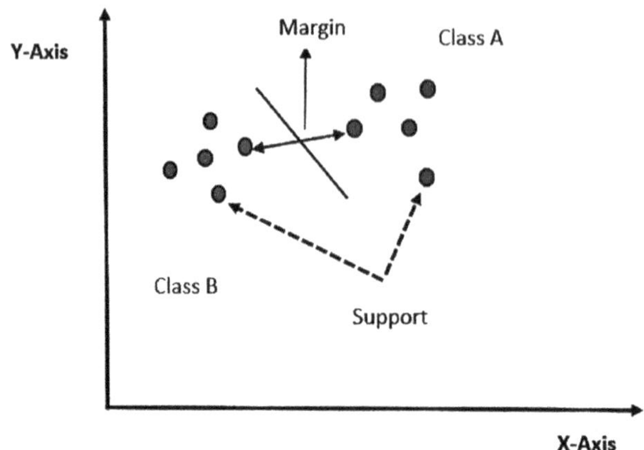

FIGURE 17.3 Support vector machine.

(sigmoid) function to classify and predict the outcome. It is mainly used for a yes or no problem or for problems that have an output of 1 or 0.

$$f(x) = 1/1 + e^{-}(x) \tag{17.1}$$

An S-shaped curve is shown to the right of Figure 17.4 which is termed a sigmoid or logistic function because the result or value of the LR should lie between the numbers 0 and 1 and cannot cross this limit. This curve helps us to map probabilities to the predicted output and these mapped values should be between 0 and 1. Also, a threshold value is specified to help us categorize the given samples of data into the different classes present.

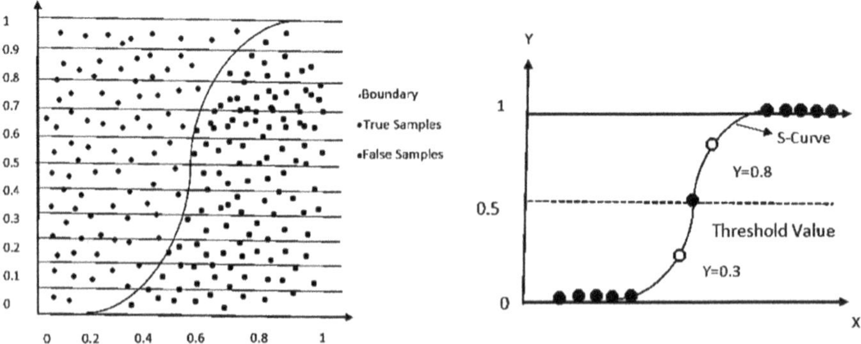

FIGURE 17.4 Logistic regression example and logistic regression sigmoid curve.

17.4.3 K-Nearest Neighbor (k-NN)

The k-NN algorithm comes under the category of supervised learning algorithms in ML; it is simple and effortless to implement and can be used to solve both regression and classification problems. K-NN studies the new data and classifies and categorizes it to the existing data categories based on similarity. K-NN is a lazy algorithm as it considers all the data for training while classification does not have any specialized training phase. Since it does not assume anything about the underlying data, we term it a non-parametric learning algorithm.

Figure 17.5 shows the two different classes A and B in the training instance or initial data. A new sample is introduced into the data which has to be classified into the two classes present. To classify the new data, we have to calculate the distance between the nearest neighbors in the training data to be classified on the one hand, and the training data on the other hand, with the help of a Euclidian method. So, based on the distance between these points we come to a conclusion as to which class the new sample introduced belongs to.

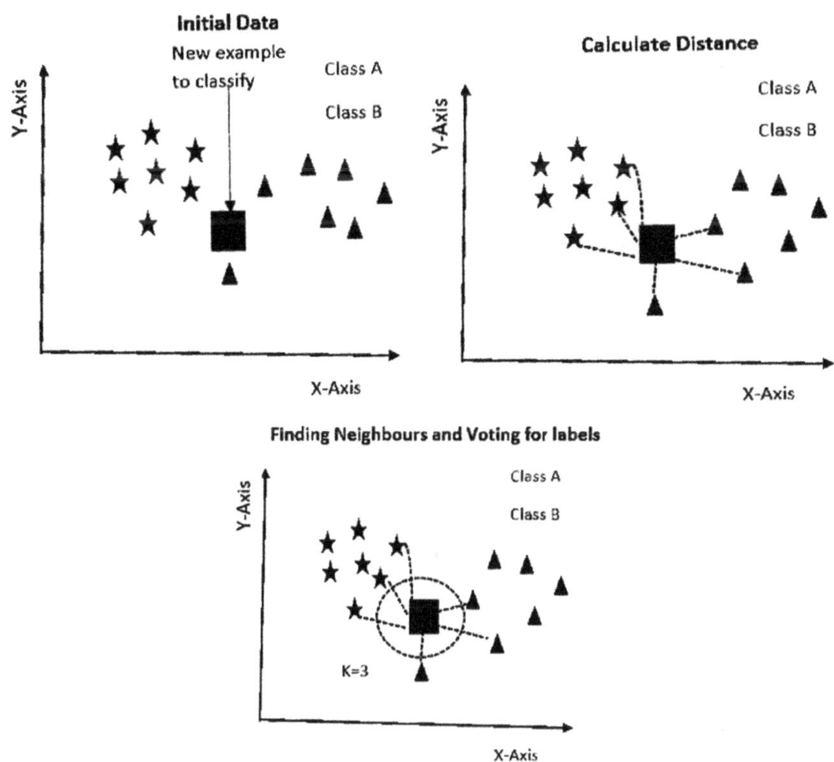

FIGURE 17.5 K-nearest neighbor process.

17.4.4 Naive Bayes

Naive Bayes is an ML calculation we can utilize to unravel classification issues. It is based on Bayes' theorem. It is one of the best yet effective ML calculations in use and finds applications in numerous industries. Naive Bayes uses Bayes' theorem and assumes that all predictors are autonomous.

This calculation may be a great fit for real-time forecasts, multi-class expectations, suggestion frameworks content classification, and sentiment analysis. A naive Bayes calculation can be constructed by making use of Gaussian, multinomial, and Bernoulli distributions.

A difference is made when calculating the back likelihood P(c|x), utilizing the earlier likelihood of course P(c), the earlier probability of predictor P(x), and the probability of the predictor given, known as the probability P(x|c). The formula to evaluate back probability is: P(c | x) = (P(x | c) * P(c))/P(x) (Figure 17.6).

17.4.5 Correlation Matrix

The correlation matrix shows the relation between all the various attributes present in the dataset and how they are related to each other, which in turn helps us to predict the outcome based on the results of the questionnaire. The co-relation values range between 0 and 1 where 0 stands for no relation between the two attributes and 1 indicates a strong relation between the two attributes (Figure 17.7).

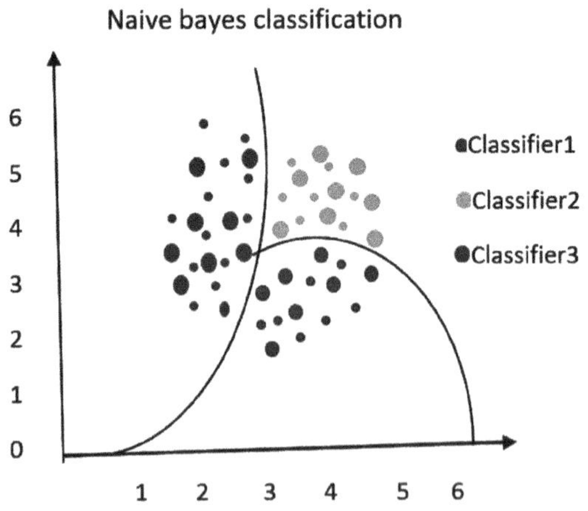

FIGURE 17.6 Naive Bayes.

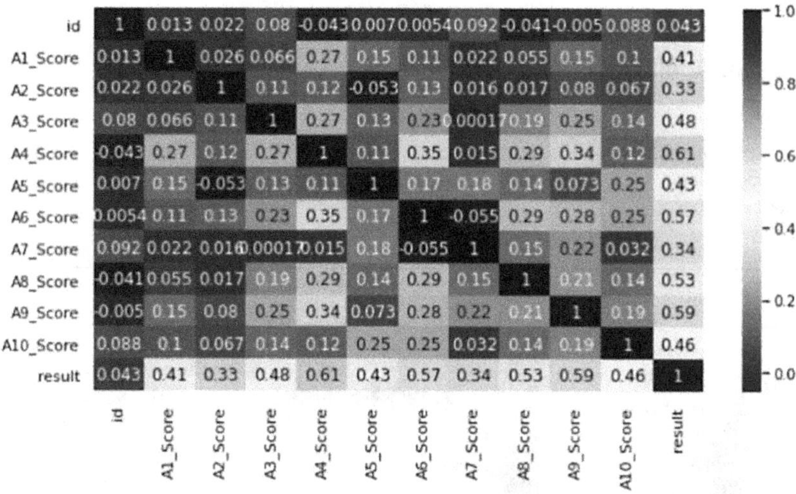

FIGURE 17.7 Correlation matrix.

17.5 RESULTS AND DISCUSSION

The outcome of our prediction model after applying the various ML algorithms helps us to analyze which algorithm is best suited for a prediction system. The results are calculated and concluded based on the accuracy of each trained model. Specificity and sensitivity were calculated using the classification and confusion matrix report, which have helped us to reach our final result (Table 17.1).

TABLE 17.1
Correlation Matrix Calculation

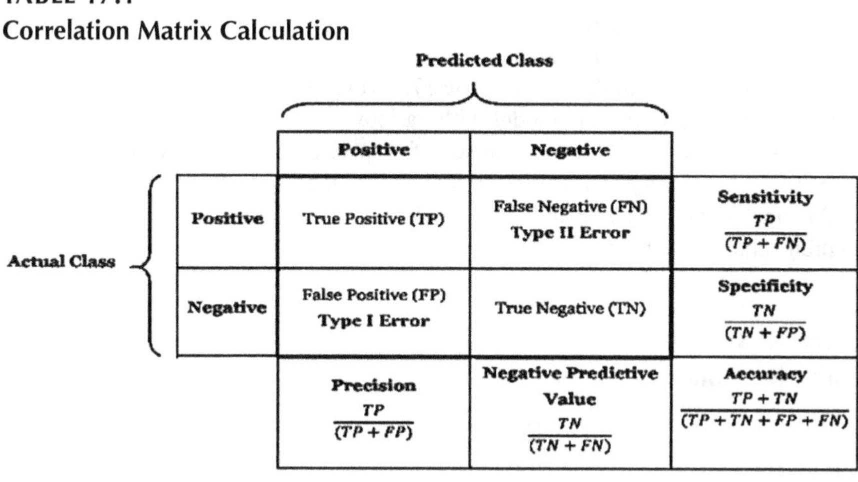

		Predicted Class: Positive	Predicted Class: Negative	
Actual Class	Positive	True Positive (TP)	False Negative (FN) Type II Error	Sensitivity $\frac{TP}{(TP+FN)}$
	Negative	False Positive (FP) Type I Error	True Negative (TN)	Specificity $\frac{TN}{(TN+FP)}$
		Precision $\frac{TP}{(TP+FP)}$	Negative Predictive Value $\frac{TN}{(TN+FN)}$	Accuracy $\frac{TP+TN}{(TP+TN+FP+FN)}$

TABLE 17.2
Results for Adult Dataset

Classifier	Accuracy	Specificity	Sensitivity
Support vector machine	96.45	0.9494	0.9575
Logistic regression	93.64	0.8945	0.9388
K-nearest neighbor	96.22	0.9696	0.9311
Naive Bayes	94.01	0.9125	0.9402

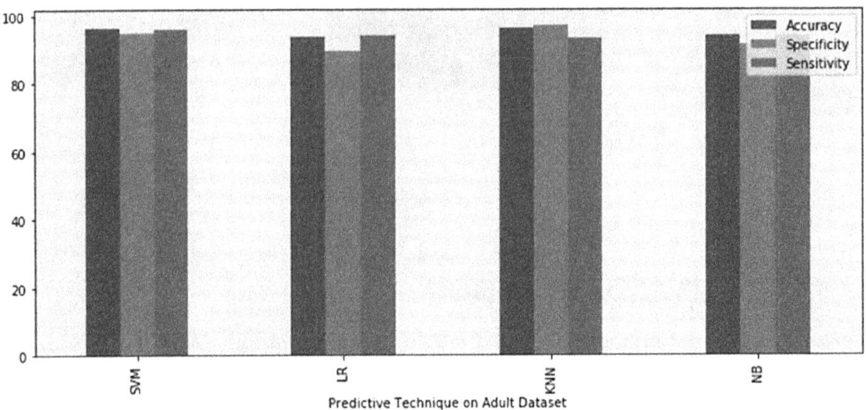

FIGURE 17.8 Graph of accuracy, specificity, and sensitivity of adult dataset.

After training the model with ASD data for adults with various ML algorithms we obtained the following results for specificity, sensitivity, and accuracy. The value of sensitivity and specificity lie between 0 and 1. After applying the various ML techniques, we arrived at the conclusion that the identification of ASD for adults achieved its highest accuracy with the SVM (Table 17.2) (Figure 17.8).

Subsequently, training the model with various ML algorithms on ASD data for children we obtained the following results for specificity, sensitivity, and accuracy. The value of sensitivity and specificity lie between 0 and 1. After applying the various ML techniques, we arrived at the conclusion that the identification of ASD for children achieved its highest accuracy with SVM (Table 17.3) (Figure 17.9).

TABLE 17.3
Results for Children Dataset

Classifier	Accuracy	Specificity	Sensitivity
Support vector machine	97.56	1.0	0.9744
Logistic regression	97.32	1.0	0.9678
K-nearest neighbor	96.22	0.9743	0.9131
Naive Bayes	94.01	0.9215	0.9564

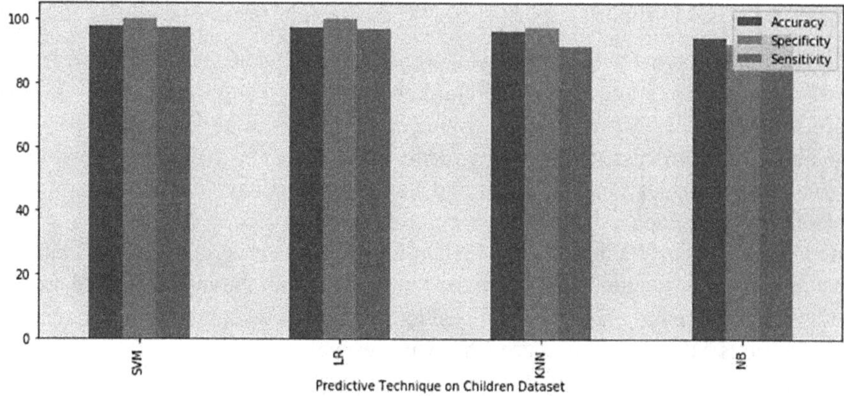

FIGURE 17.9 Graph of accuracy, specificity, and sensitivity of children dataset.

Finally, evaluating the ASD for the adolescent data for various ML algorithms we obtained the following results for specificity, sensitivity, and accuracy. The value of sensitivity and specificity lie between 0 and 1. After applying the various ML techniques, we arrived at the conclusion that the identification of ASD for adolescents achieved its highest accuracy in the SVM (Table 17.4) (Figure 17.10).

TABLE 17.4
Results for the Adolescents Dataset

Classifier	Accuracy	Specificity	Sensitivity
Support vector machine	93.51	1.0	0.7421
Logistic regression	86.32	1.0	0.8753
K-nearest neighbor	90.25	0.9743	0.8912
Naive Bayes	88.32	0.9215	0.6201

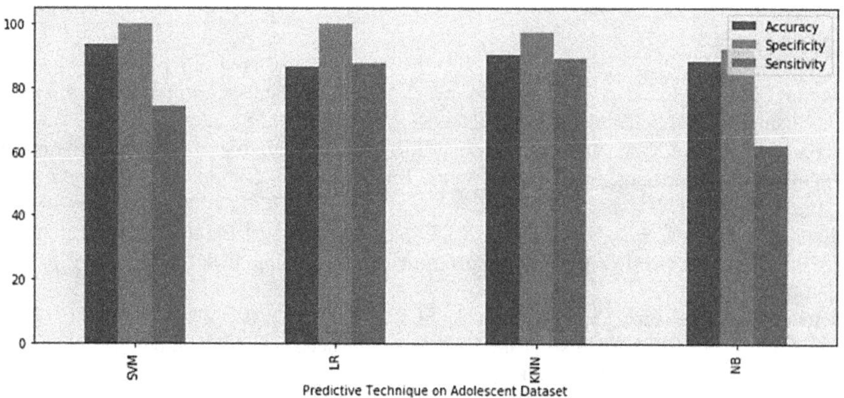

FIGURE 17.10 Graph of accuracy, specificity, and sensitivity of adolescents dataset.

17.6 CONCLUSION

This chapter has provided an effective approach to ASD detection in three different age groups (adults, adolescents, and children) achieved using the concepts of deep learning and various ML techniques. Since the diagnosis of autism is a very expensive and tedious process, the autism screening technique can guide an individual so as to be able to detect autism at the initial stage which will avoid the condition from worsening. This chapter sheds light on building a model to predict autism traits in an individual at an early stage. Each ML algorithm has its own applicability, and we have to find the best suited for our prediction model to get the most accurate results. In this research we have also tried our best to provide a good comparison between the use of various ML algorithm approaches with respect to their performance.

To analyze specificity, accuracy, and sensitivity performance evaluation, metrics were used on all the three different dataset categories. Based on the results produced by all the multiple ML algorithms used, we found that building the model with an SVM algorithm obtained the highest accuracy in comparison to the other algorithms. When we applied the SVM algorithm for adults, we obtained 96.45%; with the same algorithm on the children dataset we obtained 97.56%; and when it was applied to the adolescents dataset, we got 93.51%.

17.7 FUTURE SCOPE

The work in this chapter could further be developed into an application program that could provide people with easy and straightforward access to an ASD scan. All the user must do is answer the question set, according to their respective age group and which in turn increases the rate of early detection of autism. A potential use for the future of this research could be the study and implementation of ML and artificial intelligence models in health organizations and industry as they can provide the vast majority of patient data and health symptoms that potentially contribute to the further study of the early detection of autism. It would also be possible to develop this research in ministries of health as they have abundant data available for future development related to the early prediction of autism.

REFERENCES

Alteneiji, M. R., Alqaydi, L. M., & Tariq, M. U. (2020). Autism spectrum disorder diagnosis using optimal machine learning methods. *Autism, 11*(9), 252–260.

Büyükoflaz, F. N., & Öztürk, A. (2018, May). Early autism diagnosis of children with machine learning algorithms. In *2018 26th Signal Processing and Communications Applications Conference (SIU)* (pp. 1–4). IEEE.

Eslami, T., Raiker, J. S., & Saeed, F. (2020). Explainable and scalable machine-learning algorithms for detection of autism spectrum disorder using fMRI data. *arXiv preprint arXiv:2003.01541*.

Heinsfeld, A. S., Franco, A. R., Craddock, R. C., Buchweitz, A., & Meneguzzi, F. (2018). Identification of autism spectrum disorder using deep learning and the ABIDE dataset. *NeuroImage: Clinical, 17*, 16–23.

Omar, K. S., Mondal, P., Khan, N. S., Rizvi, M. R. K., & Islam, M. N. (2019, February). A machine learning approach to predict autism spectrum disorder. In *2019 International Conference on Electrical, Computer and Communication Engineering (ECCE)* (pp 1–6) IEEE.

Raj, S., & Masood, S. (2020). Analysis and detection of autism spectrum disorder using machine learning techniques. *Procedia Computer Science, 167*, 994–1004.

Skafidas, E., Testa, R., Zantomio, D., Chana, G., Everall, I. P., & Pantelis, C. (2014). Predicting the diagnosis of autism spectrum disorder using gene pathway analysis. *Molecular Psychiatry*, *19*(4), 504–510.

Thabtah, F. (2019). Machine learning in autistic spectrum disorder behavioral research: A review and ways forward. *Informatics for Health and Social Care*, *44*(3), 278–297.

Vaishali, R., & Sasikala, R. (2018). A machine learning based approach to classify autism with optimum behaviour sets. *International Journal of Engineering & Technology*, *7*, 18.

18 Computing Technologies for Prognosticating the Emanation of Carbon Using an ARIMA Model

Shyla, Kapil Kumar, and Vishal Bhatnagar
NSUT East Campus, India

CONTENTS

18.1 Introduction 283
 18.1.1 Formulation of the Problem 285
 18.1.2 Research Objective 285
18.2 Research Strategy 285
18.3 Related Work 286
18.4 Research Framework 288
18.5 Dataset Description 289
18.6 Forecasting Model Performance Analysis 289
 18.6.1 Autoregressive Integrated Moving Average (ARIMA) Model 290
18.7 Research Limitations 294
18.8 Conclusion and Future Scope 294
References 295

18.1 INTRODUCTION

The continuous evolution in computing technologies is resulting in an increase in carbon footprints which are harming the environment. Chen et al. (2020) found that the regulated usage of computing technologies all over the world generates a huge amount of electronic waste that is rarely recycled and is adding to environmental pollution. The relentless move towards the Internet requires consistent energy for data centers to maintain information networks, and this contributes to a large amount of carbon emanation. Companies and organizations are shifting their data centers to cooler places to deal with the problem of the enormous heat generation from computing devices. The adaptation of low-carbon computing and green computing technologies is a measure for dealing with this problem of carbon emission to a great extent.

The dissemination of societal and technological fall down can be prevented using low-carbon computing, by nullifying the production of heat waste and by the mitigation of electric energy power requirements. Riaz et al. (2020) found that energy

DOI: 10.1201/9781003146711-18

efficient technologies are designed, manufactured, and engineered to promote recyclable computing devices which are environmental friendly and economically viable. "Green use" refers to minimum electric energy utilization by computing devices and peripheral equipment and the usage of technologies in an environmentally friendly way:

- "Green design" refers to the engineering and designing of energy efficient devices, data centers, servers, printers, peripherals, and electronic devices.
- "Green manufacturing" refers to mitigating electronic waste during the manufacturing of computing devices, computers, and other equipment by organizations and manufacturers.
- "Green disposal" refers to the redeployment of existing computing devices and the recycling and disposal of electronic waste for the development of a sustainable environment.

Figure 18.1 shows the components of computing with respect to the eco-friendly usage of green technologies, the dumping of green technologies, the design of green technologies, and the manufacturing of green technologies. Computation in the context of green issues, also referred to as green information technology, is environmentally friendly computing. Qin et al. (2020) found that it is the practice of designing, engineering, using, and ultimately the placing of computing devices according to a particular strategy that minimizes the influence on the environment. This includes the

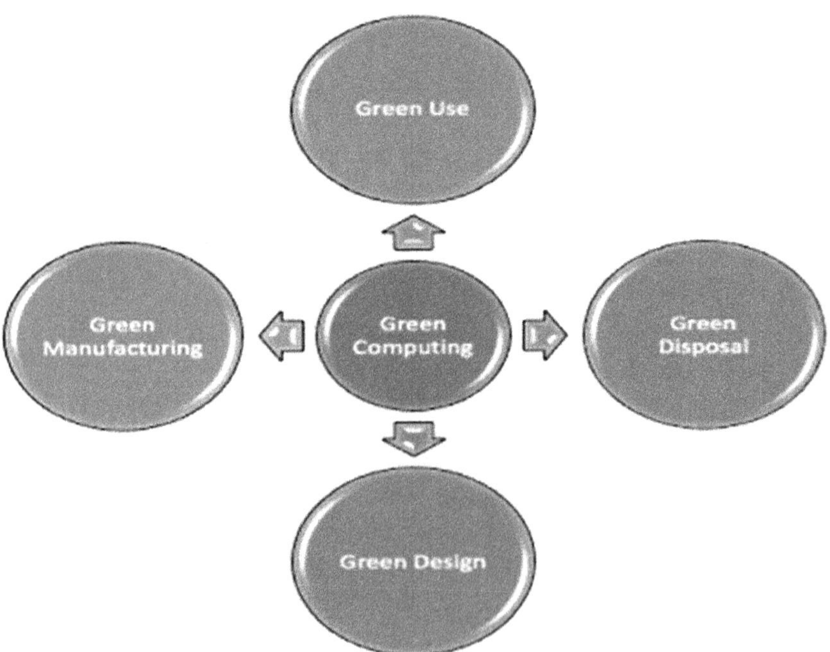

FIGURE 18.1 Components of green computing.
Source: Malik et al., 2020.

usage of energy efficient processors, servers, and other peripherals, and the proper disposal of electronic waste. It involves less energy usage and reduced waste, which results in reducing the worst impact of technological innovation on the environment. Less energy usage simply means less fossil fuel consumption and less carbon dioxide emission. The first and foremost goal is to protect the environment and make computing more environmentally sustainable so that negative consequences can be minimized. This can be achieved by implementing cloud computing. Malik et al. (2020) found that, at the time of the manufacturing of computing devices, recyclable or biodegradable materials are used for the management of electronic waste to avoid harming the environment. Less paper used will also help us to achieve the goal. The other electronic alternatives to using paper are considerable. Old machines or equipment can be replaced by new energy efficient devices, resulting in reduced heat production and other emissions; power consumption also decreases. Green information technology is cost effective too, as there is less energy consumption. It also preserves natural resources and promotes the effective utilization of existing resources.

18.1.1 Formulation of the Problem

The consistent evolution in computing technologies is causing environmental distress due to the emission of carbon and the enormous amount of electronic waste. Computing equipment generates a high amount of heat energy by consuming a large amount of electrical energy. The technological revolution creates electrical energy waste, heat energy waste, and electronic waste by the generation of carbon from energy combustion. The introduction of green information technology is the solution for the development of a sustainable environment.

18.1.2 Research Objective

- The primary objective is to determine the amount of carbon generated from various computing peripherals.
- To design a framework and methodology for forecasting carbon emanation.
- To deploy an autoregressive integrated moving average (ARIMA) model for the efficient forecasting of daily and monthly generated carbon using time series data.

The chapter is organized as follows. Section 18.2 is about our research strategy; Section 18.3 is about related work; Section 18.4 concerns the research framework; Section 18.5 describes the dataset used; Section 18.6 concerns performance analysis of the forecasting model; Section 18.7 is about the research's limitations; and Section 18.8 provides a conclusion and a look at the future scope of the work.

18.2 RESEARCH STRATEGY

The research strategy embodies the strategical representation of the proposed ideology which is trailed throughout the chapter for leading the investigation in a consecutive manner.

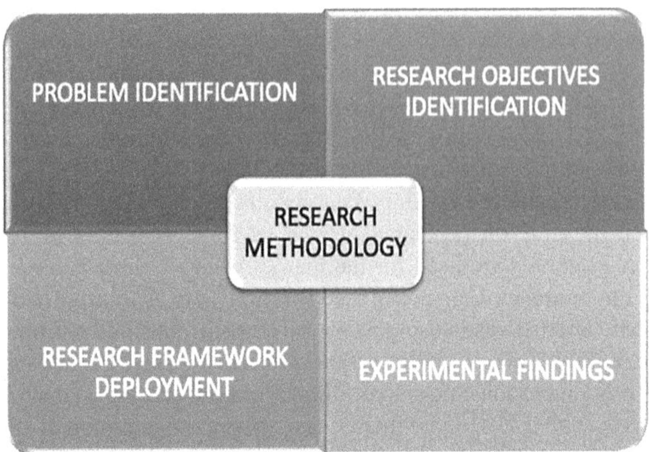

FIGURE 18.2 Research strategy.
Source: Li et al., 2020b.

Figure 18.2 illustrates the research strategy. We performed related work to analyze the prevailing research in the context of the forensic investigation of facial recognition. This highlighted limitations. The gaps define the shortcomings in prevailing work and outline further research that is required. The problem is stated by reflecting on the research gap. In this chapter we primarily report on related studies, research gaps, and identify the problem. Li et al. (2020a) found that the research framework is intended to attain goals efficiently, step toward an applicable algorithm, and that the framework is arranged using a classifier. In this paper the authors used a convolution neural network (CNN) based algorithm for the classification of data. The presentation of the algorithm is examined by researchers by processing, accuracy, recall, support, and F1 score values.

18.3 RELATED WORK

Gai et al. (2016) discussed unraveling energy left-over within the self-motivated networking milieu. The suggested framework stipulates an impending investigation with a principle and hypothetical chains. The study proposed fresh tactics for using an active distributed system based idea to attain advantages. Inspection exposed an approach used in the paper that was an active method that could permit moveable operators to find green IT within a self-motivated complex wireless atmosphere.

Kansal and Chana (2012) discussed the present weight stabilizing process executed via computation through a platform as a service, and equated them by numerous factors such as concertability, scalability, and relationship that are measured by various methods. The authors discussed these methods from the viewpoint of energy consumption and carbon release. Saha (2014) surveyed multiple significant works linked in the context of green computing and stressed the prominence of green development. The study referred to green computing that is about using computing assets in a friendly atmosphere while upholding globally computing outline. Global warming is the unceasing upswing in the temperature of the Earth's environmental system, which is causing a variety of issues. Sarkar and Misra (2016) emphasized the hypothetical modeling of the mist-figuring structure and likening its enactment with the outdated cloud computing framework. Prevailing investigation of the mechanism of mist figuring have mainly fixated on ideas of mist processing, and its effect on the IoT milieu. The effort is a significant role in the milieu, suggesting an arithmetic preparation for this novel computational example by describing its discrete mechanism and providing relative learning in terminologies that decide the provision of dormancy and power utilization. The authors modeled the architecture of fog computing and the performance was analyzed in terms of IoT applications. It was observed that the significance of fog computing is lower than cloud computing.

Lotfalipour et al. (2013) found that the economic facts of gas release and its aftermath are essential, notably with respect to the current level of growth. Consequently, the emanation of contaminated carbon dioxide was estimated precisely according to the adapted rules. Henceforth, the analysis and predicting of gas release are crucial. The intention of the study is forecasting carbon dioxide release via the mechanism of gray computing, and an integrated approach using autoregressive average by equating both tactics. The results demonstrate that the emission of carbon dioxide is rising drastically. Soomro and Sarwar (2012) found over recent years that an attentiveness in "green computing" has progressed investigation into power saving methods, from internal computing devices to industrial mechanisms. The study found in the subsequent analysis that computing cores were observed to be low. Fog computing is not the classification of cloud computing. Rather it is the next generation of IoT applications.

Qin et al. (2020) proposed a novel approach by using the hybrid algorithm that combines a self-regulated particle swarm optimization algorithm and the extreme learning machine for classification purposes. For optimization the input weights and minimum norm least-square methodology was used for lowering the service latency and energy consumption. Riaz et al. (2020) used an enhanced PSO to evaluate the output weights. The PSO is improved by using a mechanism which defines the look up behavior of the particles to determine the best response. The experiment and results show that in this framework, by using trial and error, the amount of neural operations in the processing layers does not require data variables that dynamically affect the model's accuracy and precision.

Bas et al. (2020) established a fresh innovative woolly reversion operation in the context of time series implication methods. The influence of the resource allocation graph suggests a novel implication system. The enactment is evaluated by performing the experiment with a time series dataset and a projected model. The outcomes achieved from the implication system are paralleled with some other sorted papers which are implemented based on time series data. The projected approach is performed using the method that has the best performance of all those considered.

Panigrahi and Behera (2020) planned a framework that depends on the combination of two or more algorithms. In this way the authors missed out plenty of algorithms and compared them with each other. Many time series datasets are utilized by the authors for comparative analysis and they determine the optimal performance combination of them. The results indicated, in the context of statistical patterns, that the FTSM-SVM algorithm is optimal and an advanced algorithm.

18.4 RESEARCH FRAMEWORK

The framework is the theoretical structure which is used as a systematic procedure for implementing the prediction model. Dritsaki and Dritsaki (2020) found that the initial phases of the framework include the accumulation of monthly and annually estimated carbon emission time series datasets which is then pre-processed by eliminating null values and raw information. The individual classifier is classified as a monthly average carbon emission and a yearly average carbon emission estimation. The classified dataset is used for classification using an ARIMA model. The machine learning model is represented by performance metrics. The data analysis and performance analysis is represented by the model output.

Figure 18.3 shows the research framework for designing a carbon emission forecasting model using an ARIMA model. The framework works along three phases of

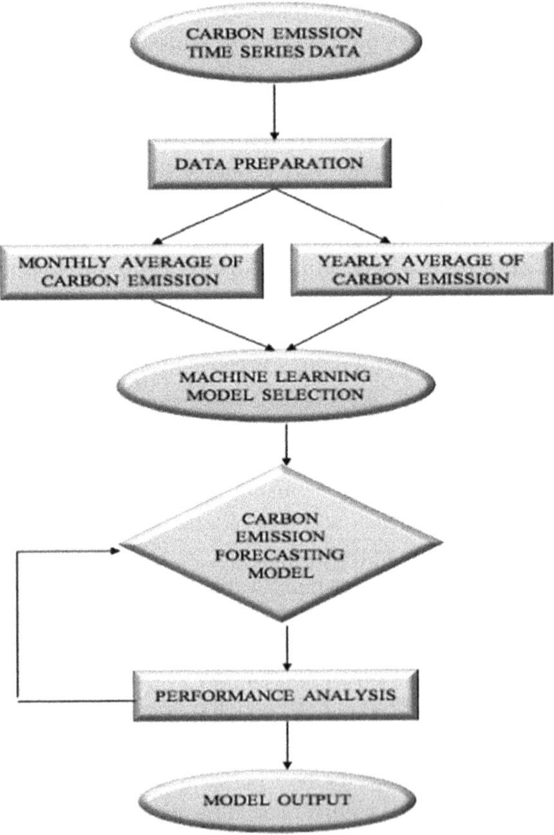

FIGURE 18.3 Research framework.
Source: Demirci, 2015.

information forecasting. Phase one involves the pre-processing of the accumulated time series data. Phase two evaluates the pre-processed data for classifier referencing and phase three includes the testing and training of data for model performance analytics.

18.5 DATASET DESCRIPTION

The monthly and annually estimated carbon emissions from computing technologies is reported in the carbon emission dataset. The dataset is obtained from the open source platform Kaggle which is of size 612.5 KB. The time series carbon emission data are collected as they can explain and enable the extraction of valuable information for future prediction, which is the aggregation of data points at constant intervals of time. The dataset is analyzed and visualized using the Python programming language with data analytics. The pre-processed analyzed data are used for testing and training a machine learning forecast model.

Figure 18.4 shows a sample of a collected dataset. The dataset consists of six columns having two integer data types with four objects and 5096 observations. The dataset is of time series data of monthly and annually estimated carbon emissions from computing devices.

18.6 FORECASTING MODEL PERFORMANCE ANALYSIS

The carbon emission data are collected and pre-processed to transform and encode them into an interpretable format for a machine learning algorithm. The features of the dataset are determined to measure the characteristics and properties of different types of embedded data. The data pre-processing phase involves feature extraction, categorization of data, quality of data assessment, aggregation of traits, sampling of traits, diminution in the dimension of data, encoding of traits, and the testing/training of the dataset.

	MSN	Value	Column_Order	Description	Unit
YYYYMM					
1973-01-01	CLEIEUS	72.076	1	Coal Electric Power Sector CO2 Emissions	Million Metric Tons of Carbon Dioxide
1973-02-01	CLEIEUS	64.442	1	Coal Electric Power Sector CO2 Emissions	Million Metric Tons of Carbon Dioxide
1973-03-01	CLEIEUS	64.084	1	Coal Electric Power Sector CO2 Emissions	Million Metric Tons of Carbon Dioxide
1973-04-01	CLEIEUS	60.842	1	Coal Electric Power Sector CO2 Emissions	Million Metric Tons of Carbon Dioxide
1973-05-01	CLEIEUS	61.798	1	Coal Electric Power Sector CO2 Emissions	Million Metric Tons of Carbon Dioxide

FIGURE 18.4 Dataset sample.

Mean sample		After filling null value mean sample	
1958-03-01	316.100000	1958-03-01	316.100000
1958-04-01	317.200000	1958-04-01	317.200000
1958-05-01	317.433333	1958-05-01	317.433333
1958-06-01	NaN	1958-06-01	315.625000
1958-07-01	315.625000	1958-07-01	315.625000
2001-08-01	369.425000	2001-08-01	369.425000
2001-09-01	367.880000	2001-09-01	367.880000
2001-10-01	368.050000	2001-10-01	368.050000
2001-11-01	369.375000	2001-11-01	369.375000
2001-12-01	371.020000	2001-12-01	371.020000

FIGURE 18.5 Sample values.

Figure 18.5 shows the mean sampled values of data. The missing and null values are eliminated by estimating mean values. The interpolation method is used to fill NaN values with the mean, median, and mode value of features.

18.6.1 Autoregressive Integrated Moving Average (ARIMA) Model

The ARIMA model is used for time series forecasting of carbon emissions from computing technologies. Lotfalipour et al. (2013) found that the model can be used for solving prediction problems for monthly and annually recorded observations. ARIMA means that, on the basis of past values of multivariate time series, future events can be forecast. The model is defined as P, D, Q, where P is the autoregressive order, Q is the moving average order, and D is the differencing to make the time series stationary (if the time series is already stationary then "D = 0").

$$\mathrm{Y}^{\mathrm{T}} = \alpha + \beta_1 Y_{T-1} + \beta_2 Y_{T-2} + \cdots + \beta_P Y_{T-P} + \epsilon_1 \tag{18.1}$$

Y depends on the lags of $Y_{T\text{-}1}$, where β_1 is the coefficient of the lag and α is the estimated intercept. E_T and $E_{T\text{-}1}$ are the errors of the autoregressive models, which are (Fatima et al., 2019):

$$\mathrm{Y}^{\mathrm{T}} = \beta_1 Y_{T-1} + \beta_2 Y_{T-2} + \cdots + \beta_0 Y_0 + \epsilon_T \tag{18.2}$$

$$\mathrm{Y}^{\mathrm{T-1}} = \beta_1 Y_{T-2} + \beta_2 Y_{T-3} + \cdots + \beta_0 Y_0 + \epsilon_{T-1} \tag{18.3}$$

The Python programming language is used for the forecasting of carbon emissions by importing Python libraries in Anaconda Jupiter.

Load the dataset

```
load --> dataset
```

The 'ms' string groups the data.

```
x --> x['co2'].resample('ms').mean()
x--> x.fillna(x.bfill())
```

Define the a, b, and c parameters.

```
a = b = c = range(0, 2)
abc-->list(iteration.items(a, b, c))
seasonal-abc --> [(x[0], x[1], x[2], 12) for x in list(iteration.items(a, b, c))]
model --> arimax(x,2, 2, 2), seasonal_array -->(2, 2, 2, 12), false ,false)
```

Fit the model

```
r-->model.fit()
r.plot_diagnostics(figuresize=(10, 10))
pred --> r.get_prediction(intialte-->pd.to_datetime('1998-01-01'), false)
x_prediction-->predion.mean
x_true -->x['1998-01-01']
```

Determine the mse.

```
mse --> ((x_predicteed - x_true) ** 2).mean()
```

Determine the forecast value and truth values of the time series

```
x_predicted --> pred_dynamic.predicted_mean
x_true --> x['1998-01-01']
```

Determine mse.

```
mse --> ((x_predicted - x_true) ** 2).mean()
```

Evaluate forecast 500 steps onwards in future

```
prediction --> result.predict(steps-->500)
```

Evaluate the confidence of forecast.

```
prediction --> c.confidence_int()
```

Figure 18.6 shows the order of differencing for making the time series stationary with autocorrelation plots for representing the estimated lag values.

Table 18.1 shows the performance metrics of the ARIMA model: the estimated values for the coefficient, standard error rate, autoregressive, and moving average. The yearly recorded carbon emission values are used for evaluating autoregressive and moving average lag values.

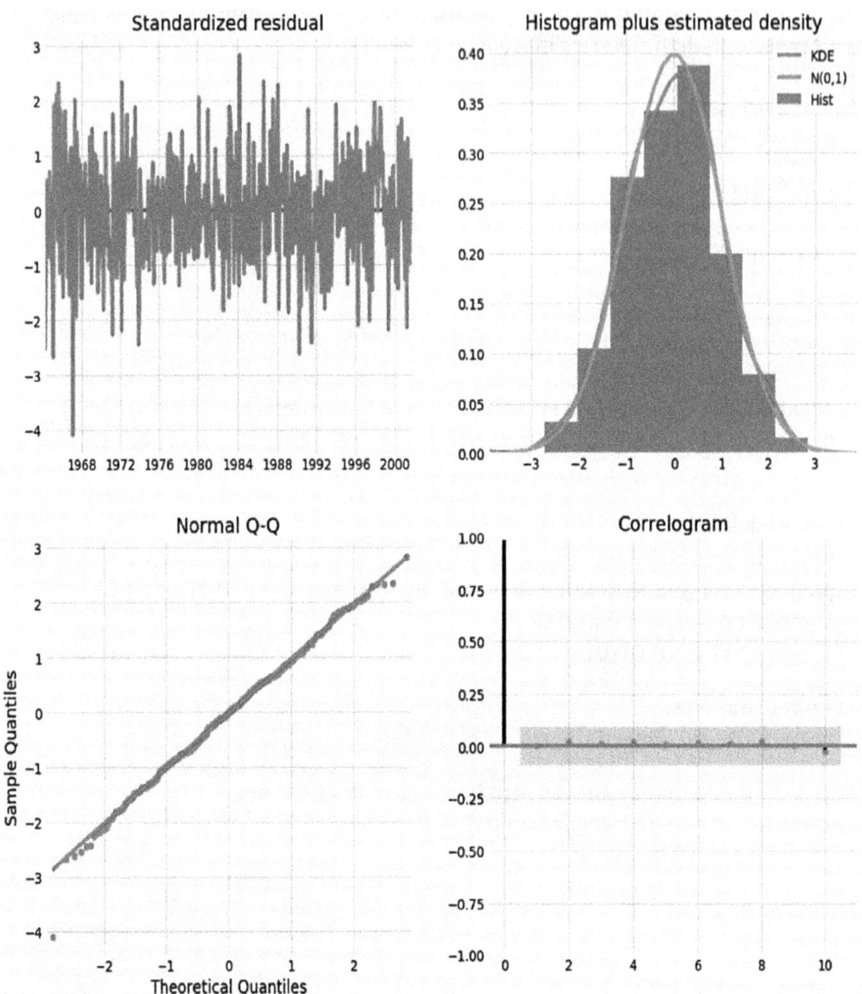

FIGURE 18.6 Order of differencing.

Figure 18.7 shows the estimated levels of carbon emission from different computing technologies. The levels are estimated values obtained from historic data from 1995 to 2001.

Figure 18.8 shows the forecast levels of carbon emission from different computing technologies. The graph shows the estimated predicted values on the basis of historic data.

TABLE 18.1
Performance Matrices for the ARIMA Model

	Coefficient	Standard Error	z	P > |z|	[0.025	0.975]
ar.L1	−0.6398	0.435	−1.472	0.141	−1.491	0.212
ar.L2	−0.4747	0.340	−1.398	0.162	−1.140	0.191
ma.L1	−0.6536	0.459	−1.425	0.154	−1.552	0.245
ma.L2	−0.1469	0.576	−0.255	0.799	−1.275	0.981
ma.L3	−0.3629	0.599	0.606	0.545	−1.537	0.811
ma.L4	0.1556	0.305	0.510	0.610	−0.442	0.753
ma.L5	−0.0876	0.185	−0.473	0.636	−0.451	0.275
ma.L6	0.0015	0.341	0.004	0.996	−0.666	0.669
ma.L7	−0.0285	0.217	−0.132	0.895	−0.453	0.396
ma.L8	0.1120	0.280	0.400	0.689	−0.437	0.661
ma.L9	0.1427	0.284	0.502	0.615	−0.414	0.700
ma.L10	−0.0593	0.370	−0.160	0.873	−0.784	0.666
ma.L11	0.0192	0.146	0.132	0.895	−0.267	0.305
ma.L12	−1.1746	0.389	-3.016	0.003	−1.938	-0.411
ma.L13	0.7445	0.501	1.485	0.138	−0.238	1.727
ma.L14	0.0988	0.578	0.171	0.864	−1.033	1.231
ma.L15	0.4149	0.482	0.861	0.389	−0.529	1.359
ma.L16	−0.1538	0.159	−0.967	0.334	−0.466	0.158
ma.L17	0.0434	0.096	0.450	0.653	−0.146	0.232
ma.L18	−0.0207	0.087	−0.239	0.811	−0.191	0.149
ma.L19	−0.0419	0.089	−0.470	0.638	−0.217	0.13
ma.L20	0.0037	0.095	0.039	0.969	−0.183	0.190
ma.L21	−0.0030	0.076	−0.040	0.968	−0.151	0.145
ar.S.L12	−0.5627	0.251	-2.246	0.025	−1.054	-0.072
ar.S.L24	−0.0559	0.046	−1.219	0.223	−0.146	0.034
ma.S.L12	−0.4626	1.959	−0.236	0.813	-4.301	3.376
ma.S.L24	−0.5418	1.083	−0.500	0.617	-2.664	1.580
Sigma 2	0.0734	0.151	0.488	0.626	−0.222	0.369

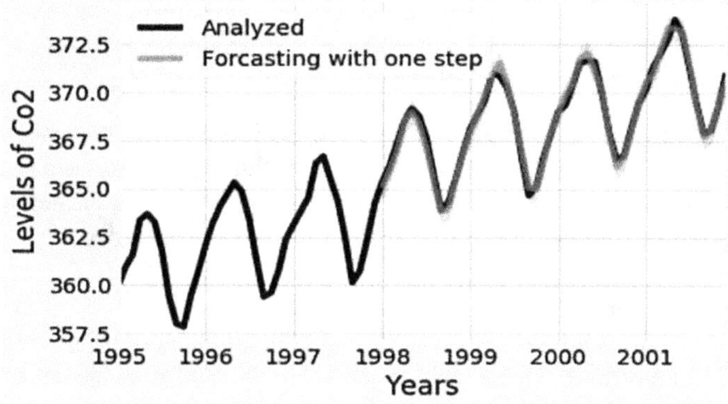

FIGURE 18.7 Estimated levels of carbon emission.

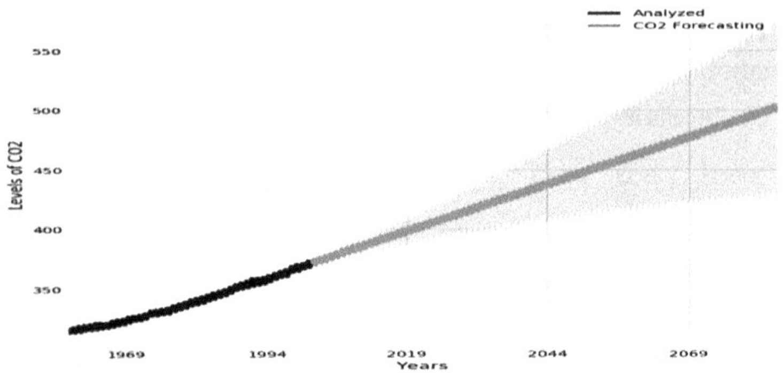

FIGURE 18.8 Forecast levels of carbon emission.

18.7 RESEARCH LIMITATIONS

The research limitations denote the range of scope of limitation experience throughout the research process. The shortcoming of research papers, articles, and journals is identified and acknowledged.

- The prior limitation is the practice of research aims, objectives, and problem statements. The problem arises when the objectives are defined too broadly and misguide the research hypothesis.
- The implementation of the data collection approach and sample size introduces research limitations as the sample size depends on the nature of the research problem and extensivity of the data collection.
- The researcher is unable to utilize the barred access software that is mandatory for conducting an experiment and proving the result.
- The lack of previous studies in the research domain due to the unavailability of open access research papers, dissertations, journals and research articles narrows the scope of the work.

18.8 CONCLUSION AND FUTURE SCOPE

The importance of green computing has been observed in the presence of carbon emitting equipment and high energy consuming technologies such as motherboards, processors, cores, and hardware machines. Green computing is dedicated to environmental science for providing energy efficient, recyclable, and decomposable technologies. In this chapter we have considered monthly/annually reported carbon emission datasets for finding the recorded carbon emission levels for previous years and forecast future carbon emission levels. The prediction of future events is determined by considering historic events using an ARIMA machine learning model. The obtained result shows that carbon emissions are periodically increasing at a higher rate which leads towards the usage of energy efficient and green computing technologies. Future work includes:

- Determining the correlation between the heat and electricity potential role for carbon emissions from different computing technologies.
- Analyzing green computing technologies and equipment to determine their efficiency regarding the environment.

REFERENCES

Bas, E., Yolcu, U., & Egrioglu, E. (2020). Intuitionistic fuzzy time series functions approach for time series forecasting. *Granular Computing*, 2, 1–11.

Chen, H., Wang, G., Li, P., Zhao, S., Yang, X., & Sibinde, D. (2020). Forecast on mid-and long-term energy and emission in Guangdong Province. *IOP Conference Series: Earth and Environmental Science, 495*(1), 012–079.

Demirci, M. (2015). A survey of machine learning applications for energy-efficient resource management in cloud computing environments. *International Conference on Machine Learning and Applications (ICMLA),* 1185–1190.

Dritsaki, M., & Dritsaki, C. (2020). Forecasting European union CO2 emissions using autoregressive integrated moving average-autoregressive conditional heteroscedasticity models. *International Journal of Energy Economics and Policy, 10*(4), 411–423.

Fatima, S., Ali, S. S., Zia, S. S., Hussain, E., Fraz, T. R., & Khan, M. S. (2019). Forecasting carbon dioxide emission of Asian countries using ARIMA and simple exponential smoothing models. *International Journal of Economic and Environmental Geology, 10*(1), 64–69.

Gai, K., Qiu, M., Zhao, H., Tao, L., & Zong, Z. (2016). Dynamic energy-aware cloudlet-based mobile cloud computing model for green computing. *Journal of Network and Computer Applications, 59*, 46–54.

Kansal, N. J., & Chana, I. (2012). Cloud load balancing techniques: A step towards green computing. *IJCSI International Journal of Computer Science Issues, 9*(1), 238–246.

Li, Z., Liu, G., & Jiang, C. (2020a). Deep representation learning with the full center loss for credit card fraud detection. *Transactions on Computational Social Systems, 7*(2), 569–579.

Li, Y., Wei,bY., & Dong, Z. (2020b). Will China achieve its ambitious goal?—Forecasting the CO2 emission intensity of China towards 2030. *Energies, 13*(11), 2924.

Lotfalipour, M. R., Falahi, M. A., & Bastam, M. (2013). Prediction of CO2 emissions in Iran using grey and ARIMA models. *International Journal of Energy Economics and Policy, 3*(3), 22–29.

Malik, A., Hussain, E., Baig, S., & Khokhar, M. F. (2020). Forecasting CO2 emissions from energy consumption in Pakistan under different scenarios: The China–Pakistan economic corridor. *Greenhouse Gases: Science and Technology, 10*(2), 380–389.

Panigrahi, S., & Behera, H. S. (2020). A study on leading machine learning techniques for high order fuzzy time series forecasting. *Engineering Applications of Artificial Intelligence, 87*, 1–11.

Qin, Q., He, H., Li, L., & He, L. Y. (2020). A novel Decomposition-Ensemble based carbon price forecasting model integrated with local polynomial prediction. *Computational Economics, 55*(4), 1249–1273.

Riaz, M., Bhatti, Z., & Mushtaq, S. (2020). Efficiency test of forecasts: An illustration for carbon emission. *Journal of Business and Social Review in Emerging Economies, 6*(2), 931–948.

Saha, B. (2014). Green computing. *International Journal of Computer Trends and Technology (IJCTT), 14*(2), 46–50.

Sarkar, S., & Misra, S. (2016). Theoretical modelling of fog computing: a green computing paradigm to support IoT applications. *Iet Networks, 5*(2), 23–29.

Soomro, T. R., & Sarwar, M. (2012). Green computing: From current to future trends. *World Academy of Science, Engineering and Technology, 63*, 538–541.

19 DWT and SVD-Based Robust Watermarking Using Differential Evolution with Adaptive Optimization

Meenal Kamlakar, Chhaya Gosavi, Vaidehee Salunkhe, Priyanka Bagul, Aishwarya Keskar, and Shweta Barge

Cummins College of Engineering for Women, Pune

CONTENTS

19.1 Introduction to Digital Watermarking 297
19.2 Background Work 298
19.3 Materials and Methods 298
  19.3.1 Discrete Wavelet Transform (DWT) 298
  19.3.2 Singular Value Decomposition (SVD) 298
  19.3.3 Differential Evolution (DE) 300
  19.3.4 Embedding Algorithm 301
  19.3.5 Extraction Algorithm 302
19.4 Results 303
  19.4.1 Imperceptibility Analysis 304
  19.4.2 Robustness Analysis 305
19.5 Concluding Remarks 309
References 310

19.1 INTRODUCTION TO DIGITAL WATERMARKING

The security of digital media on the Internet is extremely important as it is very easy nowadays to duplicate, manipulate, and distribute digital information without any loss in quality. This creates problems regarding the authentication of ownership rights. Digital watermarking is a solution for this issue. It can be interpreted as embedding ownership information in the host multimedia data. A digital watermark has three properties: robustness, imperceptibility, and security. However, there is a

DOI: 10.1201/9781003146711-19

certain tradeoff between robustness and imperceptibility. An invisible watermark has a low scaling factor and thus it has to compromise on robustness. Also a robust watermark has a high scaling factor and thus has to compromise on invisibility. The cost of embedding and extraction also increases as invisibility increases. Our aim is to achieve a balance between invisibility and robustness of the watermark using machine learning algorithms (Kapse et al., 2018).

19.2 BACKGROUND WORK

This section describes the previous work which has been done on digital watermarking by using techniques, tools, various algorithms, and their results. Vishwakarmaa and Sisaudiaa (2018) proposed a semi-blind digital watermarking technique using a discrete cosine transform (DCT), differential evolution (DE), and a kernel extreme learning machine (KELM). They applied these techniques to gray scale images. Poonam (2018) presented a combination of a discrete wavelet transform (DWT) and a singular value decomposition (SVD). Zear et al. (2017) proposed that robustness is achieved by using a fusion of DWT, SVD, and back propagation neural network (BPNN)-based image watermarking techniques which offer a good tradeoff between important factors. Jabade and Gengaje (2016) analyzed and provided a comprehensive study of possible geometric attacks on a watermarked image with mathematical modeling. Saini and Shrivastava (2015) presented a survey of image watermarking techniques in detail. Darshana Mistry (2014) evaluated the performance of DCT and DWT, where embedding of a watermark is done in DCT as well as DWT. These results are compared with spatial domain encoding. Vaishnavia and Subashinib (2014) proposed color image watermarking methods for robustness and invisibility. Table 19.1 summarizes this literature review.

19.3 MATERIALS AND METHODS

19.3.1 Discrete Wavelet Transform (DWT)

DWT is ideally the best technique for performing analysis of frequency for signal processing. DWT decomposes the original image into four frequency sub-bands, namely LL, HH, HL, and LH. LL is the low level sub-band and the remaining three are high level sub-bands. LL can be further decomposed into four sub-bands, and this can be done repetitively. The number of decompositions done on the sub-band determines the level of DWT in the image. The low frequency component (i.e., LL) contains most of the information which is sufficient to represent most of the original image.

19.3.2 Singular Value Decomposition (SVD)

SVD decomposes a real matrix M of size m and n into a product of three matrices $M = USV^T$ where U and V^T are $m \times m$ and $n \times n$ orthogonal matrices, respectively. S is an m x n diagonal matrix. The singular values of A are the diagonal elements and are nonzero. When the rank of M is r, $S = diag(\sigma 1, \sigma 2, \ldots, \sigma n)$ where $\sigma 1 \geq \sigma 2 \geq \ldots \geq \sigma r \geq \sigma r + 1 \geq \ldots \geq \sigma n = 0$.

TABLE 19.1
Summary of Literature Survey

Reference	Methods Followed	Focus	Advantages	Limitations	PSNR%
(Vishwakarmaa & Sisaudiaa, 2018)	JDE-KELM, JDCT	Prediction of coefficients using KELM nonlinear regression model, DE optimizes the multiple scaling factor	Edge of watermark remains intact adding to keeping of image	The scheme fails for rotation attack	44.9
(Poonam, 2018)	DWT-SVD, DCT	Performance analysis of an invisible watermark	Better quality , invisible and secure watermarking, secure	Use of DCT compromises on robustness	DWT-SVD : 70.46 DCT: 58.14
(Cui et al., 2018)	Spatial domain, frequency domain	Different digital watermarking techniques	NA	Does not provide balance between robustness and imperceptibility	—
(Al-Qaheri et al., 2010)	Ant colony optimization, DCT, light gradient boosting	Time consumption, error between the optimum solution and the predicted solution, watermark quality and robustness	Less execution time	Optimized techniques not used	46.38
(Kamlakar et al., 2012)	SVD	Secure and robustness of watermark, false positive issue is resolved	Secure and robust watermark sustaining most of the attacks	Small watermark image used	47.3
(Gosavi & Mali, 2017)	DCT, timestamp	Secure and robustness of watermark	Secure and robust watermark sustaining most of the attacks	—	>48
(Kamlakar Gosavi 2013)	SVD	Increase in payload for watermark, false positive issue is resolved	Secure and robust watermark sustaining most of the attacks with bigger watermark images	—	40.3

Notes: PSNR, peak signal-to-noise ratio; DE-KELM, differential evolution and kernel extreme learning machine; DCT, discrete cosine transform.

19.3.3 Differential Evolution (DE)

DE is a genetic algorithm which has been widely used in optimization studies. DE is a population-based meta-heuristic computer technology that is used to solve optimization problems. A DE algorithm needs the random selection of two individuals from an initial population. A vector difference is generated by subtraction and then another individual who was previously not chosen is again randomly considered to be added up with the vector difference. This step obtains a new individual called a donor vector. To generate the trial individual in the generated population, a target individual who is randomly chosen from the current population according to certain rules is swapped with this donor; amongst the target individual and trial individual, the one who competes with the other at the end can enter the next generation. This process goes on until the population reaches the required parameters. Figure 19.1 shows the flow diagram for differential evolution.

Figure 19.2 shows the classification of watermarking techniques done using various criteria. There are three major domains of digital watermarking. They are classified according to working domain, human perception, and type of document.

There are two types of working domain – spatial and transform. In a spatial domain watermarking can be done using either the least significant bit (LSB) or patchwork. In a transform domain watermarking can be done in three ways – by DCT, discrete Fourier transform (DFT), and DWT. Human perception can be either visible or invisible. Watermarking by type of document can be performed on either text, audio, image, or video.

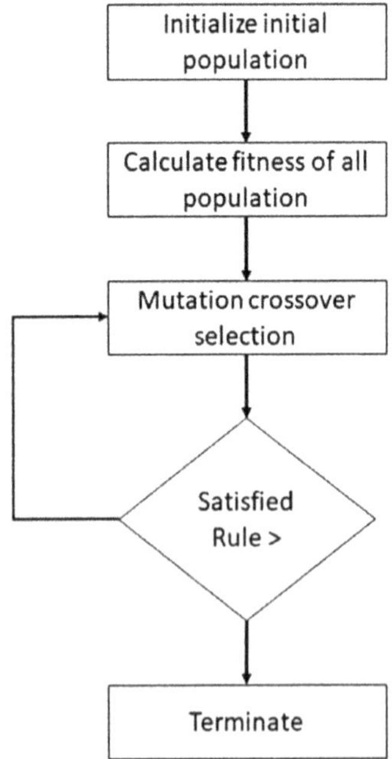

FIGURE 19.1 Flow diagram of a differential evolution algorithm.

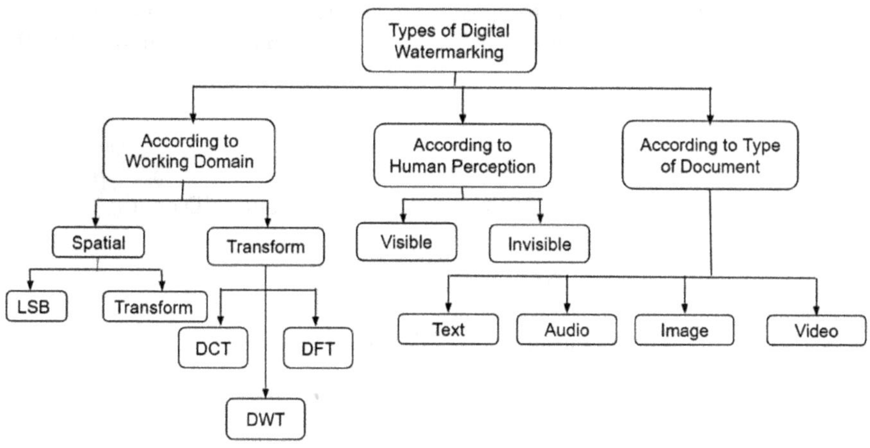

FIGURE 19.2 Types of watermarking techniques.

19.3.4 Embedding Algorithm

The watermark embedding process occurs in these steps (see also Figure 19.3):

1. Decompose the original image into the four sub-bands (LL, HL, LH, HH) by applying three-level DWT and extracting the LL sub-band.
2. Apply SVD on the LL sub-band and obtain the singular value S.

$$I = U_{LL}\ S_{LL}\ V_{LL}{}^{T}$$

3. Similarly, decompose the watermark image W into the four sub-bands (LL, HL, LH, HH) and extract the LL sub-band.

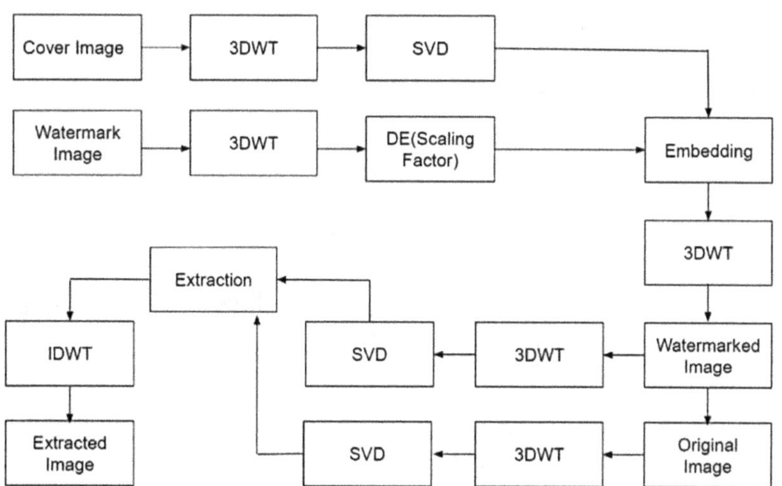

FIGURE 19.3 Flow diagram of embedding algorithm.

4. Get an optimal scaling factor q using the differential evolution algorithm.
5. Replace the singular value of the original image and the LL sub-band of W, using the scaling factor q.

$$E_{wmk} = S_{LL} + qW_{LL}$$

6. Compute the new values of SLL and then perform anti-SVD on it.

$$I_{wmk} = U_{LL}\, E_{wmk}\, V_{LL}{}^{T}$$

7. Finally, perform three-level inverse DWT to obtain the watermarked image Z.

19.3.5 Extraction Algorithm

Figure 19.4 shows a block diagram of the watermark extraction process. The steps are:

1. Using three-level DWT divide the watermarked image Z into HH, HL, LH, and LL sub-bands. Extract the LL sub-band for the extraction of the watermark.
2. Similarly, get the LL sub-band of the watermark image by performing three-level DWT.
3. Using the following equation, change the singular values of the watermarked image Z, and the original image I.

$$E_{LLL}{}^{*} = \left(Z_{LL} - S_{LL}\right)/q$$

4. Obtain the LL component of the extracted watermark W' and perform anti-SVD on it.

$$W'_{LL} = U_{wk}\, E_{LL} * V_{wk}^{T}$$

5. Perform three-level inverse DWT to extract the watermark image W'.

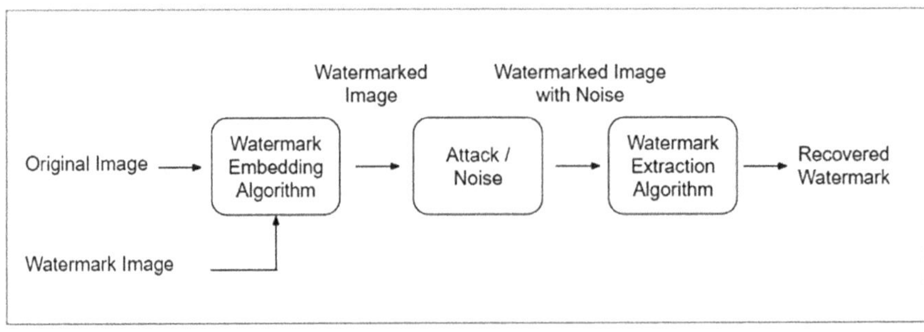

FIGURE 19.4 Flow diagram of extraction algorithm.

19.4 RESULTS

Experiments were done on the grayscale Lena and baboon cover images and leaf as watermark images whose resolution is taken as 512 × 512 pixels as shown in Figure 19.5. Snaps of the cover image and host image are shown below. The peak signal noise ratio (PSNR) and the normalized correlation (NC) were used to measure the invisibility and robustness of the proposed method. In order to test robustness, the watermarked image was subjected to various common attacks, such as:

a. Gaussian noise (GN).
b. Salt and pepper noise (SPN).
c. Rotation (RT).
d. Crop (CR).
e. Poisson noise (PN).
f. JPEG compression (JPEG).

Figure 19.6 shows watermark images embedded in the cover images.

Table 19.2 shows the PSNR and NC values of the Lena and baboon images before any attack.

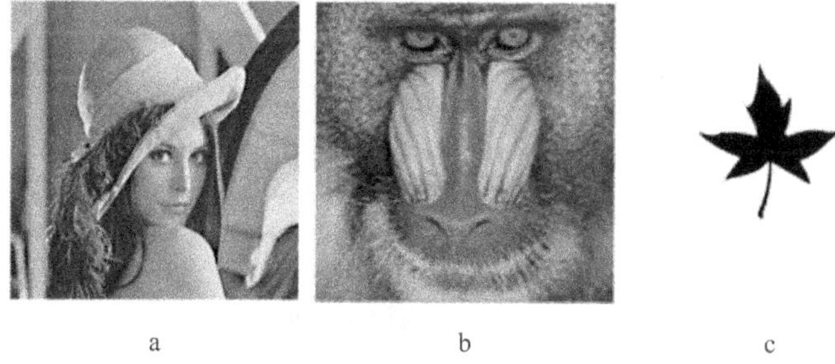

a b c

FIGURE 19.5 Cover images. (a) Lena; (b) baboon; (c) watermark image.

a

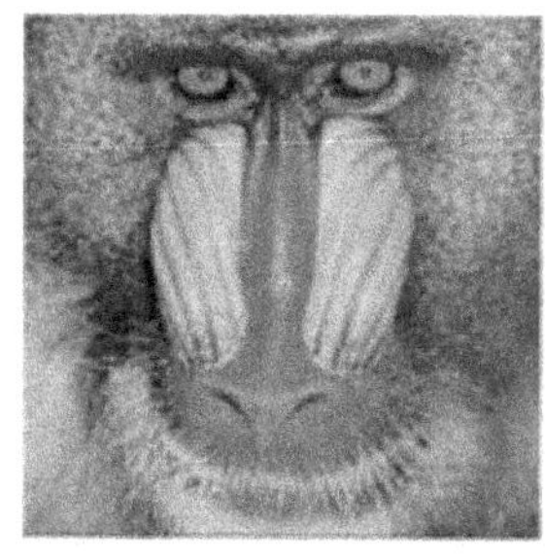

b

FIGURE 19.6 Watermarked images. (a) Lena; (b) baboon.

TABLE 19.2
Cover Images Before Attack

	PSNR	NC
Cover image Lena	55.3	0.6
Cover image baboon	53.2	0.47

19.4.1 Imperceptibility Analysis

The PSNR block computes the PSNR in decibels, between two images. PSNR is the most popular and widely used objective image quality metric; it is calculated as follows:

$$PSNR = 10log10\frac{\left(X_{MAX}\right)^2}{\left(\frac{1}{n^2}\right)\Sigma\Sigma\left(X_{ij} - W_{i,j}\right)^2}$$

where:

X is the cover image;
W is the watermarked image;
X_{MAX} is the maximum possible value of the original image X;
n is the height or weight of the image.

A high PSNR value is desired for better invisibility. If PSNR > 30 then according to the human visual system the difference between the original image and the watermarked image is negligible. Tables 19.3 and 19.4 display the results.

Similarity between the two images is measured using NC, which is used to evaluate the efficiency of watermark extraction. The higher the NC of the watermark, the more similar is the extracted watermark. Tables 19.5 and 19.6 show the NC values of the watermarked images and the extracted watermarks.

TABLE 19.3
PSNR Values for Lena

Serial no.	Type of Attack	PSNR for Cover Image	PSNR for Extracted Watermark
1.	Gaussian noise	30.8	61.6
2.	Salt and pepper	22.45	40.0
3.	Rotation attack	42.6	44.0
4.	Cropping	12.3	42.0
5.	Poisson noise	43.2	15.6
6.	Compression	48.0	47.7

TABLE 19.4
PSNR Values for the Baboon

Serial no.	Type of Attack	PSNR for Cover Image	PSNR for Extracted Watermark
1.	Gaussian noise	24.9	61.64
2.	Salt and pepper	22.6	31.6
3.	Rotation attack	42.4	37.2
4.	Cropping	13.8	22.8
5.	Poisson noise	14.06	21.3
6.	Compression	53.2	36.4

TABLE 19.5
NC Values for Lena

Serial no.	Type of Attack	NC for Cover Image	NC for Extracted Watermark
1.	Gaussian noise	0.429	0.52
2.	Salt and pepper	0.4	0.47
3.	Rotation attack	0.43	0.532
4.	Cropping	0.43	0.48
5.	Poisson noise	0.52	0.472
6.	Compression	0.43	0.47

TABLE 19.6
NC values for the baboon

Serial no.	Type of Attack	NC for Cover Image	NC for Extracted Watermark
1.	Gaussian noise	0.43	0.52
2.	Salt and pepper	0.42	0.53
3.	Rotation attack	0.43	0.53
4.	Cropping	0.43	0.53
5.	Poisson noise	0.43	0.52
6.	Compression	0.42	0.53

19.4.2 Robustness Analysis

To determine robustness, the watermarked image was subjected to various attacks. Tables 19.3–19.6 show the robustness of the given methods. The standard image of Lena goes through the six different attacks mentioned in each table. The extracted watermark images have a very high similarity to the original watermark; the difference between the two cannot be discerned by the naked eye. The proposed algorithm shows the best invisibility and robustness with respect to the original cover image and the attacks.

Figure 19.7 shows Lena's watermarked images and the baboon's watermarked images that have been tested using various kinds of attack, such as GN, RT, and CR

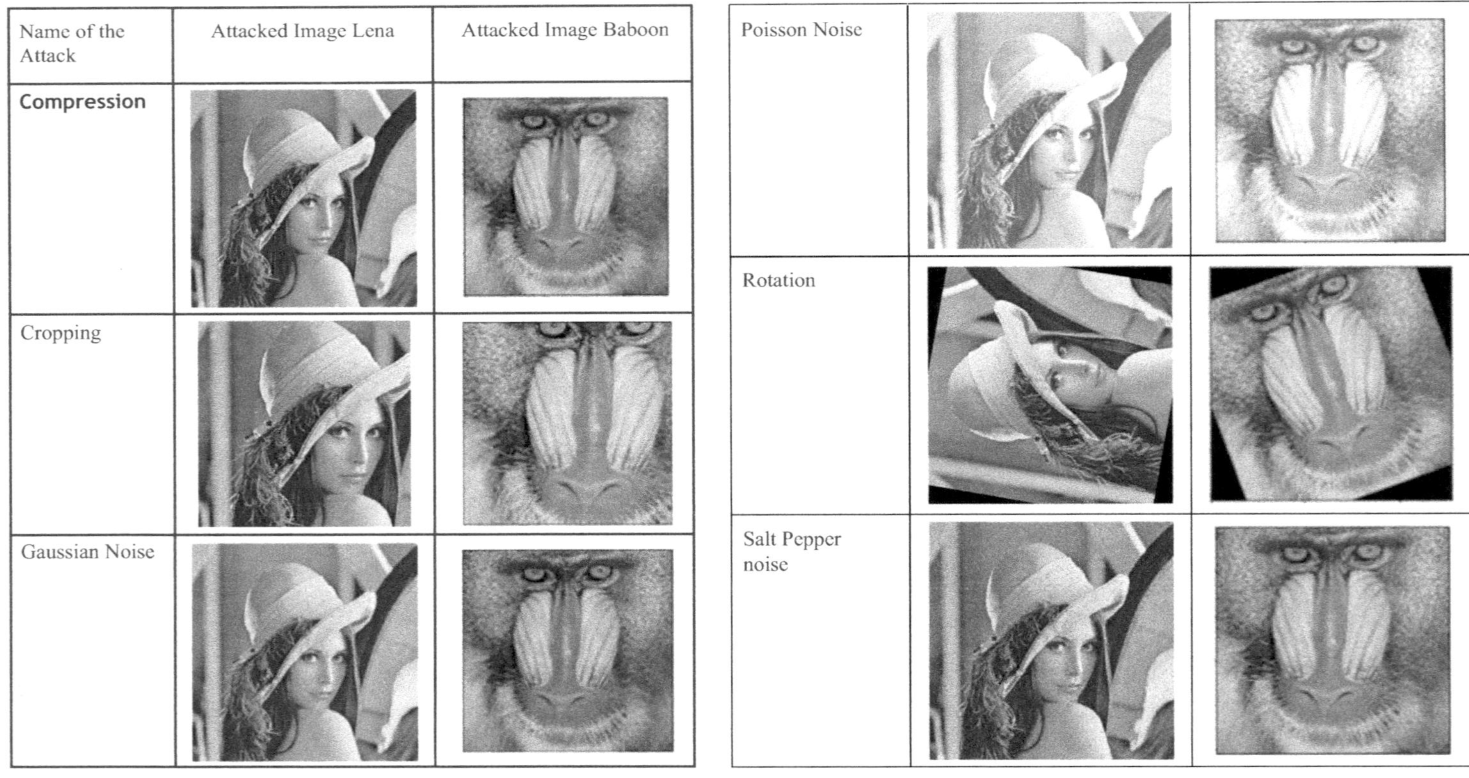

Name of the Attack
Attacked Image Lena
Attacked Image Baboon
Compression
Cropping
Gaussian Noise
Poisson Noise
Rotation
Salt Pepper noise

FIGURE 19.7 Attacked watermarked images for Lena and the baboon.

Attack	Extracted Watermark
Compression	
Cropping	
Gaussian Noise	
Poisson Noise	
Rotation	
Salt pepper noise	

FIGURE 19.8 Watermark extracted after various attacks.

Figure 19.8 shows the extracted watermark images after the test was performed on them. These extracted watermark images look similar to the original watermark image to a great extent.

The results show that the proposed method can extract the embedded watermark successfully without much tampering and with good visibility.

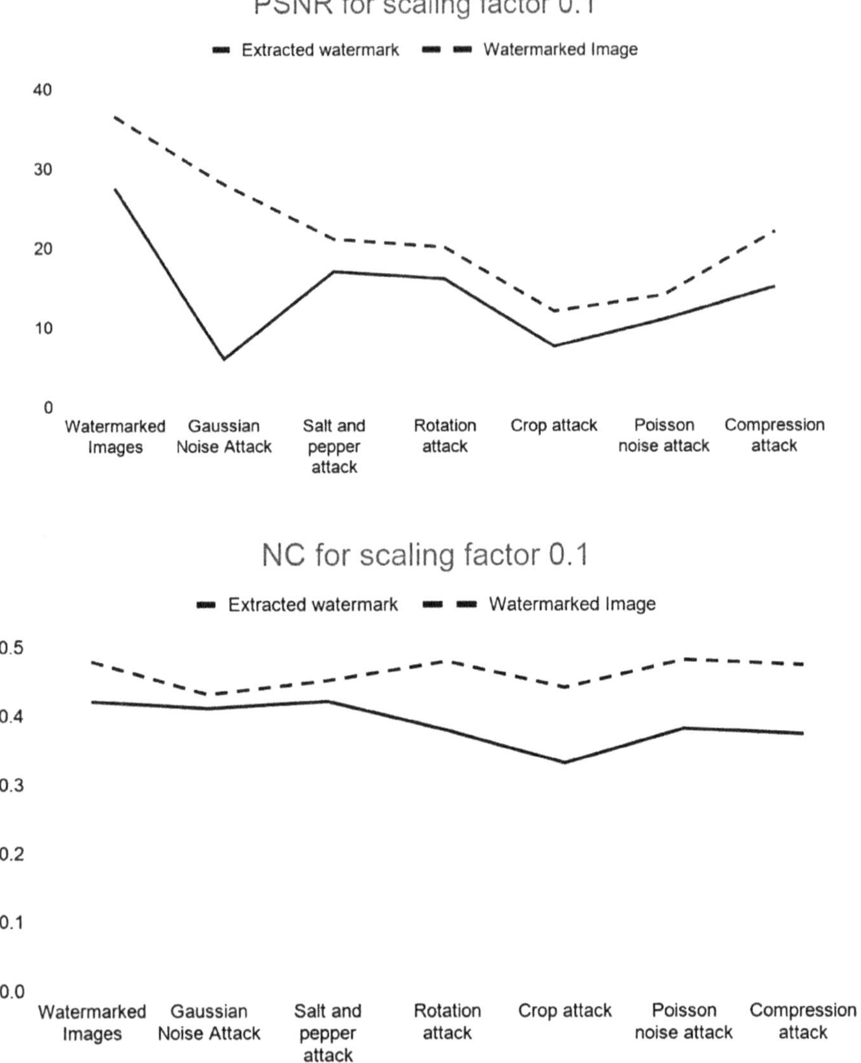

FIGURE 19.9 Graph of results before differential evolution has been applied.

Figure 19.9 shows graphs of the experimental results of watermarking when the scaling factor is set to 0.1

Figure 19.10 shows graphs of the experimental results of watermarking when the scaling factor is set using differential evolution. There is a significant rise in the values of PSNR and NC, proving that differential evolution is an efficient algorithm for finding a scaling factor.

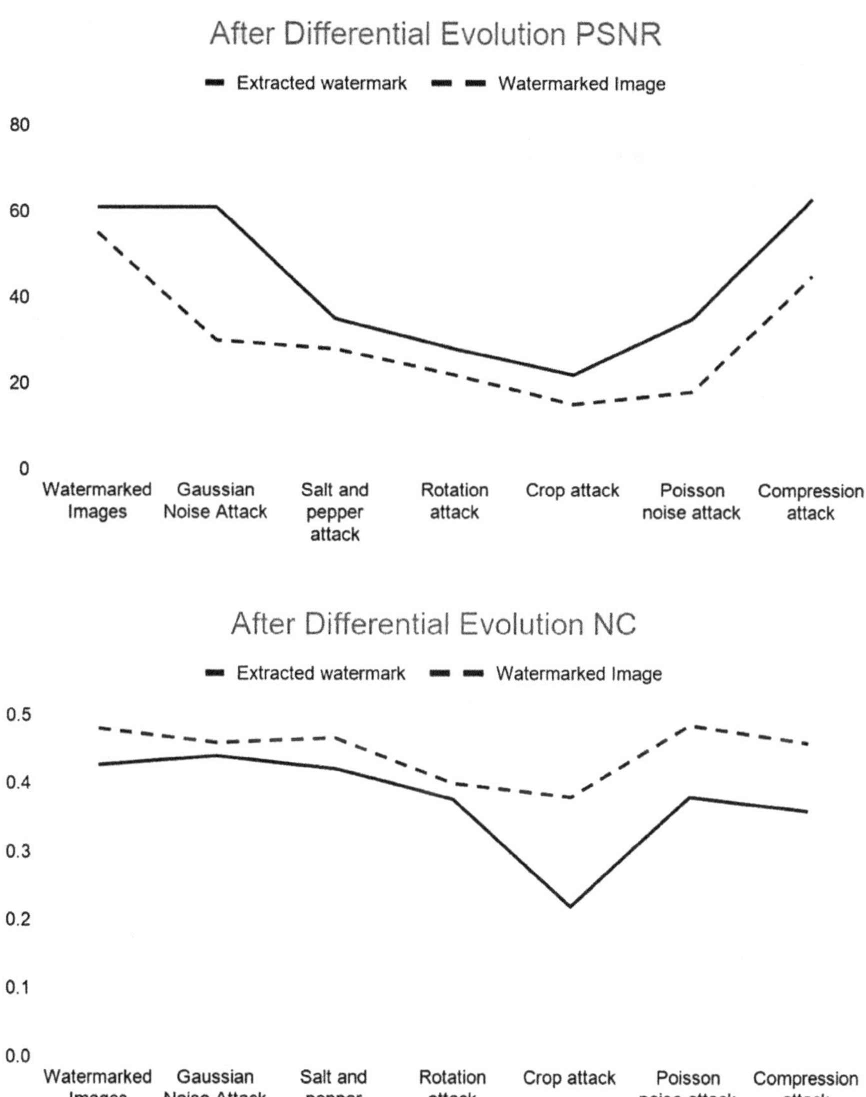

FIGURE 19.10 Graph of results after differential evolution has been applied.

19.5 CONCLUDING REMARKS

The proposed method of digital watermarking employs a wavelet domain using differential evolution and SVD. Differential evolution was used to find the appropriate scaling factor, which gives a balance between robustness and invisibility of a watermarked image. The performance of the proposed method was checked with respect to various attacks. This improved the results of PSNR and NC by 40%. The discussed algorithm provides the best imperceptibility and robustness for the cover image and attacks with the help of machine learning. It can also be used for IoT applications.

The algorithm can be extended to do watermarking on video. Timestamps can also be added to the watermarking process. Watermarks could also contain copyright information and video signatures which could further increase the probability of rightful authentication. The payload of the watermark can also be improved.

REFERENCES

Al-Qaheri, H., Mustafi, A., & Banerjee, S. (2010). Digital watermarking using ant colony optimization in fractional Fourier domain. *Journal Ubiquitous International, (JIHMSP), 1*(3), 179–189.

Cui, X., Niu, Y., Zheng, X., & Han, Y. (2018). An optimized digital watermarking algorithm in wavelet domain based on differential evolution for color image, *Plos One Journal*, 13(5): e0196306. https://doi.org/10.1371/journal.pone.0196306.

Gosavi, C. S., Mali, S. N. (2017). Secure, robust video watermarking to prevent camcorder piracy. *Indian Journal of Science and Technology, 10*(18), 1–10. https://doi.org/10.17485/ijst/2017/v10i18

Jabade, V., & Gengaje, S. (2016). Modelling of geometric attacks for digital image watermarking. *International Journal of Innovations in Engineering Research and Technology (ijiert), 3*(3), 1–8.

Kamlakar, M., & Gosavi, C. (2013). Various techniques using block based SVD in different colour channels for secure, robust and more efficient video watermarking. *International Journal of Engineering Research & Industrial Applications (IJERIA), 6*(3), 139–149.

Kamlakar, M., Gosavi, C. & Patankar, A. (2012). Secure, robust and more efficient watermarking for video using SVD for protecting rightful ownership. *International Journal of Engineering Research & Industrial Applications (IJERIA),* 5(2), 51–60.

Kamlakar, M., & Patankar, A. (2012). Single channel watermarking for video using block based SVD. *International Journal of Advances in Computing and Information Researches, 1*(2), 2277–4068.

Kapse, A. S., Belokar, S., Gorde, Y., Rane, R., & Yewtkar, S. (2018). Digital image security using digital watermarking. *International Research Journal of Engineering and Technology (IRJET), 5*(3), 163–166.

Mistry, D. (2014). Comparison of digital water marking methods. *International Journal of Computer Science Trends and Technology (IJCST), 2*(9), 2905–2909.

Poonam, S. M. A. (2018). A DWT-SVD based robust digital watermarking for digital images. *International Conference on Computational Intelligence and Data Science, 132*, 1441–1448. https://doi.org/10.1016/j.procs.2018.05.076

Saini, L. K., & Shrivastava, V. (2015). A survey of digital watermarking techniques and its applications. *International Journal of Computer Science Trends and Technology (IJCST),* 2(3) 70–73.

Vaishnavia, D., & Subashinib, T.S. (2014). Robust and invisible image watermarking in RGB color space using SVD. *International Conference on Information and Communication Technologies*. 46, 1770–1777.

Vishwakarmaa, V. P., & Sisaudiaa, V. (2018). Gray-scale image watermarking based on DE-KELM in DCT domain. *International Conference on Computational Intelligence and Data Science- ICCIDS*, 132, 1012–1020. https://doi.org/10.1016/j.procs.2018.05.017

Zear, A., Singh, A. K., & Kumar, P. (2017). Robust watermarking technique using back propagation neural network: A security protection mechanism for social applications. *International Journal of Information and Computer Security, 9*(1/2), 20–35. https://doi.org/10.1504/IJICS.2017.082837

20 A Novel Framework Based on a Machine Learning Algorithm for the Estimation of COVID-19 Cases

Rashmi Welekar, Sharvari Tapase, Shubhi Bajaj, Isha Pande, Abhishek Verma, Ashutosh Katpatal, and Vaibhav Mishra

Shri Ramdeobaba College of Engineering and Management, Nagpur

CONTENTS

20.1 Introduction 311
20.2 Literature Review 312
20.3 Principles of the SIR Model 314
20.4 Implementation of the Back-end 315
20.4.1 Data Understanding 316
20.4.2 Data Preparation 316
20.4.3 Understanding Exploratory Data Analysis (EDA) 317
20.4.4 Modeling Spread 317
20.4.5 Evaluation Walk-through 317
20.4.6 Overfitting 318
20.4.7 SIR Modeling 319
20.5 User Interface (UI) Implementation 322
20.6 Analysis 323
20.6.1 Hover-over India Map 323
20.6.2 Statewise updates 325
20.7 Results 328
References 329

20.1 INTRODUCTION

The first reported case of COVID19 was reported on 31 December 2019 in Wuhan, China. Outside China the first case reported was in Thailand on 13 January 2020.

DOI: 10.1201/9781003146711-20

After this the ongoing outbreak spread to more than 50 countries. The WHO declared this pandemic to be an emergency situation on 30 January 2020. The situation in India was also very concerning as with other countries. Because of its asymptomatic nature, it is difficult for people to predict the behavior of the virus as well as its spread. However, it is also difficult for us to know when cases will decrease.

Our aim is to predict when the number of cases will reduce in India. For this we use the susceptible-infectious-removed (SIR) model. First, we predict the decrease at the country level. We then expand our scope to the state level. We check the accuracy of our model by training it on a specific state where there already has been a visible decrease in cases.

The reason why we attempt this is because the life of the individual has come to a standstill. Our results will be advantageous to the Government of India, administrative units, as well as various frontline health workers; they can then use resources efficiently on the basis of how the reduction in cases is taking place, which may be at a slow or fast pace. Our model may also help various organizations and companies that have started manufacturing masks, PPE kits, sanitizers, or other sanitary products, so that they know when to reduce or stop production in order to avoid facing heavy losses. It will also be helpful for the ministries to decide when to enforce lockdowns or to reopen schools, colleges, and so on in any particular area. In a precise way, it becomes easier to predict when conditions will be normal again.

20.2 LITERATURE REVIEW

There are various research papers published which indicate that research on COVID-19 is going on across the entire globe. Using the linear regression model with a fast artificial neural network (FANN) for the prediction, a machine learning model for the COVID-19 pandemic in India was developed by Sujath et al. (2020), which detects rising new cases. Excess mortality in England was analyzed by Peckham (2020) to see the difference of death cases in the presence of COVID-19 and its absence.

We can also find similarities between the coronavirus and the SARS virus that was identified in 2003 and in the study by Vasantha et al. (2020) who analyzed Indian publications on SARS-CoV-2 as found in the WHO COVID-19 database. There has been a rise in the publishing of journals related to coronavirus recently, with authors belonging to Indian institutes like AIIMS and ICMR.

In "CoronaTracker: World-wide COVID-19 Outbreak Data Analysis and Prediction" (Hamzah et al., 2020), it is observed that a corona tracker was created, through which the cases of China can be tracked. To observe the effect of this virus on the mental health of people, the authors use sentiment analysis, and perform prediction with the help of a SEIR model.

Why the prediction of COVID-19 cases is so difficult is addressed by Roda et al. (2020). They consider that the SIR model works better than the other models and have predicted and stated that complex models will not guarantee more reliability or accuracy, as compared to results from simpler models.

In a similar paper, Di Girolamo and Reynders (2020) study the characteristics of the articles or reports of the first three months of the coronavirus epidemic. They searched articles having "COVID" or "COVID-19" as the keywords, and found many.

Yang et al. (2020) derived the epidemic's curve by integrating the data of population migration after and prior to 23 January 2020 and the latest version of COVID-19 data in the susceptible-exposed-infected-recovered algorithm. They used artificial intelligence trained on SARS data to determine the increase of the pandemic.

From Cooper et al. (2020), we can see that with the help of the SIR model the effects of coronavirus on various countries like India, South Korea, and the USA have been noted. Also, the difference in the number of cases is observed in the time period January 2020 to June 2020, that is, prior to and after imposing strict guidelines.

We can see from Hoque et al. (2020) that the misconceptions about the effect of weather on the virus should be rectified, especially in the largely affected regions like the USA. Using a prediction model on the USA, similar conclusions were drawn regarding India too. The relationship between weather conditions and the virus help to identify the high-risk zones in India.

Zheng et al. (2020) address how the coronavirus affects people already suffering from certain ailments. SARS-CoV-2 infects the body's cells through the AC2E cells, leading to the COVID-19 fever, leading to a near to permanent damage to the affected person's cardiovascular system.

Fang et al. (2020) show that patients suffering from diabetes and some cardiovascular diseases are at an increased rate of infection. Understanding chest CT-scan features of the coronavirus affected victims will help to detect the virus earlier and to treat the affected patients earlier. Chest CT scans show the high possibility of the presence of COVID-19 pneumonia if the person has come in contact with a COVID-19 affected person.

Remuzzi and Remuzzi (2020) was published when there was a surge in the number of cases in Italy, after the surge in the number of cases in China. Also, the exponential growth in the number in Italy could not be properly predicted and managed due to the difference in the social distancing measures and the capacity to arrange facilities, as compared to that of China.

Vasantha (2020) reports similarities between the SARS virus which was identified in 2003 and coronavirus, also mentioned in Indian journals on SARS-CoV-2 in WHO data on coronavirus. In the past few months, there has been an increase in articles related to COVID-19.

The SpO2 level (the percentage saturation of oxygen in the blood) is important in determining the presence of coronavirus in the body of a person. The SpO2 detection module was produced in response to the importance of monitoring SpO2 levels in COVID-19 patients, which was suggested by Xie et al. (2020).

Baena-Díez et al. (2020) studied two parameters: age and the socio-economic status of the affected person. They observed that the virus had impacted the low-income group of the country. They suggest that the people who already are ill or have some sort of disease should be prioritized. They also suggest a healthcare strategy for the whole population, especially in the financially affected areas.

Ali et al. (2020) provide specific and precise information on coronavirus, high level techniques for the management and prevention of spread, and treatment strategies to help spread understanding and perception amongst people.

20.3 PRINCIPLES OF THE SIR MODEL

SIR (Figure 20.1) is a mathematical epidemiology model that computes, theoretically, the number of people contaminated with an infectious disease in a static dataset of people in a particular period.

The rate of infection (Equation 20.2), S, represents the rate of spread from an infectious to a susceptible individual (Equation 20.1). The recovery rate represents how much an infection can be recovered from over time (Equation 20.3).

The base SIR model is as follows:

$$\frac{\mathrm{d}S}{\mathrm{d}t} = -\beta SI \tag{20.1}$$

$$\frac{\mathrm{d}I}{\mathrm{d}t} = \beta SI - \gamma I \tag{20.2}$$

$$\frac{\mathrm{d}R}{\mathrm{d}t} = \gamma I \tag{20.3}$$

$$N = S(t) + I(t) + R(t) \tag{20.4}$$

where:

N is the total population;
S is the total number of susceptible individuals;
I is the total number of infected individuals;
R is the total number of recovered individuals;
dS, dI, and dR delta increase/decrease.

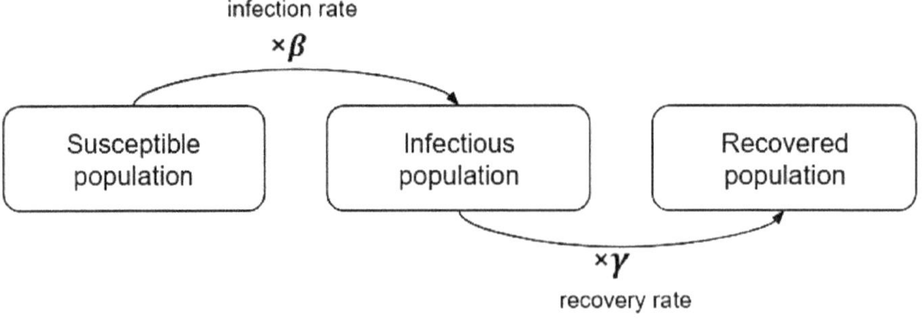

FIGURE 20.1 The SIR model concept.

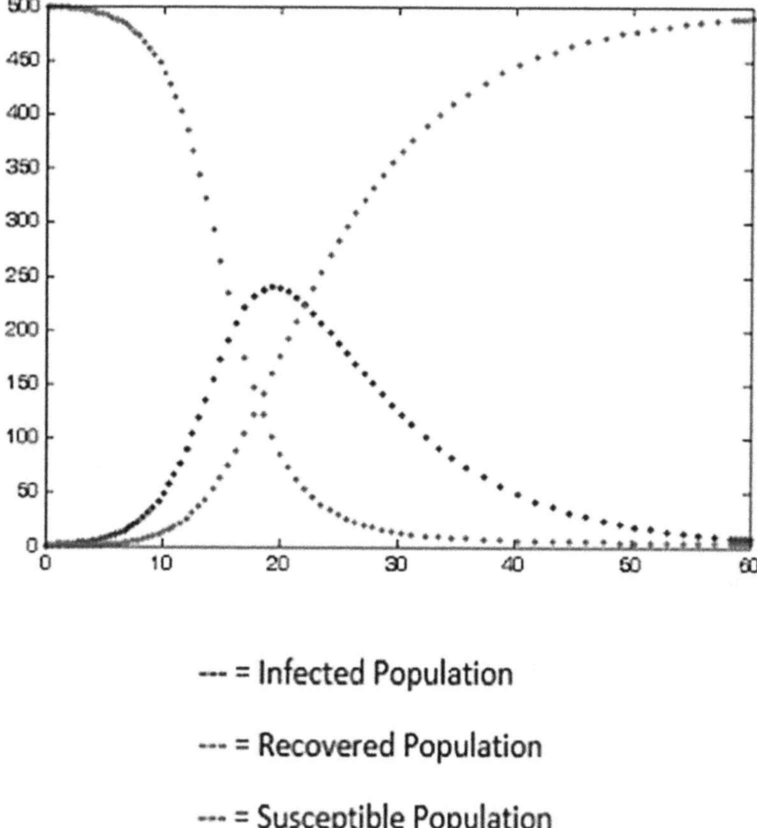

FIGURE 20.2 SIR ideal curves where the x-axis represents the number of days and the y-axis represents the population.

Here, it is assumed that the population is constant (Equation 20.4). Secondly the increased rate of the infected population is directly proportional to the contacts amongst the susceptible and infected population. This takes place at a fixed rate. A third assumption relates to the removal rate (i.e., recovery rate/death rate is constant). This gives us the graph shown in Figure 20.2.

20.4 IMPLEMENTATION OF THE BACK-END

For the implementation of the back-end we used a cross-industry standard process for data mining framework (Figure 20.3a). CRISP-DM is a process model that shows the basic approaches used by specialists. It's a highly used technique since it is very robust. The sequence of phases is highly iterative and not a stringent step-by-step approach; it also aligns well with agile development principles. The cycle shown in Figure 20.3a can also be depicted as shown in Figure 20.3b.

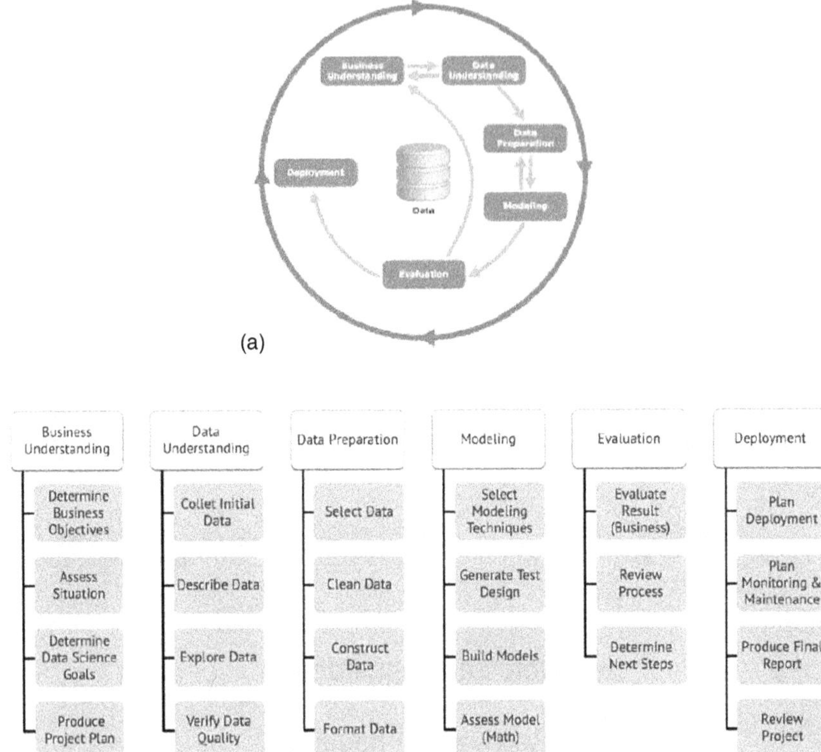

FIGURE 20.3 (a) CRISP-DM flow chart; (b) explanation of flow chart.

For implementation of our model, we used the ipy notebooks as follows.

20.4.1 Data Understanding

The data for our model was extracted from the Johns Hopkins University GitHub repository (CSSEGISandData/COVID-19). We cloned the GitHub repository and selected the important file for us. We then analyzed the structure of the data provided so that we could perform further modeling.

20.4.2 Data Preparation

The dates in the data were present as a column, but for modeling purposes we required the dates to be in rows, so we converted the data according to the required format using the pandas library. The data applied to all countries, but our main focus was India, so we extracted that data and that of a few other countries for comparison purposes.

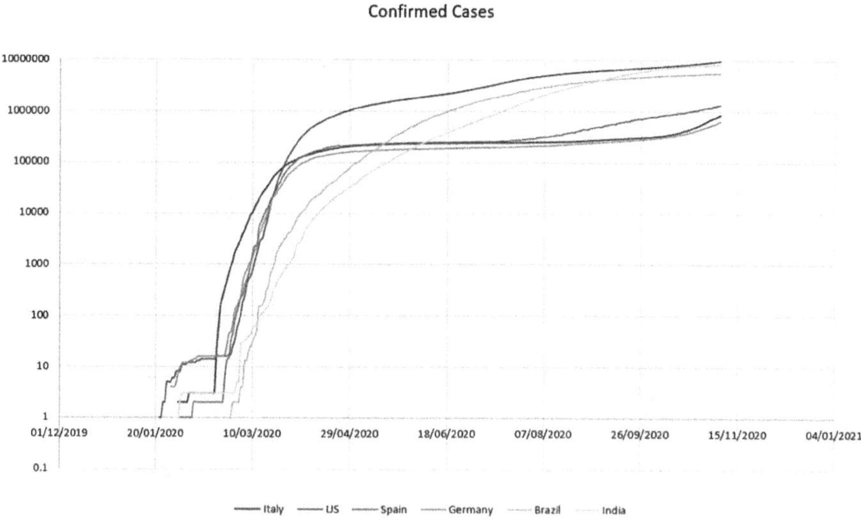

FIGURE 20.4 Confirmed cases, where the x-axis represents the dates and the y-axis represents the number of confirmed cases.

20.4.3 Understanding Exploratory Data Analysis (EDA)

Here we plotted the graph between the time and number of confirmed cases for all the extracted countries to study the trends of COVID-19 infection (Figure 20.4).

20.4.4 Modeling Spread

Now we calculated the rate at which the cases double for the countries (the doubling rate) and plotted it to obtain a visual representation (Figure 20.7). We then implemented a linear regression model, as from the doubling rate graph we saw that cases were doubling at a linear rate. But the implementation of this model yielded results that were not accurate, as observed in Figure 20.5.

The doubling rate is that at which the number of infected people increases or doubles. The decrease in the doubling rate indicates that the measures taken by government are effective. During analysis we found that the number of cases doubled every two days. Hence the doubling rate is considered to be 2. Figure 20.6 shows the doubling rate of five countries. The decrease in the doubling rate of Italy suggests that the measures taken by the respective government were efficient.

20.4.5 Evaluation Walk-through

We updated and verified the data and applied the slope calculation or doubling rate model on the worldwide data.

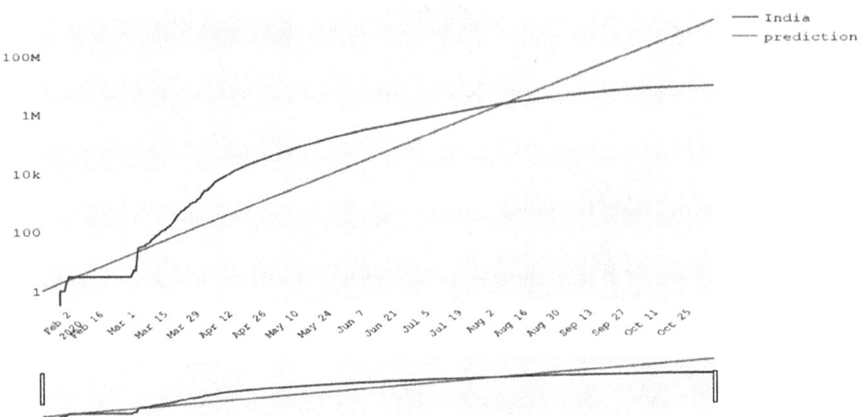

FIGURE 20.5 Linear regression curve, where the *x*-axis represents the dates and the *y*-axis the population.

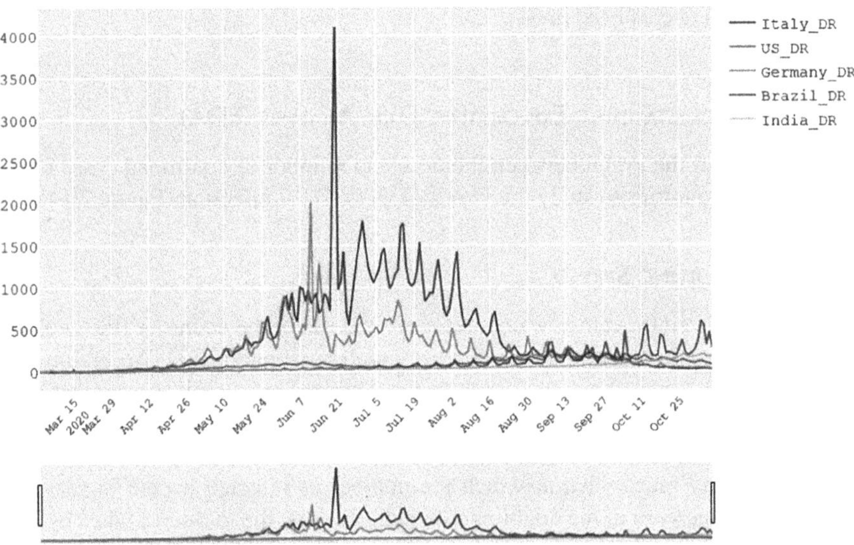

FIGURE 20.6 Doubling rate curve, where the *x*-axis represents the dates and the *y*-axis represents the doubling rate of confirmed cases.

20.4.6 Overfitting

We tried to fit the polynomial curves of different degree and calculate the mean absolute percentage error (MAPE) (Equation 20.5) values; the results obtained were not satisfactory for any polynomial. MAPE is an extension of the mean absolute error

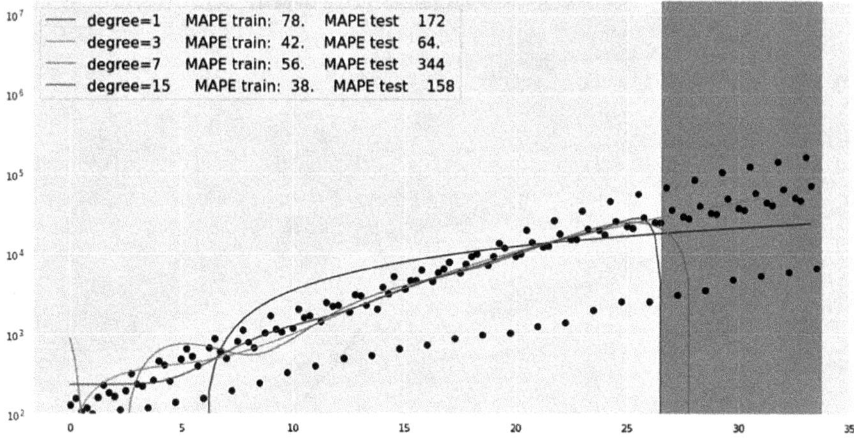

FIGURE 20.7 MAPE calculation, where the x-axis is the observed increase in cases and the y-axis is the forecast increase in cases.

(MAE) regression metric. It is used to find out how accurate the forecast is or by what percentage our forecast is off. MAPE is calculated by the following formula:

$$\text{MAPE} = \frac{1}{n}\sum_{i=1}^{n}\left|\frac{x_i - y_i}{x_i}\right| \tag{20.5}$$

where:

n is the total number of iterations;
i is the current iteration;
x_i is the actual value;
y_i is the forecast value.

The graph in Figure 20.7 indicates that the MAPE for degree 15 is 38, which is the least among other MAPEs calculated using degrees one, three, and seven, hence we go with a degree 15 equation to fold the data in 15 folds and use it for cross-validation.

20.4.7 SIR Modeling

Figure 20.8 represents the ideal curve of the SIR model. We implemented the differential equations of the SIR model (Equations 20.1, 20.2, and 20.3) and solved them using the scipy library and hence plotted the curve for the static data.

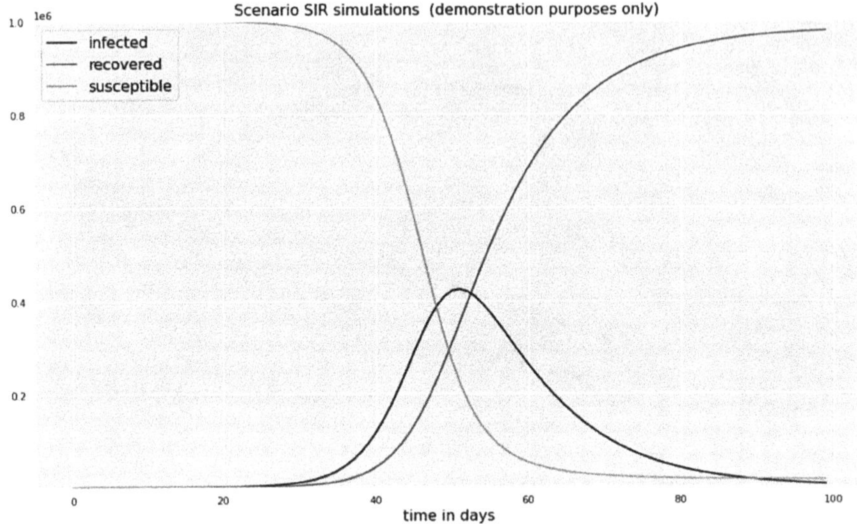

FIGURE 20.8 Ideal curve of the SIR model, where the x-axis represents the dates and the y-axis the population.

We integrated the differential equations using the following function:

CODE 20.1: HELPER FUNCTION TO USE SCIPY'S INTEGRATE. ODEINT().

```
    def fit_odeint(x, beta, gamma):
"'
helper function for the integration "'
 return integrate.odeint(SIR_model_t, (S0, I0, R0), t, args=(beta, gamma))
[:,1]      # we only would like to get dI
```

We obtained the optimal parametric values of β and ϒ as 0.1827 and 0.1125 respectively. Using these optimal values, we obtained the basic reproduction number R0 as 1.6234. The basic reproduction is the number of expected cases generated by a single active case.

Using these values, we now implemented the SIR model:

CODE 20.2(A): HELPER FUNCTION FOR SIR MODEL

```
def SIR_model_t(SIR,t,beta,gamma):
 "' Simple SIR model
  S: susceptible population
  t: time step, mandatory for integral.odeint I: infected people
```

```
R: recovered people
Beta: overall condition is that the sum of changes (differnces) sum up to 0
 dS+dI+dR=0
S+I+R= N (constant size of population)
"'
S,I,R=SIR
dS_dt=-beta*S*I/N0      #is the number of susceptible individual that
  become infected per day dI_dt=beta*S*I/N0-gamma*I
dR_dt=gamma*I
return dS_dt,dI_dt,dR_dt
```

followed by:

CODE 20.2(B): IMPLEMENTING SIR MODEL

```
SIR=np.array([S0,I0,R0])
propagation_rates=pd.DataFrame(columns={'susceptible':S0,
        'infected':I0, 'recoverd':R0})
for each_beta in pd_beta:
new_delta_vec=SIR_model(SIR,each_beta,gamma)
SIR=SIR+new_delta_vec propagation_rates=propagation_rates.
  append({'susceptible':SIR[0],
        'infected':SIR[1]*3.7,
'recovered':SIR[2]},
ignore_index=True)
```

Then, to plot the graph we used:

CODE 20.2(C): PLOTTING THE OUTPUT GRAPH

```
fig.ax1=plt.subplots(1,1)
ax1.plot(propagation_rates.index+8,propagation_rates.infected,label='infecte
  d',linewidth=3) t_phases=np.array([t_initial,t_intro_measures,t_hold,t_
  relax]).cumsum()
ax1.bar(np.arange(len(ydata)),ydata,width=0.8,label=' current infected
  India',color='r')
ax1.axvspan(0, t_phases[0], facecolor='b', alpha=0.2,label='no measures')
ax1.axvspan(t_phases[0],t_phases[1], facecolor='b', alpha=0.3,label='hard
  measures introduced') ax1.axvspan(t_phases[1],t_phases[2], facecolor='b',
  alpha=0.4,label='hold measures') ax1.axvspan(t_phases[2],t_phases[3],
  facecolor='b', alpha=0.5,label='relax measures')
```

```
ax1.axvspan(t_phases[3],len(propagation_rates.infected), facecolor='b',
  alpha=0.6,label='repead hard measures')
ax1.set_ylim(10, 100000000)
ax1.set_yscale('log')
ax1.set_title('SIR simulations',size=16)
ax1.set_xlabel('time in days',size=16) ax1.legend(loc='best', prop={'size':
16});
```

20.5 USER INTERFACE (UI) IMPLEMENTATION

For UI, we have developed a Web application using HTML, JavaScript, and CSS. This is a one-page website with a sleek design. It has five sections and a responsive menu. This web page supports three views: a desktop view, tablet view, and mobile view. This makes the website compatible with all these devices.

Firstly, we have a home screen (Figure 20.9). Whenever the website is in desktop view, it will show these sections in a bar, whereas when it is in tablet (Figure 20.10a) or mobile view (Figure 20.10b) the menu options will go to a hamburger (Figure 20.10c) by clicking on the drop-down list. A hover-over effect on the hamburger drop-down list is applied which, on hovering, displays the name of the section as a watermark on the screen.

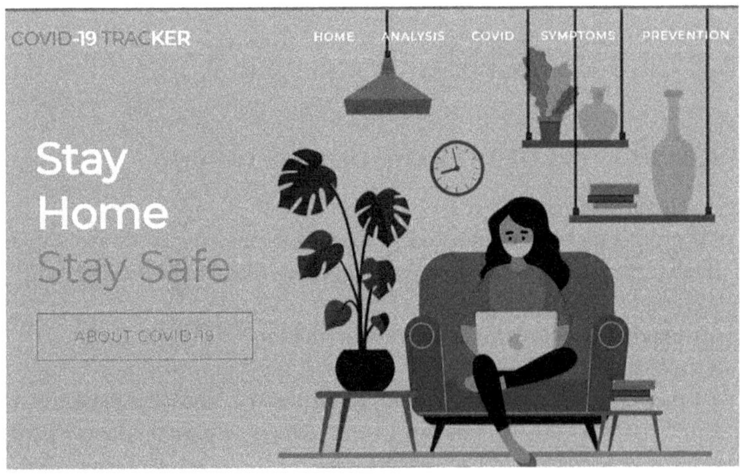

FIGURE 20.9 Home screen.

1. TabletView:

(a)

2. MobileView:

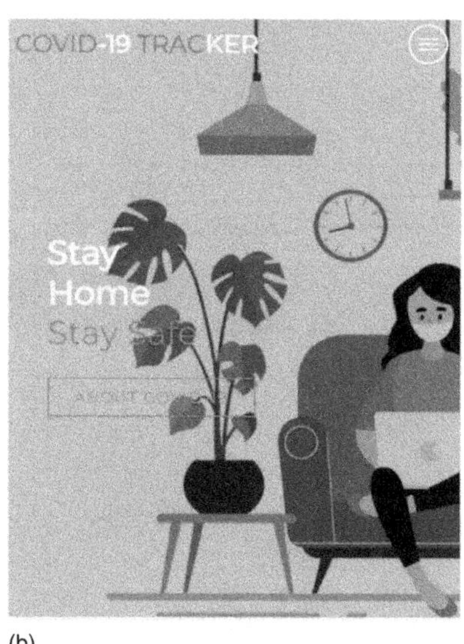

(b)

3. Hamburgermenu:

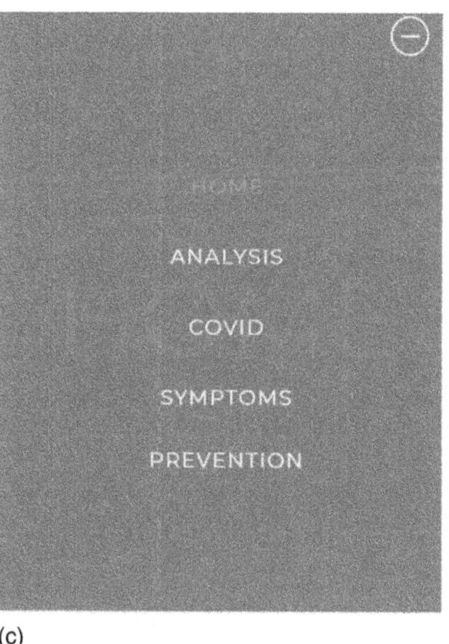

(c)

FIGURE 20.10 (a) Tablet view; (b) mobile view; (c) hamburger menu.

20.6 ANALYSIS

20.6.1 Hover-over India Map

One more way of displaying our data, so that a layperson can also understand the statistics of the COVID-19 situation in India, is a module, which consists of a map of India. On the map, if we hover over a state/union territory using our mouse, the active, confirmed, recovered, and death cases of that state/union territory are displayed in the tabs above. For example, as of now, Maharashtra (Figure 20.11a) has 43,870 active, 200,6354 confirmed (till now), 1,910,521 recovered, and 50,740 death cases. For the Andaman and Nicobar Islands, see Figure 20.11b.

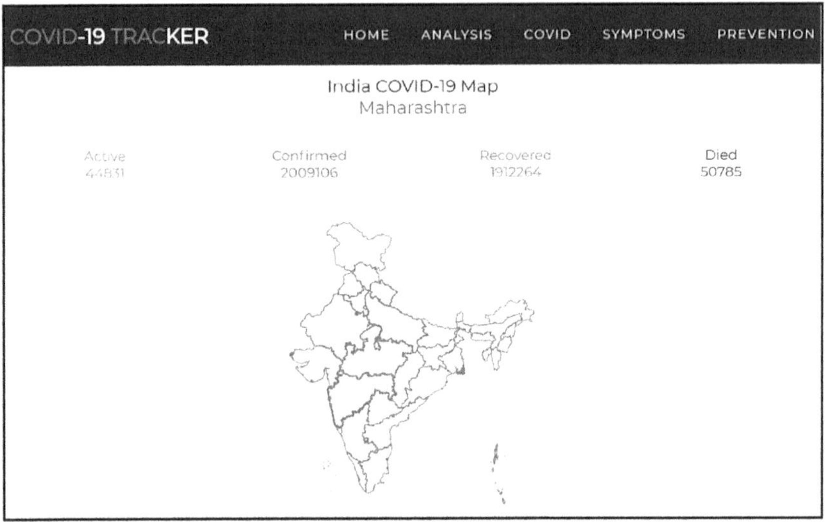

(a)

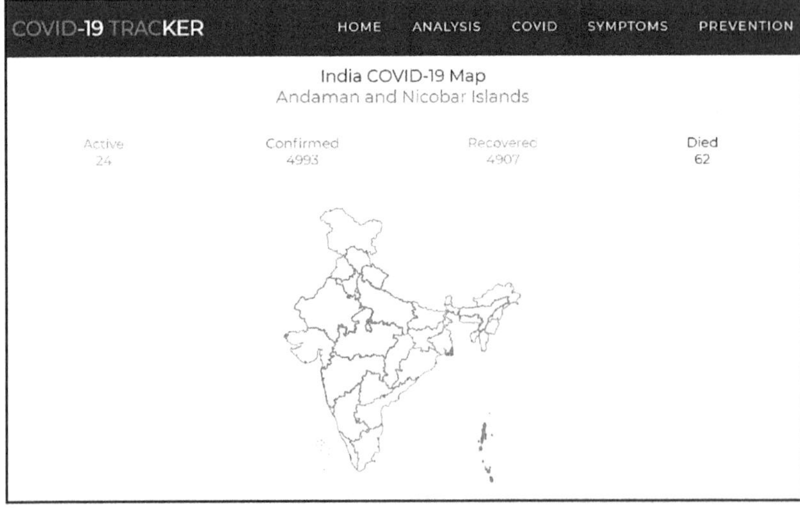

(b)

FIGURE 20.11 (a): Hover-over India map for Maharashtra; (b) hover-over India Map for the Andaman and Nicobar Islands.

This data is also fetched from the covid19india.org API. The map is taken as a scalable vector graphics (SVG) map in the background, which is taken from the application programming interface (API) charts. When we move our mouse once over it, it should retain the data of that state, and an on-mouse hover attribute fetches the information of the particular state. Each of the states have an ID which is given to each state so that when we hover over that state, by using that ID the data is fetched. There is also a path given for each boundary of the state. This on mouse out and on mouse hover is defined for all of the states and union territories of India.

20.6.2 Statewise Updates

In the analysis part, one of the modules is the tabular representation of the data for the states of India. We fetched this data from the covid19india.org API which contains the active cases, confirmed, death, and recovered cases, which we have displayed in a tabular format (Figure 20.12), creating a new row for the information for each of the states. There is also some updated news available for some of the states, which is periodically updated in the API and then brought here, (e.g., for Uttarakhand which says on **[Oct 30]: Metric of capturing the testing data has switched to "Samples**

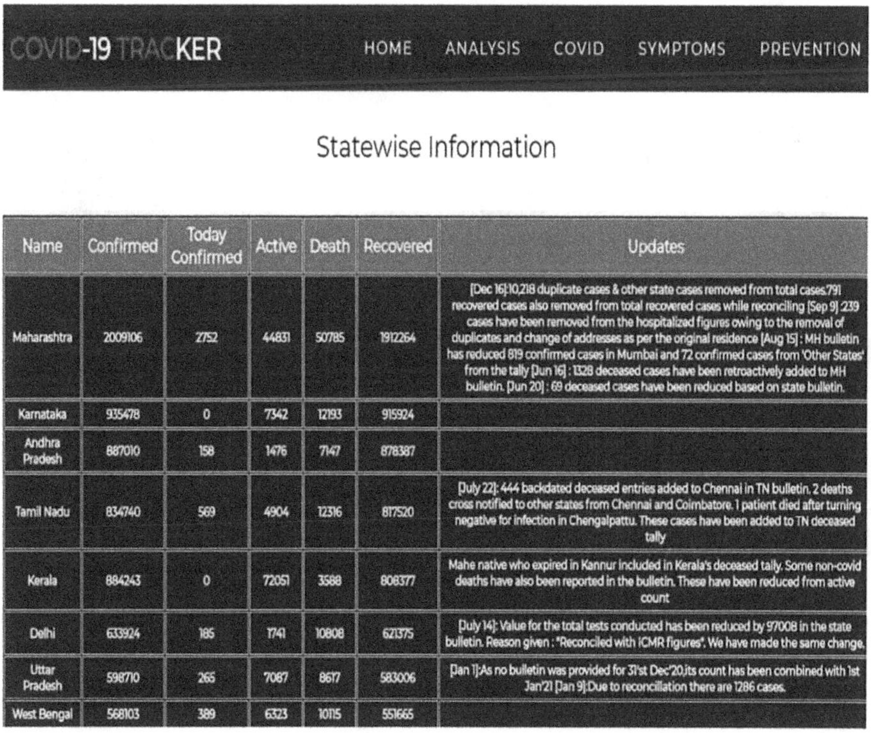

COVID-19 TRACKER HOME ANALYSIS COVID SYMPTOMS PREVENTION

Statewise Information

Name	Confirmed	Today Confirmed	Active	Death	Recovered	Updates
Maharashtra	2009106	2752	44831	50785	1912264	[Dec 16]:10,218 duplicate cases & other state cases removed from total cases.791 recovered cases also removed from total recovered cases while reconciling [Sep 9] 239 cases have been removed from the hospitalized figures owing to the removal of duplicates and change of addresses as per the original residence [Aug 15] : MH bulletin has reduced 819 confirmed cases in Mumbai and 72 confirmed cases from 'Other States' from the tally [Jun 16] : 1328 deceased cases have been retroactively added to MH bulletin. [Jun 20] : 69 deceased cases have been reduced based on state bulletin.
Karnataka	935478	0	7342	12193	915924	
Andhra Pradesh	887010	158	1476	7147	878387	
Tamil Nadu	834740	569	4904	12316	817520	[July 22]: 444 backdated deceased entries added to Chennai in TN bulletin. 2 deaths cross notified to other states from Chennai and Coimbatore. 1 patient died after turning negative for infection in Chengalpattu. These cases have been added to TN deceased tally
Kerala	884243	0	72051	3588	808377	Mahe native who expired in Kannur included in Kerala's deceased tally. Some non-covid deaths have also been reported in the bulletin. These have been reduced from active count
Delhi	633924	185	1741	10808	621375	[July 14]: Value for the total tests conducted has been reduced by 97008 in the state bulletin. Reason given : "Reconciled with ICMR figures". We have made the same change.
Uttar Pradesh	598710	265	7087	8617	583006	[Jan 1]:As no bulletin was provided for 31'st Dec'20,its count has been combined with 1st Jan'21 [Jan 9]:Due to reconciliation there are 1286 cases.
West Bengal	568103	389	6323	10115	551665	

FIGURE 20.12 Statewise updates.

Tested" from "Samples Collected"). This clarifies the differences in the statistics of some of the cases. The output images are also shown on the hover-over graph (Figure 20.13), the prediction graph shown on our website (Figure 20.14), concerning coronavirus (Figure 20.15), its prevention (Figure 20.17), and symptoms (Figure 20.16).

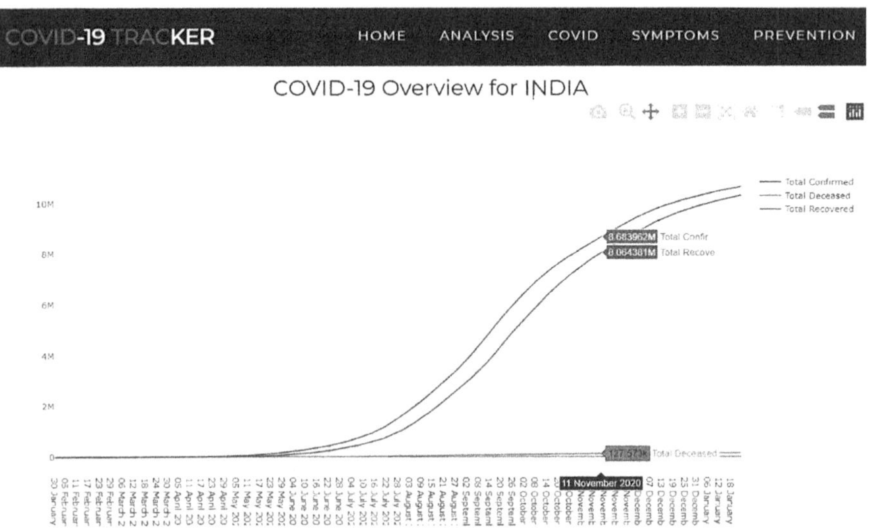

FIGURE 20.13 Hover-over line graph.

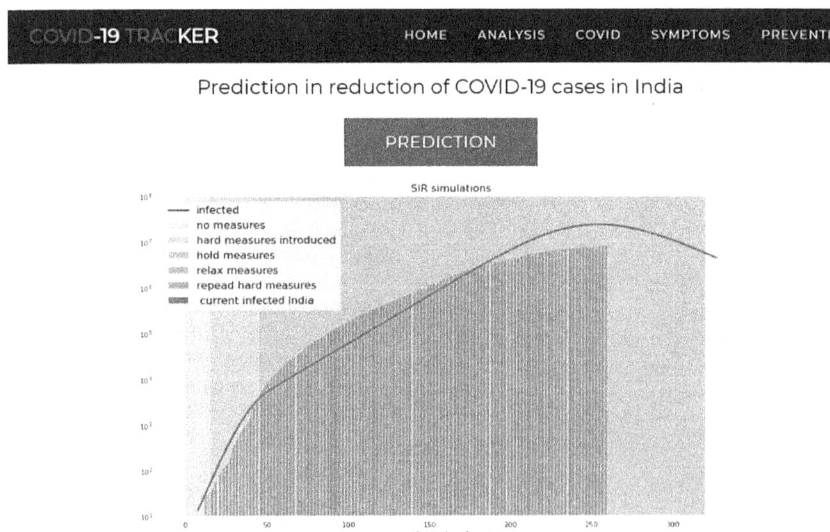

FIGURE 20.14 Prediction.

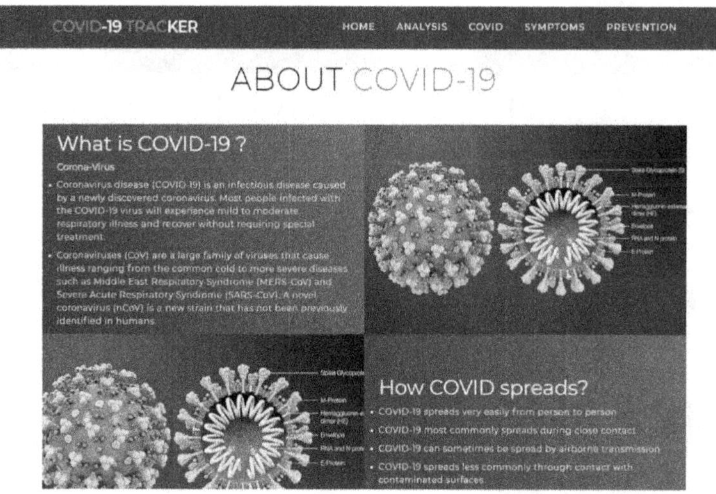

FIGURE 20.15 Information about COVID-19.

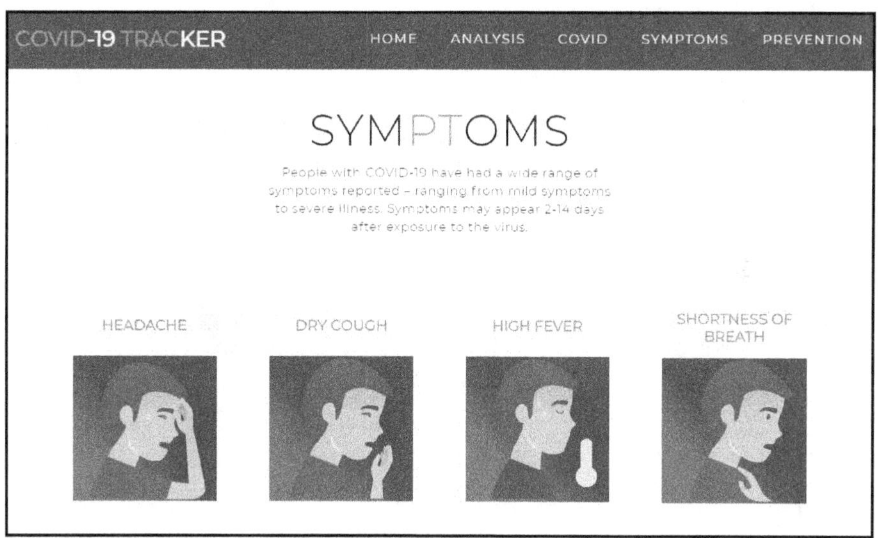

FIGURE 20.16 COVID-19 symptoms.

Much attention on the website concerns the symptoms (Figure 20.16) and prevention (Figure 20.17) section. Instead of images we have used Web animation. There are many ways to implement this, for example by including JavaScript libraries, GIFs, and embedded videos. But all these ways increase the loading period of websites. So, we have used a combination of CSS and SVG animation which have a faster load time, leaving a wonderful user experience.

FIGURE 20.17 Prevention of COVID-19.

20.7 RESULTS

After inserting data in Code 20.2(a), (b) & (c) and using Code 20.1, we obtain the result shown in Figure 20.8.

Here the line representing the infected cases (Figure 20.8) predicts the number of decreasing cases. Comparing Figure 20.18 with Figure 20.8 we find that the plotted

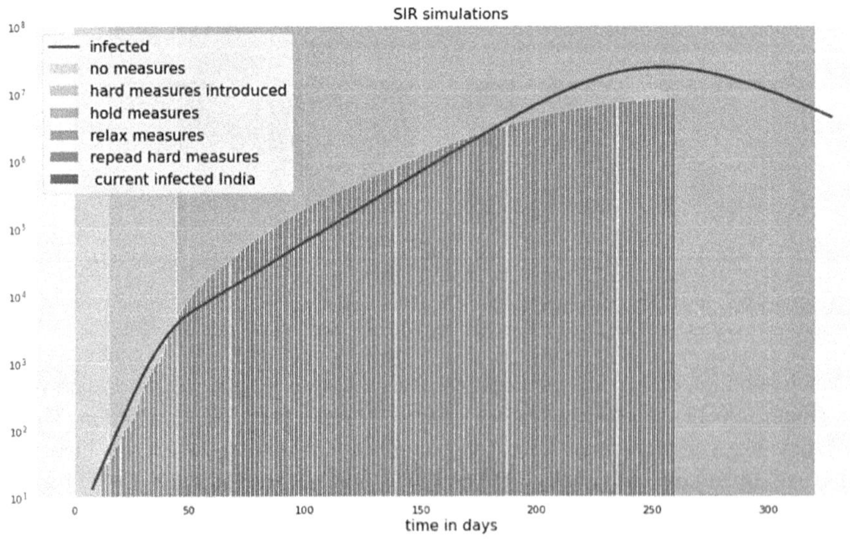

FIGURE 20.18 Obtained SIR model curve for India, where the x-axis represents the number of days and the y-axis represents the country's population.

curve is starting to take the shape of the ideal curve and hence we can say that our trained model is working as expected. The number of cases actually started decreasing in October 2020, which is about 8–9 months from the start, when the first cases appeared in India. This is also predicted in our model, where we can see the decrease in the slope of the curve at around 250–260 days (8–9 months) from the start of the pandemic. In the future, we wish to extend this model over different states of India, and then if possible to all the cities/districts of India.

REFERENCES

Ali, I., Omar M., & Harbi, Al.. (2020). COVID-19: Disease, management, treatment, and social impact. *Science of the Total Environment* 728 138861.

Baena-Díez, J.M., Barroso, M., Cordeiro-Coelho, S.I., Díaz, J.L. and Grau, M., 2020. "Impact of COVID-19 outbreak by income: Hitting hardest the most deprived." *Journal of Public Health*, 42(4), 698–703.

Cooper, I., Mondal, A., & Antonopoulos, C. G. (2020). A SIR model assumption for the spread of COVID-19 in different communities. *Chaos, Solitons & Fractals*, *139*, 110057.

Di Girolamo, N. and Reynders, R.M., 2020. "Characteristics of scientific articles on COVID-19 published during the initial 3 months of the pandemic. *Scientometrics*, 125(1), 795–812.

Fang, L., Karakiulakis, G., & Roth, M. (2020). Are patients with hypertension and diabetes mellitus at increased risk for COVID-19 infection?. *The Lancet. Respiratory Medicine*, *8*(4), 21.

Hamzah, F. B., Lau, C., Nazri, H., Ligot, D. V., Lee, G., Tan, C. L., ... & Chung, M. H. (2020). CoronaTracker: worldwide COVID-19 outbreak data analysis and prediction. *Bull World Health Organ*, *1*(32).

Hoque, M.M., Saima, U., & Shoshi, S.S., 2020. Correlation of climate factors with the COVID-19 pandemic in USA. *Biomedical Statistics and Informatics*, 5(3), 65.

Patil, S. B. (2020). Indian publications on SARS-CoV-2: A bibliometric study of WHO COVID- 19 database. *Diabetes & Metabolic Syndrome: Clinical Research & Reviews*, *14*(5), 1171–1178.

Peckham, R. (2020). COVID-19 and the anti-lessons of history. *The Lancet*, 395(10227), 850–852.

Remuzzi, A., & Remuzzi, G. (2020). COVID-19 and Italy: what next?. *The Lancet*, *395*(10231), 1225–1228.

Roda, W. C., Varughese, M. B., Han, D., & Li, M. Y. (2020). Why is it difficult to accurately predict the COVID-19 epidemic?. *Infectious Disease Modelling*, *5*, 271–281.

Sujath, R., Chatterjee, J. M., & Hassanien, A. E. (2020). A machine learning forecasting model for COVID-19 pandemic in India. *Stochastic Environmental Research and Risk Assessment*, *34*, 959–972.

Vasantha, Raju N., and S. B. Patil. (2020) Indian Publications on SARS-CoV-2: A bibliometric study of WHO COVID-19 database. *Diabetes & Metabolic Syndrome: Clinical Research & Reviews*, *14*(5), 1171–1178.

Xie, J., Covassin, N., Fan, Z., Singh, P., Gao, W., Li, G., ... & Somers, V. K. (2020, June). Association between hypoxemia and mortality in patients with COVID-19. *Mayo Clinic Proceedings,* 95(6),1138–1147.

Yang, Z., Zeng, Z., Wang, K., Wong, S. S., Liang, W., Zanin, M., ... & He, J. (2020). Modified SEIR and AI prediction of the epidemics trend of COVID-19 in China under public health interventions. *Journal of Thoracic Disease*, *12*(3), 165.

Zheng, Y. Y., Ma, Y. T., Zhang, J. Y., & Xie, X. (2020). COVID-19 and the cardiovascular system. *Nature Reviews Cardiology*, *17*(5), 259–260.

21 Assessing the Impact of Coronavirus on Pollutant Concentration

A Case Study in Malaysia

Imam Wahyu Amanullah, Sharifah Sakinah Syed Ahmad, and Emaliana Kasmuri

Universiti Teknikal Malaysia Melaka

CONTENTS

21.1 Introduction 331
21.2 Background Study 332
21.3 Proposed Methodology 332
21.4 Results and Discussion 334
21.5 Conclusion 336
21.6 Limitations and Future Work 337
Notes 338
References 338

21.1 INTRODUCTION

The coronavirus disease (Covid-19) has increased rapidly and caused numerous deaths since its beginnings in Wuhan, China in December 2019. This virus has infected more than 27 million people and caused death to 887,557 people around the world (Worldmeters, 2020). It has become a significant threat to public health. In January 2020, the World Health Organization (WHO) declared that the Covid-19 outbreak was of concern to the international health community. It has caused a variety of illnesses ranging from the common cold to fatal illness (Rasmussen et al., 2020). A person who has an unserious respiratory infection may have no symptoms. Severe contamination will result in the development of pneumonia-like symptoms that may require oxygen support. The patient who has a severe infection may develop acute respiratory distress syndrome (ARDS) (Al-Rohaimi & al Otaibi, 2020).

DOI: 10.1201/9781003146711-21

The exposure to air pollution will aggravate respiratory organ conditions (Mizen et al., 2020). In terms of Covid-19 contamination that most likely affects the respiratory organs, being in contact with air pollution might worsen the patient. Currently, several studies have been managed to assess the significance of air pollution to the cases of Covid-19 infection (Conticini et al., 2020; Ogen, 2020; Yao et al., 2020). Studies related to the newly found Covid-19 virus and air pollution are important because the effect of the Covid-19 outbreak is influenced by policies and regulations taken by local governments. Therefore, pollutant effects on Covid-19 vary among countries and regions even though similar studies have been conducted. This chapter (1) assesses the relation between air pollutants and Covid-19 patients in Malaysian regions and (2) describes the variation of air pollution concentration throughout lockdown measures in Malaysia.

21.2 BACKGROUND STUDY

Air pollution can be defined as an atmospheric condition in which contaminants and pollutant substances are present in considerably high concentrations that may influence human health or produce damaging effects to the surroundings (Vallero, 2008). The concentrations of these gases increase with the rapid development of industrial technology (Chang et al., 2020). The harmful results that air pollution brings to the human condition are making air pollution monitoring an important task. In Malaysia, particulate matter (PM) whose diameter is under 10 μm is regularly monitored by the Malaysian Meteorological Department through observation stations. Other air contaminants such as SO_2, NO_2, CO, PM10, and O_3 are also monitored by the Department of Environment (DOE) of Malaysia.

The studies have shown that the pollutants have an adverse impact on human health. The degree of adversity depends on the degree of the exposure to the pollutant. Table 21.1 summarizes the source of pollutant and its health impact.

21.3 PROPOSED METHODOLOGY

This study was designed to address the objectives that were outlined in the introduction section. Overall, we followed a data mining process that comprises five phases: (1) understanding the business objective, (2) understanding the data, (3) preparation of the data, (4) developing a model, and (5) deployment of the selected model (Miner et al., 2009). The objectives are addressed throughout the phases of the data mining process.

In business understanding, the business objectives and data mining goals are described. These business objectives are the same as the aims of the study, while the data mining goals are sets of goals to accomplish business objectives. For this study, the sets of data mining goals were: (1) acquiring suitable data for the study, (2) preprocessing the data into acceptable formats, (3) conducting a spatial and temporal reduction of the data, and (5) calculating the correlation. The data understanding process describes data acquisition, description, and quality assessment. The Covid-19 fatality data was extracted from the Malaysiakini website.[1] This website provides a

TABLE 21.1
Source of Pollutants and Health Impact

Study	Pollutant	Source	Health Impact
Xia et al., 2013	Aerosols	1. Anthropogenic activity 2. Natural phenomena	1. Deposition efficiency in respiratory tract 2. Reduction of visibility
Reumuth et al., 2019	Carbon monoxide (CO)	1. Hydrocarbon combustion	1. Health fatality consequences
Liu et al., 2019	Formaldehyde		1. Fundamental biomolecule damage
Reingruber & Pontel, 2018	Formaldehyde		1. Throat and eye irritation 2. Asthma 3. Sensitization increment 4. Sick-building syndrome 5. Increment in risk of adverse birth outcomes 6. Bronchial asthma-like symptom development
Ogen, 2020	Nitrogen dioxide (NO_2)	1. Anthropogenic activity	1. Hypertension 2. Heart and cardiovascular diseases 3. Chronic obstructive pulmonary disease 4. Lung function deficits in growth in children 5. Poor or injury of lung function 6. Diabetes 7. Inflammatory response in airways
Y. Wang et al., 2020 Yang et al., 2012 Yebin et al., 2012	Ozone (O_3)		1. Morbidity and mortality through cardiovascular disease 2. Respiratory disease
Jacobson, 2012	Sulphur dioxide (SO_2)		1. Lung damage 2. Respiratory tract damage

daily update of Covid-19 death cases for each state in Malaysia. The total cases of infection is retrieved from the Github Page[2] that gathers Covid-19 data from many sources and stores the data in a csv file. This study also used vertical airflow at 850 mb (roughly 1.5 km from mean sea level) which was accessed from the Physical Sciences Laboratory National Oceanic and Atmospheric Administration (PSL-NOAA).[3] The meteorological observational data was extracted from the Ogimet website,[4] which stores data from around the world.

Vertical airflow is represented by omega (Pascal/s). Omega is the change of pressure over time at a given atmospheric level (by pressure). Negative omega denotes upward airflow while positive omega means downward airflow. Sentinel-5P is used as the data source for pollutants. Sentinel-5 Precursor is a satellite with a low earth orbit that supplies information on air conditions, climate, and ozone and is intended to be used to monitor and forecast air quality and climate.[5] The data are accessed using the Google Earth Engine (GEE) API in Python.

We prepared the data to meet the modeling requirements. We implemented pair-wise deletion to the gathered data (Miner et al., 2009) and standardized it using z-values which are based on the mean and standard deviation.

In the modeling process, a correlation analysis was performed on the data. Through this, we could measure the strength of the relationship between two variables, resulting in a single number which is the correlation coefficient (Kanniah et al., 2020). The study period was set to 18 March to 3 May (the movement control order (MCO) period) to assess the impacts of weather and pollutant parameters on Covid-19. During this period, interstate travel was prohibited.

21.4 RESULTS AND DISCUSSION

The pollutant concentrations were divided according to the number of states in Malaysia. The pollutant concentration trends for Sabah, Johor, Selangor, and Kuala Lumpur states are selected for discussion.

Figure 21.1 shows the concentration of pollutants over selected states in Malaysia. Kuala Lumpur has the highest concentration of NO_2, followed by Sabah and Selangor. Sabah has the lowest NO_2 concentration due to low variation. The effect of the lockdown measures on a global scale have had a clear impact on NO_2 concentration, except for Sabah state for which, even before the lockdown, the concentration was low. The same result was produced by other study which stated that the NO_2 concentration was significantly reduced during the lockdown in urban areas (Kanniah et al., 2020). Sabah had the lowest NO_2 concentration compared to the others.

O_3 concentration plots for Johor, Selangor, and Kuala Lumpur almost overlap with each other. All states have their lowest concentration in mid-January 2020.

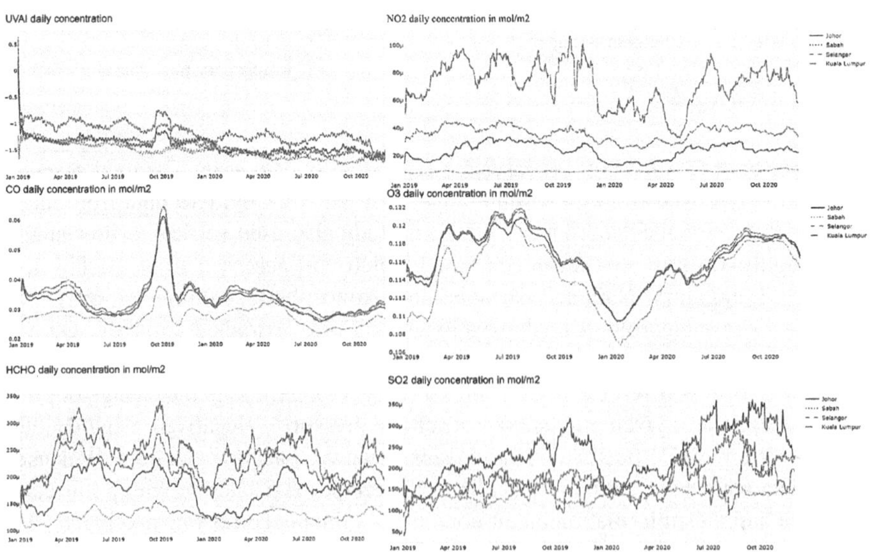

FIGURE 21.1 Smoothed daily pollutant concentration (NO_2, O_3, SO_2, ultraviolet aerosol index (UVAI), CO, and HCHO).

Concentrations then increased until mid-April 2020. Several studies have reported different results related to changes in O_3 concentration during lockdown measures. Some areas reported decreasing O_3 (Stratoulias & Nuthammachot, 2020), some reported no changes (Peralta et al., 2020), and others reported an increase (Roy et al., 2021; Sicard et al., 2020; Zhao et al., 2020). The decreasing concentration of O_3 is the effect of the reduction of anthropogenic activity. The increasing concentration can be explained by the unusual decrease in the concentration of NO_x, which results in a reduced titration of O_3 by NO.

Johor state recorded the highest concentration of SO_2, followed by Selangor and Kuala Lumpur. Sabah state again had the lowest emission compared to the others. The effects of the lockdown measures on SO_2 concentrations are not evident from the SO_2 graph in Figure 21.1. For Johor state, SO_2 concentrations were low in January; however, the trend is toward an increase. During the lockdown period, the concentration was slightly decreasing in April 2020; however, it was increasing in May, June, and July 2020. The same variation was observed in Selangor and Kuala Lumpur. Other studies have shown a slight reduction of SO_2 during the lockdown (Kanniah et al., 2020). Another study stated that the impact of the lockdown measures on SO_2 was not obvious, as there were ongoing changes in the observed levels (Wang & Li, 2020).

Kuala Lumpur has the highest UVAI compared to other states. In Selangor and Kuala Lumpur, the UVAI slightly decreased in March and April 2020. However, it was slightly increasing again in May 2020. In Johor, the UVAI increased until mid-April 2020 then decreased sharply until June 2020. Other places have reported a reduction of the aerosol during the lockdown period (Kanniah et al., 2020; Sicard et al., 2020; Stratoulias & Nuthammachot, 2020; Wang & Li, 2020). An interesting pattern occurred around September 2019. The UVAI of Kuala Lumpur, Selangor, and Johor suddenly increased. One possible reason for this is that in September 2019 there was a lot of deforestation burning in Indonesia, especially on Sumatra Island which is located near to Kuala Lumpur, Selangor, and Johor. This event caused a sudden spike in aerosol concentration (Indonesia, 2019).

The CO concentration of Sabah was marginally smaller than that of Selangor and Johor, while Kuala Lumpur's concentration was slightly higher than the others. During the lockdown period, the concentration in Johor, Selangor, and Kuala Lumpur was decreasing. In Sabah, it was increasing until early May then decreased sharply. Other studies also have reported the increasing concentration of CO (Garg et al., 2020; Kanniah et al., 2020; Stratoulias & Nuthammachot, 2020). At the end of September 2019, all states had a peak, when CO concentrations suddenly increased for a few weeks and then decreased. The reason for this might be the same as the sudden increase of the UVAI. Forest fires release large amounts of carbon into the air. Kuala Lumpur recorded the highest concentration of HCHO followed by Selangor and Johor. Sabah had the lowest concentration. During the lockdown measures, HCHO concentrations were increasing for all observed states. Another study has reported a steady concentration of HCHO during the lockdown (Ghahremanloo et al., 2021).

Correlation analysis was performed for Johor, Sabah, Selangor, and Kuala Lumpur. Of the total cases, the variables that had the highest correlation with a p-value more than 0.05 is precipitation in Johor state (–0.29). The other precipitation variable in Johor also had a negative correlation (–0.15). In Sabah state, the three weather stations had precipitation with correlations of –0.136, –0.09, and –0.03. In Selangor state, precipitation had a correlation of –0.4 (p-value less than 0.05) and –0.13. Precipitation consistently had a negative correlation with daily new cases even though the correlation was weak. These values are summarized in Table 21.2.

The next highest correlation was for the relative humidity variable in Selangor state (–0.26); the other relative humidity data in Selangor had a correlation of –0.16. In Johor state, relative humidity variables had correlations of –0.357 and –0.34, though both had a p-value of less than 0.05. In Sabah state, relative humidity variables had correlations of –0.35 (p-value less than 0.05), –0.24, and –0.23. Precipitation and relative humidity also had consistently negative correlation values, even though the correlation is considerably weak. Another variable with a consistent correlation sign is omega. In Johor, Sabah, Selangor, and Kuala Lumpur, omega variables had correlations of 0.29 (p-value less than 0.05), 0.17, 0.24, and 0.196. The highest and relative correlations for these states are listed in Table 21.3.

Precipitation and relative humidity have been shown to have a negative correlation with Covid-19 cases. Another study also generated negative correlations for precipitation and relative humidity (Zoran et al., 2020). The negative correlation of precipitation and Covid-19 cases has raised the hypothesis that rainy days motivate people to stay at home, while sunny days encourage people to go outside and hence be prone to virus exposure (Menebo, 2020).

21.5 CONCLUSION

During the lockdown measure in Malaysia some pollutants show a decreasing pattern in their concentration, inconsistent changes, and even increasing changes. NO_2 shows a decreasing trend during the lockdown, especially in urban cities. O_3 shows an

TABLE 21.2
Correlation of Precipitation for Selected States

State	Highest Correlation Value
Johor	–0.29
Sabah	–0.136, –0.09, and –0.03
Selangor	–0.4, –0.13

TABLE 21.3
Correlation Values for Humidity in Selected States

State	Highest Correlation Value	Relative Correlation Value
Selangor	–0.26	–0.16
Johor	—	–0.357
Sabah	—	–0.35, –0.24, and –0.23

increasing trend in some areas due to the lack of NO_2 titration. The trend of SO_2 is not clear. Some states reported a slight reduction, while others had no changes. The UVAI and CO show a reduction during the lockdown and peaked in September 2019 which is related to the massive wildfires in Indonesia. HCHO shows an increasing concentration.

This research has revealed the changes in the concentration of air pollutants throughout the lockdown measures which were implemented in most countries in the world. There are clear differences in some pollutants, before and during the lockdown. However, it is still difficult to say whether the changes in pollutant concentrations are due to lockdown measures only. Each pollutant has its seasonal variability. This variability is affected by the dynamics of the atmosphere which makes it difficult to explain the cause of pollutant concentration changes.

The correlation between Covid-19 death cases and weather variables is considerably weak and inconsistent. This is most likely due to no direct interaction between these two variables. The correlation between Covid-19 daily new cases and weather variables is considerably weak, but a few pollutant and weather variables do show consistent correlation signs. Precipitation and relative humidity variables have consistent negative correlation coefficients in the selected states, while the omega variable has consistent positive correlation coefficients. In the regions with high Covid-19 cases, precipitation and relative humidity have weak negative correlations. This result shows that there might be a relationship between these variables. Some hypotheses describe the relationship between precipitation and Covid-19 cases, . The first hypothesis is that rainy days encourage people to stay at home; the other hypothesis is that sunny days boost people's motivation to go outside and be exposed to the virus.

21.6 LIMITATIONS AND FUTURE WORK

Regarding the proposed methodology as well as the data used in this study, there are some limitations about describing pollutant changes and assessing their relationship with Covid-19 cases. To represent air quality over Malaysia, this study has used satellite image data. Satellite images do not directly capture air pollutant gases. Instead, they capture radiation emitted by atmospheric gases. The atmosphere is a complex system. There are many sub-systems of it that are not well quantified, resulting in errors in satellite images, which affect pollutant observation and introduce errors to the calculated concentration. Other than the quality of air pollutant data, its variability needs to be considered to better assess the changes of air pollutant concentration during lockdown measures. To assess this variability, years of pollutant data records are required. In calculating the relationship of pollutants with Covid-19 cases, the pollutant is assumed to have direct influence on such cases (pollutant concentration in a day is associated with Covid-19 cases on the same day). In fact, the role of pollutants on Covid-19 infection is not direct.

There are many improvements that could be made to this study. To get a better correlation coefficient, lagged correlation could be introduced. The relationship between Covid-19 and weather variables is not direct. Therefore, lag days could be introduced to obtain a better relationship. In normal correlation analysis, pollutant concentration on a day is associated with Covid-19 cases on that day. In lagged correlation, pollutant concentration on a day is associated with Covid-19 cases for a few days ahead, which are described as lagged days. Another way to obtain a better relationship is to use more representative pollutant data. This study has only used pollutant data that are retrieved from the Sentinel-5P satellite. Pollutant observation station data could provide a better relationship with Covid-19 cases than satellite data.

NOTES

1 https://newslab.malaysiakini.com/covid-19/en
2 https://github.com/ynshung/covid-19-malaysia
3 https://psl.noaa.gov/data/gridded/data.ncep.reanalysis.derived.pressure.html
4 https://www.ogimet.com/gsynres.phtml.en
5 https://sentinel.esa.int/web/sentinel/missions/sentinel-5p

REFERENCES

Al-Rohaimi, A. H., & al Otaibi, F. (2020). Novel SARS-CoV-2 outbreak and COVID19 disease; a systemic review on the global pandemic. *Genes & Diseases*, *7*(4), 491–501. https://doi.org/10.1016/j.gendis.2020.06.004

Chang, Y.-S., Chiao, H.-T., Abimannan, S., Huang, Y.-P., Tsai, Y.-T., & Lin, K.-M. (2020). An LSTM-based aggregated model for air pollution forecasting. *Atmospheric Pollution Research*, *11*(8), 1451–1463. https://doi.org/10.1016/j.apr.2020.05.015

Conticini, E., Frediani, B., & Caro, D. (2020). Can atmospheric pollution be considered a co-factor in extremely high level of SARS-CoV-2 lethality in Northern Italy? *Environmental Pollution*, *261*, 114465. https://doi.org/10.1016/j.envpol.2020.114465

Garg, A., Kumar, A., & Gupta, N. C. (2020). Comprehensive study on impact assessment of lockdown on overall ambient air quality amid COVID-19 in Delhi and its NCR, India. *Journal of Hazardous Materials Letters*, *2*, 100010. https://doi.org/10.1016/j.hazl.2020.100010

Ghahremanloo, M., Lops, Y., Choi, Y., & Mousavinezhad, S. (2021). Impact of the COVID-19 outbreak on air pollution levels in East Asia. *Science of the Total Environment*, *754*, 142226. https://doi.org/10.1016/j.scitotenv.2020.142226

Indonesia, B. N. (2019). *Kebakaran hutan: Titik panas berkurang drastis, tapi pemerintah harus tetap waspadai karhutla*. Retrieved 1 4, 2020, from https://www.bbc.com/indonesia/indonesia-49921090

Jacobson, M. Z. (2012). *Air pollution and global warming: History, science, and solutions*. Cambridge University Press.

Kanniah, K. D., Kamarul Zaman, N. A. F., Kaskaoutis, D. G., & Latif, M. T. (2020). COVID-19's impact on the atmospheric environment in the Southeast Asia region. *Science of the Total Environment*, *736*, 139658. https://doi.org/10.1016/j.scitotenv.2020.139658

Liu, C., Miao, X., & Li, J. (2019). Outdoor formaldehyde matters and substantially impacts indoor formaldehyde concentrations. *Building and Environment*, *158*, 145–150. https://doi.org/10.1016/j.buildenv.2019.05.007

Menebo, M. M. (2020). Temperature and precipitation associate with Covid-19 new daily cases: A correlation study between weather and Covid-19 pandemic in Oslo, Norway. *Science of the Total Environment*, *737*, 139659. https://doi.org/10.1016/j.scitotenv.2020.139659

Miner, G., Nisbet, R., & Elder, J. (2009). Handbook of statistical analysis and data mining applications. In *Handbook of statistical analysis and data mining applications*. Elsevier Inc. https://doi.org/10.1016/B978-0-12-374765-5.X0001-0

Mizen, A., Lyons, J., Milojevic, A., Doherty, R., Wilkinson, P., Carruthers, D., Akbari, A., Lake, I., Davies, G. A., al Sallakh, M., Fry, R., Dearden, L., & Rodgers, S. E. (2020). Impact of air pollution on educational attainment for respiratory health treated students: A cross sectional data linkage study. *Health & Place*, *63*, 102355. https://doi.org/10.1016/j.healthplace.2020.102355

Ogen, Y. (2020). Assessing nitrogen dioxide (NO2) levels as a contributing factor to coronavirus (COVID-19) fatality. *Science of the Total Environment*, *726*, 138605. https://doi.org/10.1016/j.scitotenv.2020.138605

Peralta, O., Ortínez-Alvarez, A., Torres-Jardón, R., Suárez-Lastra, M., Castro, T., & Ruíz-Suárez, L. G. (2020). Ozone over Mexico city during the COVID-19 pandemic. *Science of the Total Environment*, *761*, 143183. https://doi.org/10.1016/j.scitotenv.2020.143183

Rasmussen, S. A., Smulian, J. C., Lednicky, J. A., Wen, T. S., & Jamieson, D. J. (2020). Coronavirus disease 2019 (COVID-19) and pregnancy: What obstetricians need to know. *American Journal of Obstetrics and Gynecology*, *222*(5), 415–426. https://doi.org/10.1016/j.ajog.2020.02.017

Reingruber, H., & Pontel, L. B. (2018). Formaldehyde metabolism and its impact on human health. *Current Opinion in Toxicology*, *9*, 28–34. https://doi.org/10.1016/j.cotox.2018.07.001

Reumuth, G., Alharbi, Z., Houschyar, K. S., Kim, B.-S., Siemers, F., Fuchs, P. C., & Grieb, G. (2019). Carbon monoxide intoxication: What we know. *Burns*, *45*(3), 526–530. https://doi.org/10.1016/j.burns.2018.07.006

Roy, S., Saha, M., Dhar, B., Pandit, S., & Nasrin, R. (2021). Geospatial analysis of COVID-19 lockdown effects on air quality in the South and Southeast Asian region. *Science of the Total Environment*, *756*, 144009. https://doi.org/10.1016/j.scitotenv.2020.144009

Sicard, P., de Marco, A., Agathokleous, E., Feng, Z., Xu, X., Paoletti, E., Rodriguez, J. J. D., & Calatayud, V. (2020). Amplified ozone pollution in cities during the COVID-19 lockdown. *Science of the Total Environment*, *735*, 139542. https://doi.org/10.1016/j.scitotenv.2020.139542

Stratoulias, D., & Nuthammachot, N. (2020). Air quality development during the COVID-19 pandemic over a medium-sized urban area in Thailand. *Science of the Total Environment*, *746*, 141320. https://doi.org/10.1016/j.scitotenv.2020.141320

Vallero, D. A. (2008). Fundamentals of air pollution. In *Fundamentals of air pollution*. Elsevier. https://doi.org/10.1016/B978-0-12-373615-4.X5000-6

Wang, Q., & Li, S. (2020). Nonlinear impact of COVID-19 on pollutions from January 1 to October 30 – Evidence from Wuhan, New York, Milan, Madrid, Bandra, London, Tokyo and Mexico City. *Sustainable Cities and Society*, *65*, 102629. https://doi.org/10.1016/j.scs.2020.102629

Wang, Y., Wild, O., Chen, X., Wu, Q., Gao, M., Chen, H., Qi, Y., & Wang, Z. (2020). Health impacts of long-term ozone exposure in China over 2013–2017. *Environment International*, *144*, 106030. https://doi.org/10.1016/j.envint.2020.106030

Worldmeters. (2020). *Coronavirus Update (Live)*. Retrieved 7 9, 2020, from https://www.worldometers.info/coronavirus/

Xia, M., Yanjun, M., Renjie, C., Zhijun, Z., Bingheng, C., & Haidong, K. (2013). Size-fractionated particle number concentrations and daily mortality in a Chinese city. *Environmental Health Perspectives*, *121*(10), 1174–1178. https://doi.org/10.1289/ehp.1206398

Yang, C., Yang, H., Guo, S., Wang, Z., Xu, X., Duan, X., & Kan, H. (2012). Alternative ozone metrics and daily mortality in Suzhou: The China air pollution and health effects study (CAPES). *Science of the Total Environment*, *426*, 83–89. https://doi.org/10.1016/j.scitotenv.2012.03.036

Yao, Y., Pan, J., Liu, Z., Meng, X., Wang, W., Kan, H., & Wang, W. (2020). Temporal association between particulate matter pollution and case fatality rate of COVID-19 in Wuhan. *Environmental Research*, *189*, 109941. https://doi.org/10.1016/j.envres.2020.109941

Yebin, T., Wei, H., Xiaoliang, H., Liuju, Z., Shou-En, L., Yi, L., Lingzhen, D., Yuanhang, Z., & Tong, Z. (2012). Estimated acute effects of ambient ozone and nitrogen dioxide on mortality in the pearl river delta of southern China. *Environmental Health Perspectives*, *120*(3), 393–398. https://doi.org/10.1289/ehp.1103715

Zhao, F., Liu, C., Cai, Z., Liu, X., Bak, J., Kim, J., Hu, Q., Xia, C., Zhang, C., Sun, Y., Wang, W., & Liu, J. (2020). Ozone profile retrievals from TROPOMI: Implication for the variation of tropospheric ozone during the outbreak of COVID-19 in China. *Science of the Total Environment*, 142886. https://doi.org/10.1016/j.scitotenv.2020.142886

Zoran, M. A., Savastru, R. S., Savastru, D. M., & Tautan, M. N. (2020). Assessing the relationship between ground levels of ozone (O3) and nitrogen dioxide (NO2) with coronavirus (COVID-19) in Milan, Italy. *Science of the Total Environment*, *740*, 140005. https://doi.org/10.1016/j.scitotenv.2020.140005

22 A Comprehensive Review of SLAM Techniques

Karthi Mohan, Surya Bharath Achalla, and Arpit Jain
University of Petroleum and Energy Studies, Dehradun

CONTENTS

22.1 Introduction 342
22.2 SLAM and Its Methodologies 345
  22.2.1 Relative Position Measurement 345
  22.2.2 Absolute Position Measurement 345
  22.2.3 Encoders 346
22.3 Mapping the Environment and Navigation 347
  22.3.1 Mapping 348
  22.3.2 Localization 349
    22.3.2.1 Vehicle Model 350
    22.3.2.2 Sensor Model 351
    22.3.2.3 State Estimation 352
    22.3.2.4 Support for SLAM Using a Kalman Filter 352
    22.3.2.5 Particle Filter 354
22.4 Navigation 355
  22.4.1 Definition of the SLAM Problem 356
  22.4.2 Homogeneous Coordinates 356
  22.4.3 Homogeneous Taking Euclidian into Consideration 356
  22.4.4 Transformation 357
  22.4.5 Translation: (Three Parameters) ➔ (Three Translations) 357
  22.4.6 Rotation: (Three Parameters) ➔ (Three Rotations) 357
  22.4.7 Rotation Matrices 358
  22.4.8 Rigid Body Transformation: (Six Parameters) ➔ (Three Translations + Three Rotations) 358
  22.4.9 Similarity Transformation: (Seven Parameters) ➔ (Three Translations + Three Rotations + One Scale) 358
  22.4.10 Affine Transformation: (Twelve Parameters) ➔ (Three Translations + Three Rotations + Three Scales + Three Spheres) 358
22.5 Simulation 359
22.6 Hardware Requirements and Implementation in the Real World 360

DOI: 10.1201/9781003146711-22

22.6.1 Single Board Computers 361
22.6.2 Inertial Measurement Unit (IMU) 362
22.6.2.1 Odometry Model of Robot with Wheel Encoders 364
22.6.3 Basic Principle of the Accelerometer 365
22.6.4 Controllers 366
22.7 Limitations 366
22.8 Future Scope 367
22.9 Conclusion 367
References 367

22.1 INTRODUCTION

For a robot to attain a complete autonomous mode, it needs the ability to localize itself on a map or to construct a map of any surroundings with the support of computer vision. This method is called robot mapping (Thrun, 2008). Mapping indicates the movement of the stereo sensor, or any mapping device on the robot, through the environment by modeling this environment in terms of the robot's perception. Robot mapping works on methods such as state estimation, localization, mapping, navigation motion planning, and simultaneous localization and mapping (SLAM) (Behnke & Veloso, 2013).

A state is a set of quantities used to define a robot's motion, such as position, orientation, and velocity. State estimation deals with the estimating of objects poses through sensor information (Khairuddin, Talib, & Haron, 2015). Localization is a process of obtaining the robot's location in its environment. Mapping refers to building a map of any environment obtained by the robot's pose in relation to obstacles around it (Valencia & Cetto, 2018). Navigation of a robot is the process of estimating its position in its frame of reference and planning its path to its next position. Motion planning or path planning refers to the computation of all the possible sequences of configurations to navigate a robot from one point to another in the environment. SLAM is a method for computing the robot's pose and building a map of its surroundings simultaneously (Chatterjee & Rakshit, 2012).

SLAM generates a map of any environment by navigating the robot through the workspace. A robot must know its position in an area when it has to perform a specific job or reach a goal state (Dissanayake, Newman, Clark, Durrant-Whyte, & Csorba, 2001). SLAM uses algorithms to recognize, analyze, and process information using sensor values as inputs to navigate around an unknown or new environment or when mapping the same. SLAM estimates the pose of the robot and helps it to learn through a map, just like a person trying to find his way around an unknown place by looking around and recognizing the obstacles around him to obtain his relationship with obstacles by estimating his path through the area to reach his goal state. A human brain processes the information and controls its motion in a new environment by visual perception of the information it has on the surface and obstacles it encounters around itself and builds a map of the environment by an assessment based either on experience or simultaneous analysis of the information at a point and acting upon it. The likely fully autonomous robot uses SLAM to create a map of the environment by perceiving information simultaneously and acting to reach its goal state. Just as a human brain processes and plans for a trajectory by assessing the next steps of the motion, so a robot must localize its position in a new environment through visual cues and assess its next position by estimating waypoints to travel along by avoiding obstacles and optimizing its path and trajectory. To achieve this the robot must be efficient in its odometry to enable it to estimate well its position in the environment. The robot's pose in respect of each obstacle present in the space where it is placed is mapped at each instant and remapped until all the errors are minimized and modified by the algorithm (Barfoot, 2017).

SLAM is essential to building the map of any environment that needs to be explored, as well as obtaining the exact position of the robot, even in an unexplored locality, by simultaneously mapping and localizing itself. This helps in making a robot fully autonomous by enabling it to locate itself and map its environment by relative observations. The main advantage of slam would be navigation without needing to provide prior information about the environment or any topological information, making it fully autonomous. SLAM can be applied in various environments, making it autonomous, to ease the process of the application. SLAM provides a solution to manned navigation systems or any geophysical navigation systems by autonomously localizing and mapping in any unknown environment, irrespective of the location. SLAM discusses applications in various scenarios, such as: household application in an autonomous vacuum cleaner, a lawnmower for the garden; aerial surveillance with an unmanned aerial vehicle (UAV) for 3D mapping and topological views; underwater applications for reef monitoring; underground applications for mining; and space applications for terrain mapping for localization on Mars rovers (Alan & Pritsker, 1982).

Mapping can also enhance the state estimation of the robot's pose to minimize errors. Information on a map of the environment can facilitate a robot to perform path planning, to define its path in reaching its goal state, and to enable visualization for the operator. Using SLAM at each instance to localize the position of obstacles and the robot can be used to relocate the obstacles so as to have clear information on the map and avoid collisions with nearby objects (Suzuki, Amano, Hashizume, & Suzuki, 2011). The map is obtained by the robot using a grid-based mapping technique, commonly known as Gmapping or grid mapping (Wurm, Stachniss, & Grisetti, 2010).

SLAM challenges are classified as:

- Volumetric vs. feature-based SLAM.
- Topologic vs. geometric maps.
- Known vs. unknown correspondence.
- Static vs. dynamic environments.
- Small vs. large uncertainty.
- Active vs. passive SLAM.
- Anytime and any space SLAM.
- Solo robot vs multi-robot SLAM. (Ainsworth, Sternberg, Raczy, & Butcher, 2016; Fazli & Kleeman, 2006; Carlson, et al., 2014)

There are five driving levels according to the standards levels of automation (SAE) when human participation is involved in travel inside or outside the vehicle:

- Level 0: No automation.
- Level 1: Driving Assistance (e.g., adaptive cruise control (ACC), lane keeping assistance).
- Level 2: Partial driving automation – controls both longitudinal and lateral control (e.g., GM super cruise, Nissan ProPILOT assist).
- Level 3: Longitudinal control, lateral control, object and event detection and response (OEDR); includes automated object and event detection and response (e.g., Audi A8 sedan).
- Level 4: Longitudinal, lateral, OEDR, and fallback; no steering wheel (e.g., Waymo may release such a car).
- Level 5: Longitudinal control, lateral control, OEDR, fallback, no steering wheel, and unlimited operational design domain (ODD).

Figure 22.1 shows the visualization of a robot while performing SLAM. The tool used for this simulation is MATLAB (Matrix Laboratory). Exteroceptive and interoceptive sensors are used for this whole process. We will discuss these sensors and mechanisms in a later part of the chapter.

In Section 22.2, we will discuss SLAM and its methodologies, and discuss wheel encoders. In Section 22.3, we will discuss mapping and localization and its techniques with a robot model. In Section 22.4, we will look at navigation; and in

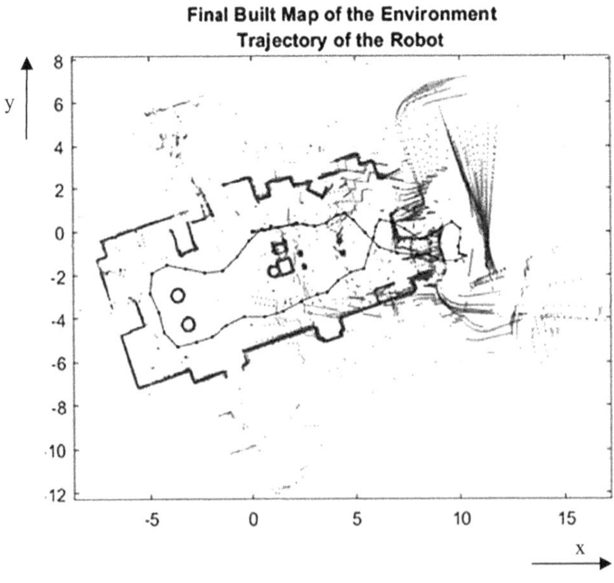

FIGURE 22.1 A preview of the mapping of an unknown environment with the necessary hardware requirements using MATLAB.

Sections 22.5 and 22.6, we examine simulation and the hardware requirements for an autonomous mobile robot (AMR). Limitations, future scope, and conclusions are discussed respectively in Sections 22.7, 22.8, and 22.9.

22.2 SLAM AND ITS METHODOLOGIES

SLAM methodologies are approaches that use probability theory to represent the uncertainties in a robot's motion. A SLAM problem works on a solution to obtain a map of any environment and provides a path to the robot from observations made on the environment and its position orientation and velocity (Hsiao & Wang, 2011).

There has been no certain methodology to classify whether localization takes place first or mapping. A map is required for localization, and estimation of poses is essential for mapping. The mapping of an environment is done through the localization of the robot's pose and obstacles in the environment. Also, localization determines the pose of the robot within a pre-built map. The localization of the mobile robot requires sensory information that will be taken as input in terms of its pose. A robot's pose or an obstacle can be obtained through sensor detection where a position acquisition can be a relative position measurement or an absolute position measurement (Guivant, Nebot, Nieto, & Masson, 2004).

22.2.1 Relative Position Measurement

The relative position of a robot can be measured using different methods and measuring devices, such as an odometer or an inertial measurement unit for inertial navigation.

Odometry methods use encoders to record the wheel rotation measurements with the mobile robot's movement to detect its present location about the origin's coordinates.

Inertial measurement unit (IMU) sensors are used to monitor the robot's linear and rotational acceleration with respect to the plane or terrain. With the information from the sensor, the robot's relative position that is obtained can be integrated to obtain speed and distance traveled from the reference point to localize its position precisely (Biswas & Veloso, 2013).

22.2.2 Absolute Position Measurement

Absolute position measurement uses sensors that are based on range finding beacons or vision-based sensors to acquire the position from the robot to the target. They use typical range-finding sensors, commonly optical-based sensors (i.e., laser range finders that estimate the range of the obstacle or boundary by estimating the segment difference between the light wave sent and the one that rebounds). Range finding beacons use techniques such as Wi-Fi localization using graph-based Wi-Fi maps from various signal strengths across the field. This method estimates the optimal location of the robot's position by use of the probability function. The linear interpolation on the graph is used to obtain the mean and standard deviation of Wi-Fi through approximations.

Robots use Global Positioning System (GPS) modules to acquire information on the position from the travel time of radio signals sent to and received from the satellite. However, indoor environments will have poor accuracy when using GPS modules.

Absolute position measurements are independent of the location estimates from previous iterations, as they are derived directly from a single measurement rather than by integrating a sequence of measurements. Hence, errors in position can be minimized easily as compared to relative position measurement techniques (Liu, Zhang, Gu, & Ren, 2013).

22.2.3 Encoders

An encoder is a simple device which is used to measure the distance covered by the mobile robot in space. The main objective of encoders is to measure the mobile robot from its initial state to its goal state with exact quantities, such as 0 meters (start position) and 100 meters (target position), along with precision. In commercially available motors, most are integrated with a Hall sensor to measure the distance covered by the electric vehicle. A similar kind of mechanism is used in gasoline vehicles. But here, we are going to use only the encoders for mobile robots. There are motors integrated with encoders and there are motors which do not have the features of encoders; some have the facility to have an encoder added to it. There are four different varieties of encoders available:

- Mechanical.
- Optical.
- Magnetic.
- Electromagnetic induction. (Niku, 2001a)

In this chapter, we are diving into optical encoders, where a light source, like an LED, is placed on one side which allows a beam of light to pass through the other side via an encoder disk to the photosensitive sensor (Figure 22.2). If there is any change in the angular position of the disk, then the light will pass through it and be counted as steps. The steps are counted by the light received at the phototransistor; if it receives the light then it is HIGH, if it does not receive the light then it is LOW. A positive count provides the steps each time the robot moves from one place to the other. There are some main factors we need to consider before writing the code for the encoders of the mobile robots:

- Wheel diameter.
- Gear ratio.
- Line encoder.
- Base motor count per revolution (CPR) or pulse per revolution (PPR).
- CPR or PPR at the output of the motor shaft.
- Number of slots in the encoders. (Brown & Schneider, 1987)

FIGURE 22.2 A rotary incremental encoder disk.

FIGURE 22.3 Rotary encoder sensor module, MH Sensor Series.

The below mentioned sensor can work within the range 3.3 to 5.0 V, and there are four pins that can be externally seen on it. The positive terminal of a voltage common collector (VCC) represents +5 V or +3.3 V, the negative ground A0 represents the analog output, and D0 represents digital output. Figures 22.2 and Figure 22.3 are used together; the encoder wheel is mounted on the shaft of the motor, and on the other shaft the robot's wheel is mounted. The rotary encoder sensor is placed in such a way that the whole encoder wheel comes in between the photosensitive sensor.

22.3 MAPPING THE ENVIRONMENT AND NAVIGATION

Creating a map of any unknown space is one of the challenging tasks in the autonomous navigation of the robot or vehicle. Hence the robot needs to map the environment, localize the position, and avoid obstacles as well. This is how a smart navigating robot or vehicle functions, based on their features, as shown in Figure 22.4. The consistency of mapping and localizing the robot is the main task here because step by step the workflow continues and follows a sequence of orders to achieve the goal.

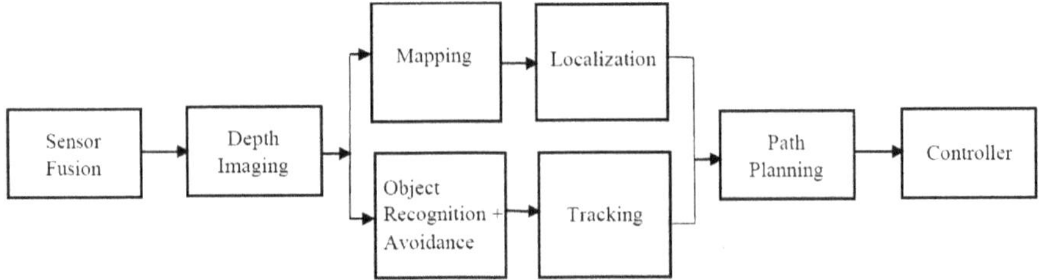

FIGURE 22.4 Simple block diagram of complete working SLAM technique.

Data acquisition (DA) also plays a major role in navigation: if the robot moves in a known environment, then it knows the coordinates of all the points in the environment, so it only has to compute the distance between the initial and goal state (Durrant-Whyte & Bailey, 2006).

22.3.1 Mapping

Gmapping is one of the traditional methods published in Open SLAM in ROS packages. The Gmapping package offers laser-based SLAM, a ROS node called slam_gmapping. They are commonly implemented on differential drive robots in which odometry is used to define the position and Gmapping with ROS supports packages that are implemented on existing models such as Turtlebot3 using 3D visualization software packages like Rviz; the process is run into a remote PC for monitoring, as visualized in Figure 22.5. Once a mobile robot is placed in any space without map data as such, the robot is controlled to move around each corner of the environment defining its position each time it reaches the corners or an obstacle when using a real robot (Ocando, Certad, Alvarado, & Terrones, 2017). If a simulation version of a turtle bot with ROS packages is used, the teleoperation of keyboard commands or joypad packages can be used to move the robot in the environment. The details of the map data obtained through this operation are obtained on a grid map. SLAM produces a 2D occupancy grid map and the pixels of the map show the data in three different colors, thus classifying the areas in which the robot can or cannot move. These are represented in white for a free area in which a robot can move; an occupied area is displayed in black to show the obstacles or the area the robot cannot move in; and an unknown area is shown in gray where the robot must remap to obtain information. The map obtained by moving the robot along the corners of the environment is saved using the robot's map saver nodes and a telemetry node, which are used to obtain the angular velocities of the robot; the Gmapping nodes then draw a map using the LiDAR scan data which can be further used in the process of navigation (Wang, Liang, & Guan, 2011). There are many other mapping techniques available based on the hardware components because cameras, LiDAR, and depth cameras are also used for mapping the environment, whether in 2D or 3D.

The 2D-LiDAR-based mapping techniques are:

- Gmapping.
- Hector.

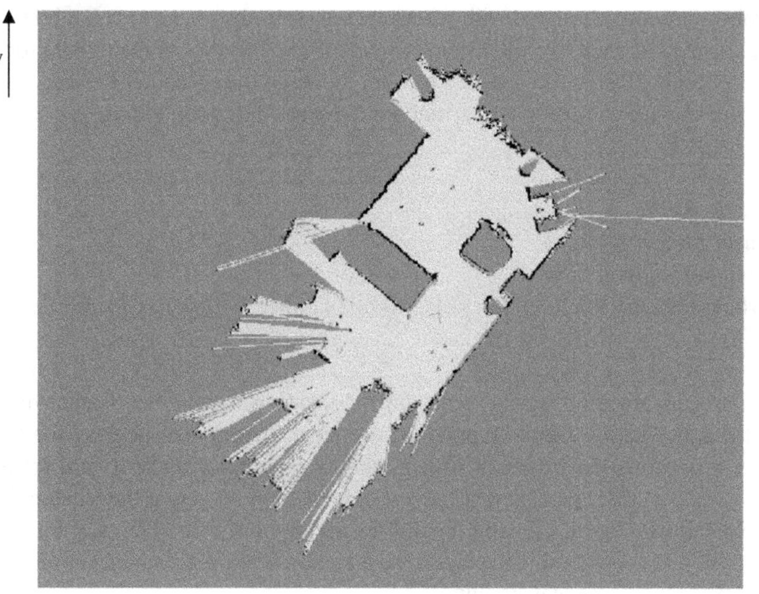

FIGURE 22.5 A preview of the mapping of an unknown environment with the necessary hardware requirements using ROS.

- SLAM.
- Cartographer.

Mapping based on a monocular camera:

- LSD SLAM (large scale direct monocular SLAM).
- ORB SLAM.
- Direct sparse odometery (DSO).

Depth camera mapping techniques:

- ZEDfu.
- Real time appearance based mapping (RTAB map).
- ORB SLAM.
- Stereo parallel tracking and mapping (S-PTAM).

22.3.2 Localization

Localization is the estimation of the mobile robot's location in the state space. Localization is the operation of performing a task, where the robot is situated with respect to the environment. GPS is also used for localization, but in terms of latitude and longitude, which is applicable and generated uniquely over a worldwide range.

Mobile robots are also equipped with wheel encoders, so they have to count how far they have traveled and the overall distance covered. The location of the mobile robot is updated periodically, or the distance moved, by using the extended Kalman filter (EKF). There are a few techniques carried out to locate the robot with respect to the environment:

- Vehicle model.
- Sensor model.
- Extended Kalman filter.
- Particle filter. (Alencastre-Miranda, Munoz-Gomez, Murrieta-Cid, & Monroy, 2006)

In other words, localization means specifying the location of the robot in an environment. If the robot is place in a mapped environment, then it will be easy for it to localize its pose in the environment. When outdoor conditions make it hard for the robot to localize in that environment, we use a GPS module, using the trilateration method. Sometimes due to climatic conditions and interference GPS may lose the connection, so it is very feasible to place a local marker or beacon in the environment to localize the robot using checkpoints that it passes.

22.3.2.1 Vehicle Model

The arithmetic models represent the action of the mobile robot and the sensors integrated on it, which are the key features that play a vital role in localization once the pose of the robot is updated. The kinematic model represents the motion of the robot as a rejoinder to the driving actions. Here, we are going to investigate a unicycle mobile robot design and discuss its kinetics, angular velocity, and forward/backward velocity. It is a two-wheeled robot with dedicated motors to each wheel, as shown in Figure 22.6

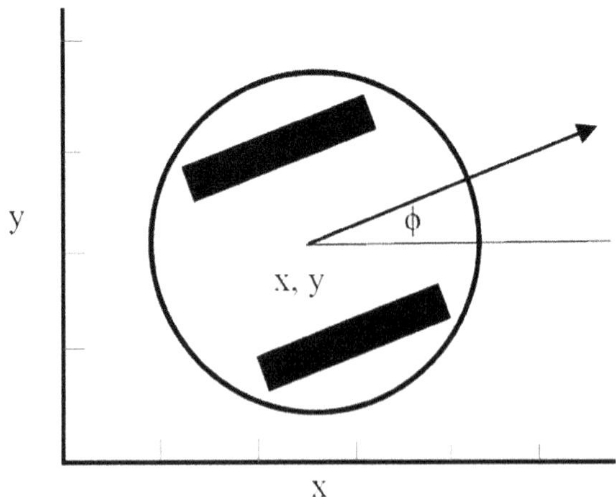

FIGURE 22.6 A differential drive robot with two motors which can operate in two different planes.

The kinematics regulating the motion of the differential drive mobile robot are provided by:

$$x(t) = \left(v(t) + \delta v(t)\cos(\vartheta)t\right) \tag{22.1}$$

$$y(t) = \left(v(t) + \delta v(t)\sin(\vartheta)t\right) \tag{22.2}$$

$$\vartheta(t) = w(t) + \delta w(t) \tag{22.3}$$

where: the coordinates x(t) and y(t) represent the position of the center of the mobile robot at time variant (t); orientation is denoted by ϑ(t), which is the angle between the heading and x-axis of the global coordinate of the frame; x(t) denotes the derivatives of $x^{\cdot}(t)$ with respect to time t. The forward velocity v(t) and angular velocity w(t) are important control inputs of the mobile robot. δv(t) and δw(t) are the values which, in the difference between the intended control and actual control values (noises), are assumed to be zero in Gaussian. The sampling time ΔT and the Euler method results are:

$$x^{\cdot}_{k+1} = x_k + \left(v_k + \delta v_k\right)\Delta T\cos\left(\vartheta_k\right) \tag{22.4}$$

$$y^{\cdot}_{k+1} = y_k + \left(v_k + \delta v_k\right)\Delta T\sin\left(\vartheta_k\right) \tag{22.5}$$

$$\vartheta^{\cdot}_k + 1 = \vartheta_k + \left(w_k + \delta w_k\right)\Delta T \tag{22.6}$$

where: x_k, y_k, ϑ_k are the location of the mobile robot with respect to time k; v_k is the velocity with respect to the time k; w_k is the angular velocity with respect to k; and δw_k is the discrete-time and angular velocity noises, respectively.

22.3.2.2 Sensor Model

Assume an environment that is packed with N landmarks at known position (xL, yL), i = 1…N. In the sensor model, the mobile robot is equipped with a laser sensor, which emits light at a certain frequency and waits for the reflection when it falls on an object. If there is no reflection then there's no object nearby; it can also define the location and distance of objects from the robot. The relative angle to the object in respect of the mobile robot and the world axis coordinates is given as input to the robot for reference.

$$r_{k+1} = \sqrt{\left(x_L - x^{\cdot}_{k+1}\right)^2 + \left(y_L - y^{\cdot}_{k+1}\right)^2}\, w_r \tag{22.7}$$

$$\varphi_{k+1} = \operatorname{atan}\left(\frac{y_L - y^{\cdot}_{k+1}}{x_L - x^{\cdot}_{k+1}}\right) - \vartheta^{\cdot}_{k+1} + w_\theta \tag{22.8}$$

where: w_r and w_θ are the zero-mean Gaussian observation noises.

Most sensors used for obtaining range and bearing measurements to landmarks are ultrasonic sensors and laser range finders (LRFs). φ_{k+1} is the sensor model and is able to observe the bearing for camera movement. For the objects observed by the sensors mounted on the robot, if the reflection is received by the sensor then it is denoted by the term d_{max}, which means the distance between the robot and the object is at a maximum and not nearby.

For an overly complicated task like SLAM to be applied, a map needs to be build based on the pose. To do so, pose estimation is essential, without any uncertainties as to the sensory information. Hence, various localization methods have been proposed to overcome these uncertainties by the use of a Kalman filter (Indiveri, 2002).

22.3.2.3 State Estimation

- Estimate the state of the mobile robot (x) of a system, given observation-like sensors (z) and control-like motors (u).
- Goal p (x | z, u). (Cook & Zhang, 2020)

There are two important situations where the EKF method is not a good choice for localization of the robot. The first is when the environment is taken by an occupancy grid, which provides the information of all the grids, whether occupied or free. This process is carried out using a laser scanner, which emits light at a certain frequency and waits for the reflection when it falls on an object; if it does not reflect then there is no object in the next grid; it can also define the distance of the object from the robot. The range measurement observed by the robot depends upon the environment and its own location, although it is not feasible to find an analytical observation model (Queiroz, Cai, & Feemster, 2019). Using the state estimation condition only allows the planning to be tracked, whether the robot has moved from point A to B or not. This also registers what path the robot goes along and how long it will take for it to reach the goal state, which is based on the linear and angular velocity of the robot. In the ROS software, these two parameters are denoted in meters per second (m/s). This also gives feedback connection to the local and global planners, which define the path for the robot. The static objects are also localized using the particle filter, so that the robot can avoid the objects easily and travel on a safe and secure path to reach its goal state.

22.3.2.4 Support for SLAM Using a Kalman Filter

A Kalman filter (KF) is a recursive filter. In control theory it is also branded as a linear quadratic estimation that uses iterative measurements calculated over time, including noise (that is observed from the surroundings) and imprecisions and provides an estimation of unknown variables by estimating the joint probability distribution to track the status of the robot and obstacles. KF is based on Bayes probability theory, which assumes the model and predicts the present state from earlier states. It also estimates the error between the predicted values and the previous steps obtained by the measuring instruments used to update the estimation to achieve accurate state values, as visualized in Figure 22.7 (Andreasson, Treptow, & Duckett, 2005).

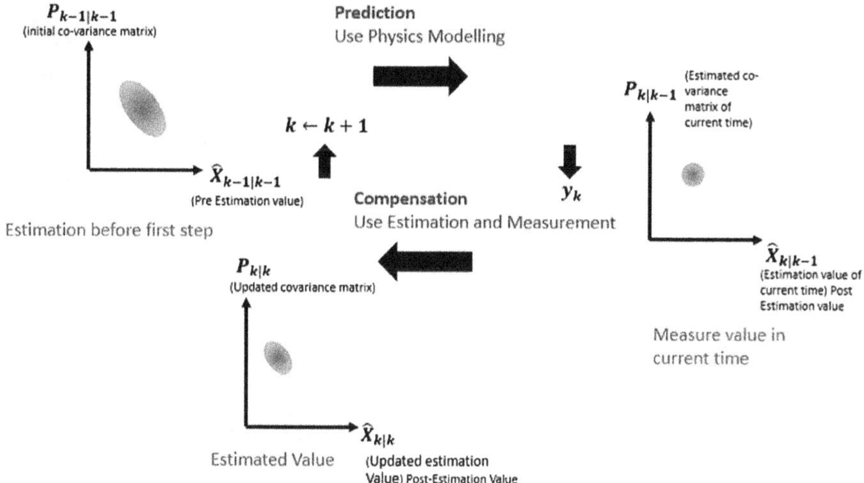

FIGURE 22.7 Kalman filter estimates the state of the system and variants, and updates the state after considering the system state y_k at k times step.

A KF completes a goal by using the linear projections, while a particle filter uses the sequential Monte Carlo localization (MCL). Both are different techniques but show significant outcomes. The better filter is discussed in the conclusion and is justified with a proper proof and statistics.

The accuracy of the model can be improved by repeating the Kalman filtering.

A Bayes filter is the base for the KF and extended Kalman filter (EKF) for mapping and localizing the robot.

Recursive Bayes filter 1:

$$\mathrm{bel}(x_t) = p(x_t \,|\, z_{1:t}\, u_{1:t}) \tag{22.9}$$

Recursive Bayes filter 2: (Bayes' Rule)

$$\mathrm{bel}(x_t) = \eta\, p(z_t\,|, x_t|, z_{1:t-1}|, u_{1:t})\, p(u_t \,|\, z_{1:t-1}\, u_{1:t}) \tag{22.10}$$

Recursive Bayes filter 3: (Markov Assumption)

$$\mathrm{bel}(x_t) = \eta\, p(z_t \,|\, x_t)\, p(x_t, z_{1:t-1}, u_{1:t}) \tag{22.11}$$

Recursive Bayes filter 4: (Law of total probability)

$$\mathrm{bel}(x_t) = \eta\, p(z_t \,|\, x_t) \int_{x_{t-1}} p(x_t \,|\, x_{t-1}, z_{1:t-1}, u_{1:t})\, p(x_{t-1}, z_{1:t-1}, u_{1:t})\, dx_{t-1} \tag{22.12}$$

Recursive Bayes filter 5: (Markov Assumption)

$$\text{bel}(x_t) = \eta p(z_t | x_t) \int_{x_{t-1}} p(x_t | x_{t-1}, u_t) p(x_{t-1}, z_{1:t-1}, u_{1:t}) dx_{t-1} \tag{22.13}$$

Recursive Bayes filter 6: (Markov Assumption)

$$\text{bel}(x_t) = \eta p(z_t | x_t) \int_{x_{t-1}} p(x_t | x_{t-1}, u_t) p(x_{t-1} |, z_{1:t-1} |, u_{1:t-1}) dx_{t-1} \tag{22.14}$$

Recursive Bayes filter 7: (Recursive term)

$$\text{bel}(x_t) = \eta p(z_t | x_t) \int_{x_{t-1}} p(x_t | x_{t-1}, u_t) \text{bel}(x_{t-1}) dx_{t-1} \tag{22.15}$$

Prediction Step:

$$\overline{\text{bel}}(x_t) = \int p(xt | ut, xt - 1) \text{bel}(x_{t-1}) dx_{t-1} \tag{22.16}$$

Correction Step:
Sensor or observation model (Bennewitz, Stachniss, Burgard, & Behnke, 2006; Nistic`o & Hebbel, 2009; Stachniss, 2009; Mottaghi & Vaughan, 2006; Giorgio, Stachniss, & Burgard, 2007)

$$\text{bel}(x_t) = \eta p(z_t | x_t) \overline{\text{bel}}(x_t) \tag{22.17}$$

22.3.2.5 Particle Filter

The particle filter is used in object tracking, and is popular in MCL. The KF is accurate for linear systems that have Gaussian noise. Mostly the non-linear systems used in common are robots and sensors. Therefore, a particle filter is used to predict all the way through a simulation based on the trial-and-error method. The estimated values obtained from the probability distribution (PD) in the system are signified as particles. The particle filter, otherwise called the sequential Monte Carlo (SMC) method, performs pose estimation by estimating the error from the input information. When a particle filter is implemented while using SLAM, the robot's current position is estimated through range sensors and odometry. The particles or samples describe the uncertain pose of the robots. The particles are then moved to another assessed present, dependent on the portable robot's model and probabilities. The weightage of every particle is estimated by the genuine estimated value and the noise is step-by-step decreased to assess an exact position.

The technique for estimation of the posture of the robot is one of updating the conveyance of the particles that uncovers the probability of the portable robot on the X–Y coordinate plane fixated on the estimated sensor value. Every molecule of a portable robot is a portrayal of the estimated pose of the robot, communicated as x, y, i, and the weight of every particle as:

Particle = pose (x, y, i). (Garcia, Montiel, Castillo, Sepu´lveda, & Melin, 2009)

The particle is estimated by the following procedure:

- Step1: Initialization. The robot's initial pose is unknown in a new space. Along these lines, the particles are nonchalantly organized inside a reach where the pose can be obtained with N particles, with every one of its underlying particle weights as 1/N delivering a whole of particle weight to 1.
- Step 2: Forecast based on system model. This depicts the motion of the portable robot; it moves each particle according to the odometry data and noise.
- Step 3: Update based on measured sensor information. Based on the measured sensor information, the probability of every particle is calculated, and its weight is then revised.
- Step 4: Pose estimation. The pose and weight of each particle is utilized to estimate the pose of the robot by means of mean, median, and maximum weight values.
- Step 5: Resampling. Resampling is the step of engendering new particles to eliminate less weighted particles and produce new particles to balance the number of particles N.

When estimating the position, if the number of particles is enough, the particle filter can be more precise than EKF obtained through the KF. However, when the number of particles is not sufficient, a SLAM-based Rao-Blackwelled particle 336 ROS robot programming filter (RBPF) is used which utilizes both the particle and Kalman filter (Grisetti, Stachniss, & Burgard, 2005, 2007).

22.4 NAVIGATION

The mobile robot navigation scope is an approach that lets you steer the mobile robot to achieve a goal state in the state space with static and dynamic obstacles in a good and harmless way by avoiding all these obstacles. There are two essential tasks involved in navigation: environment perception and path following. The idea of mission refers to the realization of a set of navigation and operational goals; in this sense, the mobile robot should have an architecture able to coordinate the onboard components: the sensorial system, movement, and operational control, to attain properly the different aims detailed in the mission with an efficiency that can be supported either in indoor or outdoor environments (Niku, 2001b). Commonly, global planning methods supplemented with local methods are used for indoor missions since the surroundings are partially known; for outdoor applications, local planning methods are also suitable, with global planning methods a counterpart, because of the adequate information of the surroundings (Garcia, Montiel, Castillo, Sepu´lveda, & Melin, 2009).

The navigation problems of a mobile robot can be split up into four subproblems:

- World perception: The mobile robot senses the world it represents in features.

- Path planning: The mobile robot uses the features to generate an ordered progression of objective points that it should attain.
- Path generation: The goal is to achieve a path through the series of objective points.
- Path tracking: The mobile robot controls the path it is to follow. (Triebel & Burgard, 2005)

22.4.1 Definition of the SLAM Problem

The main objective and the solution for the SLAM problem is to map an unknown environment and travel in that environment avoiding obstacles and following the shortest distance between the two points or multiple points based on the application.

The robot control variables are:

$$u_1 : T = \{u_1, u_2, u_3, \ldots, u_T\}$$

The observations are:

$$z_1 : T = \{z_1, z_2, z_3, \ldots, z_T\}$$

where:

u_1 ➔ is the command to drive/stop for 1 meter (or) turn 45° to the left/right side;
z_1 ➔ laser scanner for sensing and knowing/mapping the environment.

Requirements:

- Map of any/specific environment (m).
- Path of mobile robot ➔ x_0: T = {x_0, x_1, x_2, . . ., x_T}.

22.4.2 Homogeneous Coordinates

The representation x of a geometric object is homogeneous if x and λx represent the same object for $\lambda \neq 0$.

22.4.3 Homogeneous Taking Euclidian into Consideration

$$X = \begin{bmatrix} u \\ v \\ w \end{bmatrix} = \begin{bmatrix} wx \\ wy \\ w \end{bmatrix} = \begin{bmatrix} x \\ y \\ 1 \end{bmatrix} \quad X = \begin{bmatrix} x \\ y \end{bmatrix}$$

$$\begin{bmatrix} u \\ v \\ w \end{bmatrix} = \begin{bmatrix} \frac{u}{w} \\ \frac{v}{w} \\ 1 \end{bmatrix} \rightarrow \begin{bmatrix} \frac{u}{w} \\ \frac{v}{w} \end{bmatrix} = \begin{bmatrix} x \\ y \end{bmatrix} \text{ (dividing by the last component).}$$

22.4.4 Transformation

A projective transformation is an invertible linear mapping:

$$X' = Mx.$$

22.4.5 Translation: (Three Parameters) ➔ (Three Translations)

If a frame moves in the space with no adjustment in its direction, the change is a pure translation. For this situation, the directional unit vectors remain similar and consequently are not modified:

$$M = \lambda \begin{bmatrix} I & t \\ Ot & 1 \end{bmatrix}$$

$$I = \begin{bmatrix} 1 & 0 & 0 \\ 0 & 1 & 0 \\ 0 & 0 & 1 \end{bmatrix}$$

$$t = \begin{bmatrix} tx \\ ty \\ tz \end{bmatrix}$$

$$O^T = \begin{bmatrix} 0 \\ 0 \\ 0 \end{bmatrix}$$

22.4.6 Rotation: (Three Parameters) ➔ (Three Rotations)

To unravel the derivation of rotations about an axis, first assume that the edge is at the inception of the reference outline and corresponds to it. Later, grow the outcomes to different rotations just as if they were combinations of turns:

$$M = \lambda \begin{bmatrix} R & 0 \\ 0 & 1 \end{bmatrix}$$

22.4.7 Rotation Matrices

For two dimensions:

$$R(\theta) = \begin{bmatrix} \cos\theta & -\sin\theta \\ \sin\theta & \cos\theta \end{bmatrix}$$

For three dimensions:

$$Rx(\theta) = \begin{bmatrix} 1 & 0 & 0 \\ 0 & \cos\theta & -\sin\theta \\ 0 & \sin\theta & \cos\theta \end{bmatrix}$$

$$Ry(\theta) = \begin{bmatrix} \cos\theta & 0 & \sin\theta \\ 0 & 1 & 0 \\ -\sin\theta & 0 & \cos\theta \end{bmatrix} \quad Rz(\theta) = \begin{bmatrix} \cos\theta & -\sin\theta & 0 \\ \sin\theta & \cos\theta & 0 \\ 0 & 0 & 1 \end{bmatrix}$$

22.4.8 Rigid Body Transformation: (Six Parameters) → (Three Translations + Three Rotations)

$$M = \lambda \begin{bmatrix} I & t \\ Ot & 1 \end{bmatrix}$$

22.4.9 Similarity Transformation: (Seven Parameters) → (Three Translations + Three Rotations + One Scale)

$$M = \lambda \begin{bmatrix} mR & t \\ Ot & 1 \end{bmatrix}$$

22.4.10 Affine Transformation: (Twelve Parameters) → (Three Translations + Three Rotations + Three Scales + Three Spheres) (Mautz & Tilch, 2011)

$$M = \lambda \begin{bmatrix} A & t \\ Ot & 1 \end{bmatrix}.$$

22.5 SIMULATION

Discovering maps with the mobile robot is a high-dimensional state estimation challenge that entails a simultaneous solution to the question of how to renew the map given sensory input and an estimate of the pose of the mobile robot for the given map.

The following map is built in ROS using TurtleBot 3, which is the exact simulation of the real-life mobile robot. TurtleBot 3 is built with a single board computer (SBC) with a Raspberry Pi 4 as the processor and OpenCR integrated with the IMU which acts as the controller for the robot and the 360-degree servo motors use wheel movements and an encoder measure for the distance. In the customized version, most of the users change the controller to Arduino and attach an external IMU because it is cheaper than OpenCR and effective in motion. The servo motors are replaced with the encoded DC motor. The environment is mapped using a LiDAR (Light Detection and Ranging), the light being discharged from a continuously firing laser. The light travels to the ground or any objects around it and reflects an image of the things, such as buildings, tree, and cars. The reflected light energy returns to the LiDAR sensor where it is recorded.

To map any environment, the mobile robot should explore the place by moving to each nook and corner. This also applies to humans, as when we go to a new place and explore it and get to know how things are placed and how spacious it is. From humans, it is taken to the robot for exploring the place and being helpful in performing the SLAM. Even though a new obstacle is placed after mapping and is on the path of the goal state, the robot will avoid the obstacle and reach the goal state. If the object is removed from the path afterward, the mobile robot will sense that and move according to its goal state without colliding with any new obstacles and still follow the planned path to reach the goal state with reduced distance.

In this simulation, we implemented Gmapping which is a highly efficient Rao-Blackwellized particle filter to train the grid maps using sensors like ultrasonic ones and an LRF. We conducted this simulation because it is more efficient when compared to other techniques. Even in localization there are few disadvantages to EKF when compared to the particle filter. Therefore, Figure 22.8 shows the predefined mapping of an environment in the simulation software (Koubâa, 2019).

The path planning problem is also well known as NP-hardness (Chen & Quan, 2008), where the complexity increases with the degrees of freedom (DOF) of the robot. While in simulation, the differential drive robot is a ground vehicle, so it is reduced to 2D space (Bačík, et al., 2017). In Figures 22.8–22.11, we can see that the mobile robot moves from the initial state to the goal state with the shortest distance traveled and selects the node points based on its location. There are only a few software packages available for simulation with full access; among them is a powerful tool called ROS. ROS is a plastic framework for writing robot software. The main ROS client libraries are equipped for a Unix-like system, primarily because of their dependence on a large collection of open-source software. There are a set of tools in ROS for taping from and playing back to ROS topics. This is anticipated to be high performing and to avoid deserialization and reserialization of the message (Babaians, Tamiz, Sarfi, Mogoei, & Mahrabi, 2018).

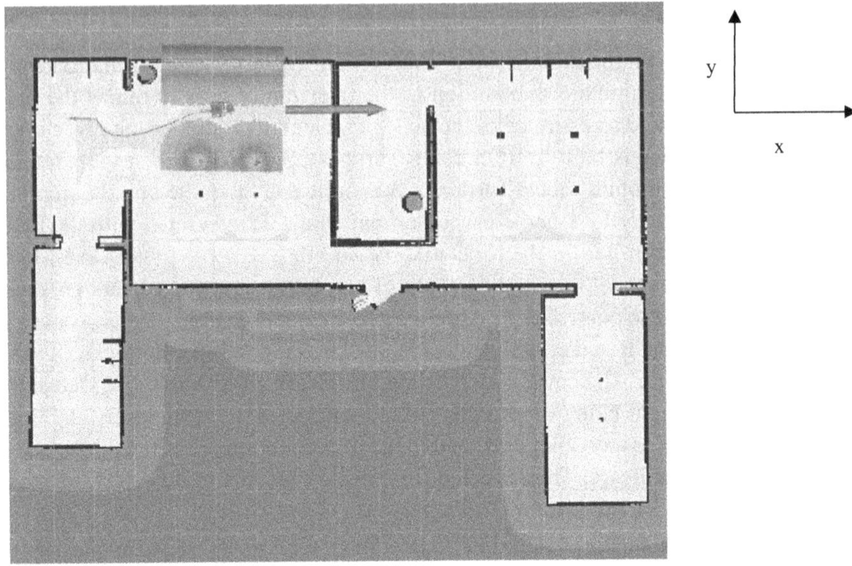

FIGURE 22.8 Mapping the TurtleBot 3 house in a robot operating system.

Notes: The large green arrow represents the direction the mobile robot faces towards and its current position; the small red arrow defines the goal state and pose.

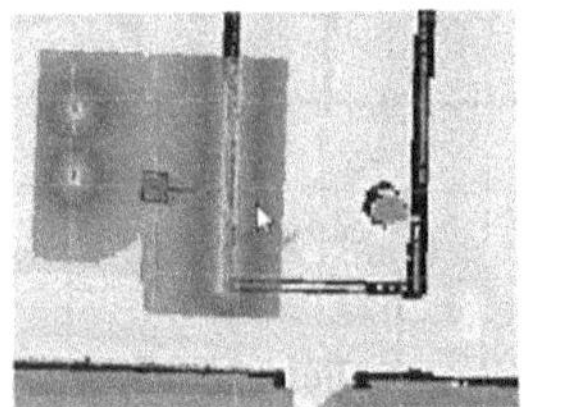

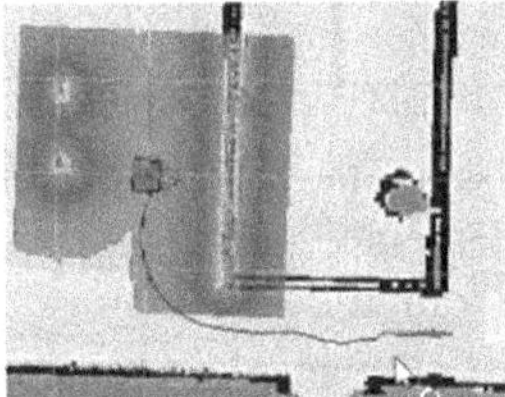

FIGURE 22.9 The mobile robot is positioned at a place in the house.

Note: The small red arrow denotes the pose of the mobile robot.

22.6 HARDWARE REQUIREMENTS AND IMPLEMENTATION IN THE REAL WORLD

There are certain components that should be fused together to build autonomous vehicles or robots because heavy intensive computational values need to be calculated for the application it will use. Figure 22.10 shows a basic block diagram designed for indoor autonomous robots. If the application is different, the robot needs to ride around the outdoor environment; then the parameters will increase and

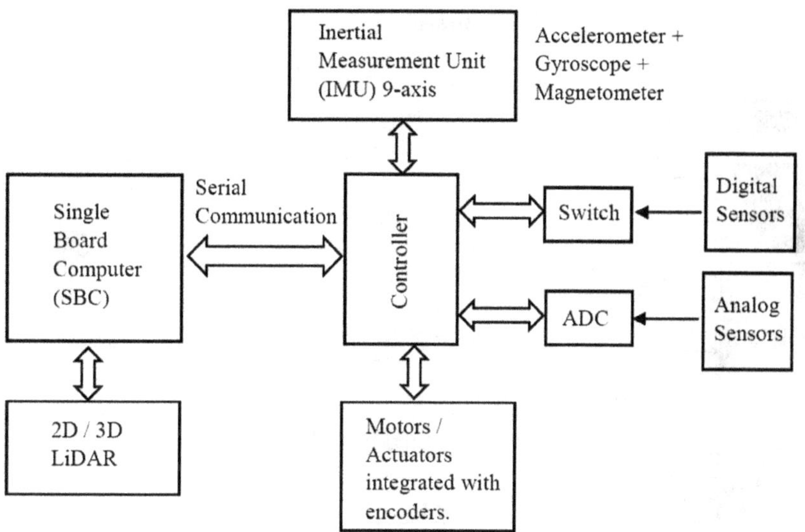

FIGURE 22.10 Basic requirement of an autonomous robot and how parts are integrated.

there will be more functions to calculate; simultaneously complexity will also increase. A simple hardware mapping is shown in Figure 22.10.

22.6.1 Single Board Computers

As we all heard in childhood, the "CPU is the heart of the computer"; likewise for an autonomous robot there are too many complexities that can be solved by a controller; but a processor can handle all these tasks and perform them without any issue. All self-driving and most smart cars have an in-built SBC. This is because the acceleration of the processor helps the machine to perform more and heavier tasks, which cannot be done by any of the controllers, due to their reduced processing capacity. Based on the requirement of the mobile robot, hardware is customized so that we can easily list it from high specification to low level. Among the high range computing devices, Nvidia is a much-preferred device for deep learning and self-driving cars and is able to push the limit of computationally intensive applications. Due to its heavy hardware compatibility with other external devices and sensors, it is widely preferred for computation and acts as a brain for the system. From school to many commercial and industrial-grade products, Raspberry Pi is used due to its cost-effectiveness and decent computation unit. Although it is introduced for educational purposes later, it is used for mobile robots (Peng, Zhang, Liu, Asari, & Loomis, 2019). The SBC is the brain of autonomous vehicles: they need to process a lot of data and calculate the risk factors. A LiDAR sensor can be attached to the SBC,

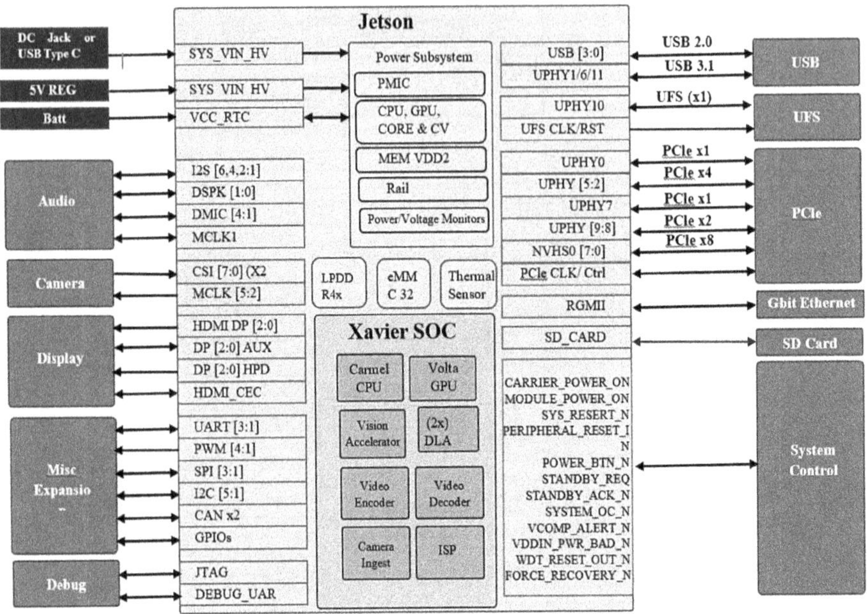

FIGURE 22.11 The rich set of high-speed input/output ports of Nvidia AGX Xavier.

because of its high-computational output. SBCs are most like normal computers with high processing power, but they operate and communicate via a meta-operating system using a complete operating system. A few widely used SBCs were mentioned above for computational purposes. There are other exteroceptive sensors that need to be attached to the SBC: (a) a camera, for computer vision; (b) a stereo-camera, for depth imaging; (c) RADAR, for robust object detection and relative speed detection; and (d) SONAR, short range and all-weather distance measurement at low cost. We have mentioned the architecture of the Nvidia AGX Xavier, which is used in many self-driving cars from Nvidia itself. Its architecture is shown in Figure 22.11.

22.6.2 Inertial Measurement Unit (IMU)

An IMU is a very important unit for all true autonomous robots. The development of micro-electromechanical systems (MEMS) has made it viable to fabricate cheaply a single chip. The main objective of the IMU is to define the position of the robot, the direction it is facing, and how much it is elevated above the ground. An IMU is a cluster of accelerometer, gyroscope, and magnetometer, where each sensor is used for a different purpose with the goal of localizing the robot and providing its pose (position + orientation) (Skog & Handel, 2006). Based on the purpose, a few other sensors are also attached. The MEMS sensor has made it possible to build a low-cost global navigation satellite system (GNSS) aided by an inertial navigation system (INS) for monitoring the robot's movement and localizing it (Skog, 2006).

In the phase of programming, we need to consider a few parameters before initiating the ports of the IMU. This is because the accelerometer, gyroscope, and magnetometer have different addresses in hexadecimal and decimal. Here, we will consider the MPU-6000 variant and see how to trigger the registers of the sensors. There are certain addresses necessary for accessing registers in the accelerometer, gyroscope, and magnetometer. Tables 22.1 and 22.2 represent the registers in hexadecimal format. The mechanism and working principle are visualized in Figure 22.12. These sort of sensors fall under a MEMS concept as we discussed earlier in this section.

Since its functionality is different compared to these two sensors, the magnetometer does not operate like an accelerometer or a gyroscope. In the gyroscope, the accelerometer senses the forward velocity and any vibration, and measures how stable the structure is from a point and at a certain angle. The functionality of the accelerometer and gyroscope is the same around the world. But it varies from place to place for the magnetometer, since its magnetic attraction value can vary in various locations. We use GPS for navigation in cell phones, and we can clearly see a blue color facing the way we travel from one position to another, which shows because of the magnetometer values and how they vary by changing the angle (Specification, 1998).

Sensors such as the accelerometer, gyroscope, and magnetometer can now be used in smart phones. When technology advances and evolves, a revolution enters the commercial market, and educates everybody. Smart phones were the first revolution, and it was believed that self-driving cars would be the second revolution. Scientists have incorporated several sensors for a smart or palm-held computer, but the lives of

TABLE 22.1
Register Mapping of Accelerometer in the MPU-6000 Series

Serial No.	Hexadecimal Address	Decimal Address	Register Name
1	3B	59	ACCEL_XOUT_H
2	3C	60	ACCEL_XOUT_L
3	3D	61	ACCEL_YOUT_H
4	3E	62	ACCEL_YOUT_L
5	3F	63	ACCEL_ZOUT_H
6	40	64	ACCEL_ZOUT_L

TABLE 22.2
Register Mapping of Gyroscope in the MPU-6000 Series

Serial No.	Hexadecimal Address	Decimal Address	Register Name
1	43	67	GYRO_XOUT_H
2	44	68	GYRO _XOUT_L
3	45	69	GYRO _YOUT_H
4	46	70	GYRO _YOUT_L
5	47	71	GYRO _ZOUT_H
6	48	72	GYRO _ZOUT_L

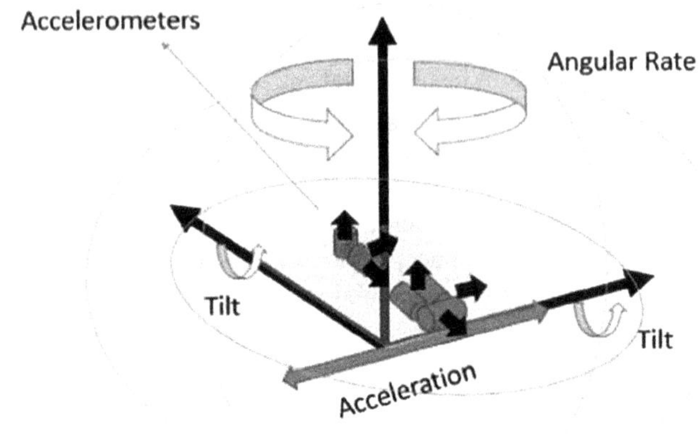

FIGURE 22.12 Insight of inertial measurement unit.

people in autonomous vehicles are a trade-off, so you need to be alert and careful about the environment. Normally, the field of view of a human driver is 120 degrees; if these types of sensors are combined then the estimate of vision and condition would be 360 degrees, which is three times greater than humans and eliminates errors. Users can communicate these sensor values with other vehicles and share the state estimation values with them. It would be easy to run autonomous vehicles with zero accidents in a connected network and remove 92% of human error (http://www.magnetic-declination.com/).

22.6.2.1 Odometry Model of Robot with Wheel Encoders

By integrating visual odometry with traditional wheel odometry and other sensor devices, a posture evaluation produced in this way can be enhanced. This can be accomplished through various methods of flag preparation known as the combination of sensors or sensor fusion. The KF is particularly useful for integrating signals from various sensors and eliminating localization errors that occur due to numerous components such as sensor clamor, quantization, imperfect display planning, and robot moves from $(\overline{x}, \overline{y}, \overline{\theta})$ to $(\overline{x}', \overline{y}', \overline{\theta}')$ sensor inclination or drift as shown in Figure 22.13. Typically, in the static position or the home position or pre-motion position, these values $(\overline{x}, \overline{y}, \overline{\theta})$ are stated; once the vehicle is moved from its initial position to the target position, then the modified values are represented by $(\overline{x}', \overline{y}', \overline{\theta}')$. It is also appropriate at low speeds to look only at kinematic vehicle models. Dynamic modeling is more involved, but over a large operating area, vehicle behavior is more accurate. When creating the vehicle kinematic model, there are some words to be recalled: ENU – Earth North UP; ECEF – Earth centered Earth fixed; and odometry information.

Wheel encoders are utilized to measure the precise speed of each wheel and can give data for odometry and localization. The angular speed of the left and right wheel

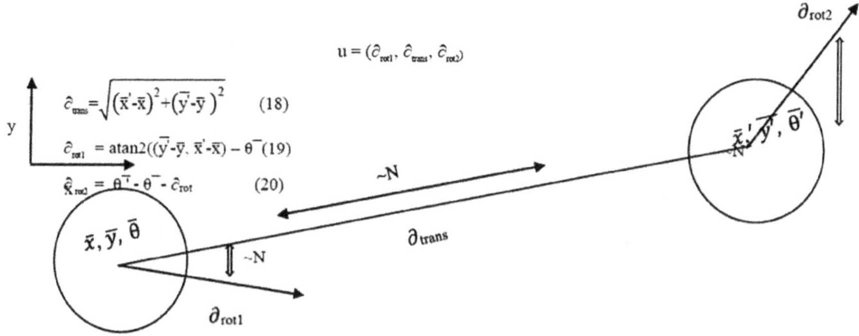

FIGURE 22.13 Robot moving from the initial position to the goal state with a change in pose.

is given by ω_R and ω_L, individually. These estimations are changed into the direct speed of the wheels, given by:

$$v_L = \omega_L \text{ x r} \tag{22.18}$$

$$v_R = \omega_R \text{ x r} \tag{22.19}$$

where r ϵ R^+ is the constant radius of the wheel. Denoting the constant distance between the wheels b ϵ R^+, the linear and rotational velocity of the robot is given by (Chen, Lee, & DeBra, 2012):

$$v_x = \left(\left(v_L + v_R\right)/2\right) \tag{22.20}$$

$$\omega_z = \left(\left(v_L - v_R\right)/b\right) \tag{22.21}$$

22.6.3 Basic Principle of the Accelerometer

The accelerometer is a precision instrument intended to measure a part of the vector amount called specific constrain f. It is characterized within the condition:

$$f = a - g \tag{22.22}$$

To acquire the inertial acceleration of the mobile robot, we need to calculate g with the use of:

$$a = f + g \tag{22.23}$$

a ➔ inertial acceleration
g ➔ gravitational force attraction per unit mass (Shen, Tick, & Gans, 2011).

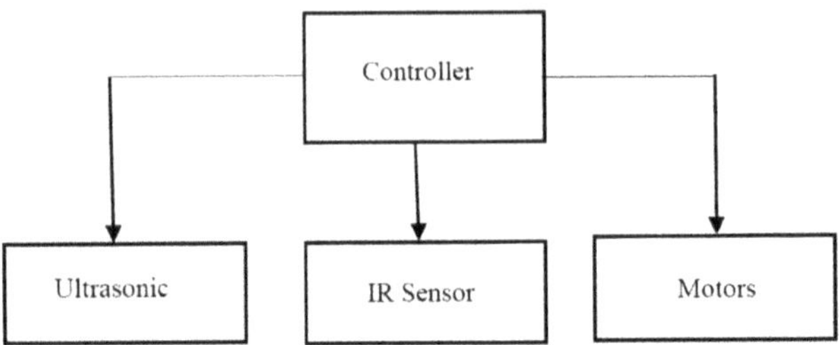

FIGURE 22.14 Block diagram of controller part of an autonomous robot.

22.6.4 Controllers

There is a wide range of controllers used for autonomous vehicle navigation systems; some developers also use customized controllers for the system to actuate motors. Controllers like OpenCR have an inbuilt IMU in it for better calibration and quicker receiving and transmitting of signals than an external device (Guizzo & Ackerman, 2017). Most of the developers prefer the Arduino and external IMU because it is cheaper than OpenCR and as effective. Similarly, the motors also change according to the environment and amount of torque required, based on the application of the robot. The controllers will be connected to the SBC via a USB cable which provides the in and out data transmission, so based on other sensor values a robot can move easily in the space. Figure 22.14 represents how the controller is connected to the exteroceptive sensors and motors.

22.7 LIMITATIONS

SLAM has several limitations in terms of efficiency with the use of computer vision as it is found to be successful only in a map of a building or an indoor environment. Mapping will become complex with the change in dynamics of the robot's environment. It is limited by inconsistent sensor measurements and updating of the state estimation. A KF is based on a Gaussian assumption and it is slow in maps with larger dimensions. Localization requires robust features and data association as sometimes it needs to merge multiple maps. It has high computational costs, and it is unstable in an environment with large dimensions. Outdoor environment mapping is hard for the robot because we need to calculate more parameters than in indoor conditions, like wind resistance, friction, climatic conditions, and ups and downs in the road. In indoor conditions, the climate will be the same all day due to its lighting conditions, but outside the climate will vary, which leads to false values of the sensors that are trying to learn about the environment. This climatic condition, due to this period's limited technology, may lead to the fatality of passengers or pedestrians due to errors in the driving robot. Wind resistance and friction can be managed and eliminated with training and machine learning of various environments and repeated training to gain more feasibility. Some places on roads are damaged and when the vehicle approaches any plot hole or manhole it should slow down for the passengers' convenience.

22.8 FUTURE SCOPE

The implementation of localization and mapping have endless possibilities for solving numerous challenges in terms of navigation and perception for autonomous robots, wherever it needs to find its current position, build the map of the environment in the case of an unknown environment, or constantly update the states of estimation in the case of dynamic environments. SLAM is used in the navigation tools in autonomous vehicles, aerial vehicles, and industrial robots for mapping the environment and estimating the states. It can devise a solution to create mechanisms to compute and summarize statistical data from past behavior. The future of SLAM is likely to focus on unresolved challenges in improving robust performance, robot perception, advancements in the construction of maps with large dimensions, high level computation, awareness about the environment, integration of multiple maps, task-oriented research on updating and refreshing information, and defining when to access at a larger scale in robotic systems.

22.9 CONCLUSION

In extensive Monte Carlo simulations, it has been verified that FEJ-EKF is more precise and more stable than both the standard EKF and robot-centric mapping (Castellanos, Neira, & Tard´os, 2004). All these techniques consider inputs given by humans for certain constraints, like wheelbase, width, diameter, and friction, based on this localization. These areas need to be concentrated and improved, more for feasibility. 3D sensors require more computational power because more nodes running altogether occupy memory. Optimization of the techniques and algorithm need to be done for less computational power and high feasibility. A particle filter uses more particles for localization because more particles of memory will be also occupied. With the crossover technique, these can be reduced: there are algorithms built for this, but they are not robust enough to do the job. With the help of AI, we can expect these techniques to come into the picture in the near future.

REFERENCES

Ainsworth, D., Sternberg, M. J., Raczy, C., & Butcher, S. A. (2016). k-SLAM: Accurate and ultra-fast taxonomic classification and gene identification for large metagenomic data sets. *Nucleic Acids Research*, 45, 1649–1656.

Alan, A., & Pritsker, B. (1982). Applications of SLAM. *IIE Transactions*, 70–77.

Alencastre-Miranda, M., Munoz-Gomez, L., Murrieta-Cid, R., & Monroy, R. (2006). local reference frames vs. global reference frame for mobile robot localization and path planning. *IEEE Xplore*, 309–318.

Andreasson, H., Treptow, A., & Duckett, T. (2005). *Localization for mobile robots using panoramic vision, local features and particle filter.* IEEE Xplore.

Babaians, E., Tamiz, M., Sarfi, Y., Mogoei, A., & Mahrabi, E. (2018). ROS2Unity3D; High-Performance Plugin to Interface ROS with Unity3d engine. *IEEE Xplore*, 59–64.

Bačík, J., Ďurovský, F., Biroš, M., Kyslan, K., Perduková, D., & Padmanaban, S. (2017). Pathfinder–development of automated guided vehicle for hospital logistics. *IEEE Xplore*, 26892–26900.

Barfoot, T. D. (2017). *State estimation for robotics.* Cambridge University Press.

Behnke, S., & Veloso, M. M. (2013). *RoboCup 2013: robot world cup XVII.* Springer.

Bennewitz, M., Stachniss, C., Burgard, W., & Behnke, S. (2006). *Metric localization with scale-invariant visual features using a single perspective camera.* Springer-Verlag 195–209.

Biswas, J., & Veloso, M. (2013). Multi-sensor mobile robot localization for diverse environments. In *Robot Soccer World Cup*. Springer, Berlin, Heidelberg, 468–479.

Brown, R. H., & Schneider, S. C. (1987). Velocity observations from discrete position encoders. *In Signal Acquisition and Processing - International Society for Optics and Photonics.*, 1111–1118.

Carlson, J. D., Mittek, M., Parkinson, S. A., Sathler, P., Bayne, D., Psota, E. T., ... Bonasera, S. J. (2014). Smart watch RSSI localization and refinement for behavioral classification using laser-SLAM for mapping and fingerprinting. *IEEE Xplore*, 2173–2176.

Castellanos, J. A., Neira, J., & Tard´os, J. D. (2004). *Limits to the consistency of EKF-based slam.* Elsevier, 716–721.

Chatterjee, A., Rakshit, A., & Nirmal Singh, N. (2012). *Vision based autonomous robot navigation: Algorithms and implementations.* Springer.

Chen, J. H., Lee, S. C., & DeBra, D. B. (2012). Gyroscope free strapdown inertial measurement unit by six linear accelerometers. *Journal of Guidance Control and Dynamics*, 286–290.

Chen, B., & Quan, G. (2008). NP-Hard problems of learning from examples. *IEEE Xplore*, 182–186.

Cook, G., & Zhang, F. (2020). *Mobile robots: Navigation, control and sensing, surface robots and AUVs (Second Edition).* John Wiley & Sons.

Dissanayake, M. G., Newman, P., Clark, S., Durrant-Whyte, H. F., & Csorba, M. (2001). A solution to the simultaneous localization and map building (SLAM) problem. *IEEE*, 229–241.

Durrant-Whyte, H., & Bailey, T. (2006). Simultaneous localization and mapping: part I. *IEEE Xplore*, 99–110.

Fazli, S., & Kleeman, L. (2006). Simultaneous landmark classification, localization and map building for an advanced sonar ring. *Robotica*, 283–296.

Garcia, P. M., Montiel, O., Castillo, O., Sepu´ lveda, R., & Melin, P. (2009). Path planning for autonomous mobile robot navigation with ant colony optimization and fuzzy cost function evaluation. Elsevier, 1102–1110.

Giorgio, G., Stachniss, C., & Burgard, W. (2007). Improved techniques for grid mapping with Rao-blackwellized particle filters. *IEEE transactions on Robotics*, 34–46.

Grisetti, G., Stachniss, C., & Burgard, W. (2005). Improving grid-based SLAM with Rao-blackwellized particle filters by adaptive proposals and selective resampling. *IEEE Xplore*, 2432–2437.

Grisetti, G., Stachniss, C., & Burgard, W. (2007). *Improved techniques for grid mapping with rao-blackwellized particle filters. IEEE Xplore.*

Guivant, J., Nebot, E., Nieto, J., & Masson, F. (2004). Navigation and mapping in large unstructured environments. *The Interantional Journal of Robotics Research*, 449–472.

Guizzo, E., & Ackerman, E. (2017). The turtleBot3 teacher [Resources_Hands On]. *IEEE*, 19–20.

Hsiao, C. H., & Wang, C. C. (2011). Achieving undelayed initialization in monocular SLAM with generalized objects using velocity estimate-based classification. *IEEE Xplore*, 4060–4066.

Indiveri, G. (2002). *An introduction to wheeled mobile robot kinematics and dynamics. Slides for lecture given at RoboCamp, Paderborn (Germany).*

Khairuddin, A. R., Talib, M. S., & Haron, H. (2015). Review on simultaneous localization and mapping (SLAM). *IEEE*, 85–90.

Koubâa, A. (2019). *Robot operating system (ROS).* Springer.

Liu, T., Zhang, W., Gu, J., & Ren, H. (2013). A Laser Radar based mobile robot localization method. *IEEE Xplore*, 2511–2514.

Mautz, R., & Tilch, S. (2011). *Survey of optical indoor positioning systems. IEEE Xplore.*

Mottaghi, R., & Vaughan, R. (2006). An integrated particle filter and potential field method for cooperative robot target tracking. *IEEE Xplore*, 1342–1347.

Niku, S. B. (2001a). *Introduction to robotics: Analysis, systems, applications.* Semantic Scholar.

Niku, S. B. (2001b). *Introduction to robotics: Analysis, systems, applications* (Vol. 7). New Jersey: Princeton Hall.

Nistic`o, W., & Hebbel, M. (2009). *Particle filter with temporal smoothing for mobile robot vision-based localization.* Springer-Verlag Berlin Heidelberg, 167–180.

Ocando, M. G., Certad, N., Alvarado, S., & Terrones, Á. (2017). Autonomous 2D SLAM and 3D mapping of an environment using a single 2D LIDAR and ROS. *IEEE Xplore*, 1–6.

Peng, T., Zhang, D., Liu, R., Asari, V. K., & Loomis, J. S. (2019). Evaluating the power efficiency of visual SLAM on embedded GPU systems. *IEEE*, 117–121.

Queiroz, M. d., Cai, X., & Feemster, M. (2019). *Formation control of multi-agent systems: A graph rigidity approach.* John Wiley & Sons.

Shen, J., Tick, D., & Gans, N. (2011). Localization through fusion of discrete and continuous epipolar geometry with wheel and IMU odometry. *IEEE*, 1292–1298.

Skog, I., & Handel, P. (2006). Calibration of a mems inertial measurement unit. *XVII IMEKO World Congress*, 1–6.

Skog, I., Schumacher, A., & Handel, P. (2006). A versatile PC-based platform for inertial navigation. *IEEE*, 262–265.

Specification, U.S.B. (1998). *Revision 1.1.* In SBS-Implementers Forum.

Stachniss, C. (2009). *Mapping and localization in non-static environments.* Springer, 161–175.

Suzuki, T., Amano, Y., Hashizume, T., & Suzuki, S. (2011). 3D terrain reconstruction by small unmanned aerial vehicle using SIFT-based monocular SLAM, 292–301.

Thrun, S. (2008). *Handbook on robotics.* Springer.

Triebel, R., & Burgard, W. (2005). Improving simultaneous mapping and localization in 3D using global constraints. *AAAI*, 1330–1335.

Valencia, R., & Cetto, J. A. (2018). *Mapping, planning and exploration with Pose SLAM.* Springer.

Wang, Y., Liang, A., & Guan, H. (2011). Frontier-based multi-robot map exploration using particle swarm optimization. *IEEE Xplore*, 1–6.

Wurm, K. M., Stachniss, C., & Grisetti, G. (2010). Bridging the gap between feature- and grid-based SLAM. *Elsevier*, 58, 140–148.

23 A Novel Evolutionary Computation Method for Securing the Data in Wireless Networks

Inderpreet Kaur, Vibha Tripathi, and Nidhi
Galgotia College of Engineering and Technology

CONTENTS

23.1 Introduction 372
23.2 Ant Colony Optimization 372
    23.2.1 Biological Analogy 372
    23.2.2 ACO Algorithm 374
23.3 Artificial Bee Colony Optimization 375
    23.3.1 Bee Foraging Behavior 375
    23.3.2 Modified Approach 376
    23.3.3 Algorithm 377
23.4 Cuckoo Approach 377
    23.4.1 Introduction 377
    23.4.2 Cuckoo Reproductive Method 378
    23.4.3 The Lévy Flights Mechanism 378
    23.4.4 Algorithm 378
23.5 Particle Swarm Optimization (PSO) 380
    23.5.1 Introduction 380
    23.5.2 The History of Particle Swarm Optimization 381
    23.5.3 Algorithm 381
23.6 Comparative Study of PSO and CS 383
    23.6.1 Experiments 383
23.7 Comparative Study between Ant Colony Optimization and Cuckoo Search Optimization 384
23.8 Comparison Between ACO, ABC, PSO, and Cuckoo Search 386
23.9 Implementation of Cuckoo Search Algorithm 386
23.10 Conclusion 386
Bibliography 387

DOI: 10.1201/9781003146711-23

23.1 INTRODUCTION

The development in wireless communication became the cause of many elementary changes in data, networks, and telecommunication, and has made unified networks come to life. By going cordless, personal communication, wireless LANs, mobile networks, and mobile systems fulfill the possibilities of completely distributed computing and communications anytime and anywhere. This become more popular every passing minute due to their user-friendliness. Consumers are no longer reliant on cables, so moving from one place to another is so easy, while not worrying about getting disconnected from the cable. The ease of movability is one of the great characteristics of wireless networks that make them attractive.

Yet aside from this advantage, the protection of the transmitted data is a major problem that accompanies wireless communication. Due to the ease of unauthorized access to the system, wireless networks are more insecure than a wired network. The need for more efficient protection measures emerged with the rising technology.

In the field of computer security, techniques taken from the field of artificial intelligence, especially evolutionary computation, are constantly becoming more and more present, both in network/host security and in the very challenging field of cryptology. For example, in the design and study of a variety of modern cryptographic primitives, ranging from pseudo-random number generators to block ciphers, in the cryptanalysis of state-of-the-art cryptosystems, and in the identification of network attack patterns, to name a few, several algorithms have been suggested in recent years that take advantage of methods focused in evolutionary computation.

An evolutionary algorithm is a stochastic search algorithm dependent on populations. The basic concept is to produce iteratively random variance within population individuals, which represents the candidate's solution to the problem, and to select the most suitable candidates for the task in hand. These algorithms have, to date, been mainly used in medicine and physics, and in recent years their use in computer security has emerged.

This chapter briefly explains not all but only four of such mechanisms with the intention to use their modified form for the security of data in wireless networks. The first part of the chapter deals with ant colony optimization (ACO), the second deals with artificial bee colony optimization (ABC), the third with the Cuckoo Approach, and lastly the fourth deals with particle swarm optimization.

23.2 ANT COLONY OPTIMIZATION

23.2.1 Biological Analogy

Proof-of-concept implementations are initially based on research on a new metaheuristic for optimization. It is only after the practical interest of the method has been shown through experimental work that researchers attempt to deepen their understanding of the working of the method not only through more and more complex experiments, but also through an effort to construct a theory. It is important to address questions such as "how and why the method works" because finding a response can help to improve its applicability. Optimization of the ant colony, which was implemented in the early 1990s as a modern technique to solve challenging problems of combinatorial optimization, is currently at this stage of its life cycle.

Ants leave a trail of a chemical named pheromone on the ground while walking. This pheromone is smelled by other ants. A path is constructed by this pheromone from nest to food source. Other ants searching food, use this trail to find the food. Ants can determine the shortest way from nest to food by these trails. Deneubourg et al. (1993) created an experiment to study and research ants' behavior in determining the quickest way to food. They used a variable length double bridge to connect a food source and a nest of ants. They did many experiments with a longer path and a shorter path.

At the start of every experiment, ants begin with a random motion toward the food, but a shorter path is taken ultimately. This outcome is explained below. At the beginning, zero ants traveled, so the chemical pheromone present everywhere was equally giving the same probability to any route. The ants who chose the shortest way reach the food source in less time and return to the nest before the ants who took the longer way. To go back for the next trip, the ants decide to follow one path, due to the high pheromone level on that shorter way. The pheromone level on the shorter way begins to build up faster, diverting the entire colony to go along this path.

Ant systems and various types of ACO algorithms use digital or artificial methods for implementation. Ants are stimulated by using digital artificial ants (agents). Artificial digital pheromones are created and used as pheromone trails and a graph is used instead of a double bridge or path. Exclusive powers and features are used to implement the restrictions and no defects are provided. Ants use it to solve realistic issues. The amount of chemical set down by artificial agent ants is fixed proportional to the quality and efficiency of the obtained solution – something similar to natural ant behavior (Figure 23.1).

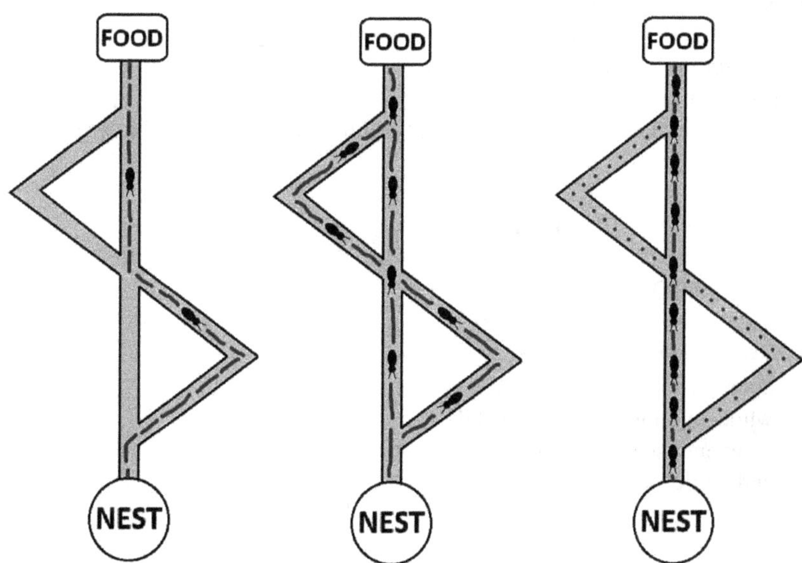

FIGURE 23.1 Ant colony optimization.

This is a procedure which is required to resolve problems on optimization that are complex combinatorial. ACO is an algorithm that adheres to the actions of actual ant colonies as they hunt for food and return to the nest again. Similar to the biological approach, artificial ants that are simple agents communicate with each other, indirectly, with the help of artificial pheromones inside a colony. In a colony of agents, called ants (artificial), there is indirect communication performed by using (artificial) pheromone lines. Pheromones act as distributed information as well as digital information which is used by ants to find solutions to a problem. Adapting to the problem while executing the algorithm as their search experience changes is the difference between an ACO algorithm and other construction algorithms.

23.2.2 ACO Algorithm

Metaheuristic approaches are extended to complicated problems with a difficult computational point of view when it becomes non-effective by using standard numerical methods. The challenges in everyday life require exponential amounts of evaluations and the one alternative, when the issue is big enough with large calculations, is to apply metaheuristic ways to obtain an optimal solution in a fair amount of time by the ACO algorithm by Marco Dorigo (Dorigo et al. 1996). Later, changes are made specifically to the rules for pheromone upgrading. In ACO algorithms, artificial ants mimic the behavior of an ant. The issue is characterized by using a graph. The alternatives for the solution are described in a graph by paths which are then searched for a shorter path that fits the constraints given.

ACO algorithm specifications are:

- An adequate graph for the topic.
- Reasonable positioning of pheromones on the nodes or on the arcs of the map.
- Reasonable heuristic feature of the dependent problem, which guides the ants to reinforce solutions.
- Laws for updating pheromones.
- Probability law of change, which determines how to use, in the semi-solution, new nodes.

Structure of the ACO algorithm:

```
Initialize number of agents;
Initialize the ACO parameters;
while end condition not reached do
    for j = 0 to number of agents
        agent j chooses starting node;
        while solution is not reached do
            agent j select higher probability node;
        end while
    end for
    update pheromone values;
end while
```

Probability of transformation *pi, j,* when selecting node *j,* if present node is *i,* a heuristic knowledge product, *ni, j,* this move was correlated with the trail level of pheromone *ti, j,* where *i, j=1.... n.*

$$p_{i,\ j} = \frac{T_{i,j}^{a}\ \ \eta_{i,j}^{b}}{\sum k \notin unused\ \ T_{i,k}^{a}\ \ \eta_{i,k}^{b}} \tag{23.1}$$

where *unused* is a collection of the graph's unused nodes.

If the value of the heuristic is greater, a node becomes more important; or when associated pheromones are higher to begin with, the degree of pheromones is the same for all, which is assigned a less positive constant: $t0$, $0 < t0 < 1$ value. Ants at the conclusion of every iteration update the values of the pheromones. The degree of pheronomes is updated as per the ACO algorithm. The principal law for the pheromone trail in updating is:

$$\tau \mathrm{i,j} \leftarrow \rho \tau \mathrm{i,j} + \Delta \tau \mathrm{i,j} \tag{23.2}$$

p reduces the pheromone's value, like the reduction of the *g* pheromone level with time.

At *i, j* is a fresh pheromone introduced, and is related to the solution's efficiency, which is determined from the amount of the target function of the ant-produced solution.

An agent begins to produce a solution from any random part of the question-stimulated graph. A random beginning is a diversification of searching. Comparatively few spontaneous beginners can be used due to the random start than various population base metaheuristics. The heuristic data reflect the previous understanding of the question, which is used to enhance the ants' management. The pheromone is a worldwide awareness, and seeks the optimal solution for the ants. The pheromone is a device for concentrating the hunt for the right options so far.

23.3 ARTIFICIAL BEE COLONY OPTIMIZATION

23.3.1 Bee Foraging Behavior

A new kind of swarm intelligence algorithm proposed by Karaboga and Basturk (2007) is the artificial bee colony (ABC) algorithm; it simulates the intelligent actions of honey bees, and bees conduct various nectar collecting activities according to their respective division of labor to understand the source of knowledge sharing and exchange. It has gained considerable recognition and analysis from many academics because of its basic form, less constraints, and efficient implementation. At present, in many areas, such as the neural network, filter design, parameter optimization, and optimization techniques, the ABC algorithm has been successfully applied.

The ABC algorithm also suffers from an early maturity flaw, comparable to other swarm intelligence algorithms. Some scholars use confusion to set up the population to enhance the diversity and nonlinearity of the swarm for this reason. In order to prevent early maturity, the allocation process of the bidding match and ranking is proposed to reduce the impact of super-individuals in the population. The mutation operators are usually integrated into the ABC algorithm in order to enable the algorithm to step out of a local extremum.

According to Chen et al. (2009), in nature, bees have three separate functions when they gather nectar: they are working, onlooker, and scout bees. All three categories work as a team together to accomplish the collection of nectar. Moreover, the elementary conduct of bees can be classified into two groups, namely recruiting bees for collection of nectar and declaring a source not fit.

This optimization algorithm, based upon population, uses an iterative approach to attain a global maximum or minimum.

The entire search range of bees in this algorithm reflects the solution space problem for optimization. A random source site is a deterministic solution to the issue of optimization and the volume of nectar from the food source is considered a fitness value which is further used to define the consistency of solutions. Hence the process of gathering nectar is also the method of mapping the solution.

23.3.2 Modified Approach

In this model, the community of bees contain three clusters: employed, onlooker, and scout. The first half of this community is formed by working, while the next part contains the onlookers. Only one bee is allotted to one food source. To put it another way, the amount of bees working is equivalent to the total number of sources of food around the hive. When the food source of any employed bee is exhausted, it turns into a scout bee.

Give the scouts the initial sources of food:

REITERATE

Send the employed bees to the sources to assess the volume of nectar
On the basis of the preference of the onlooker bees, find out the probability value of each source.
Stop the mechanism of mining of the sources abandoned by the bees
Send the scout bees towards the quest area, arbitrarily, to locate new sources for food.
Maintain a record of best food sources discovered so far.

UNTIL (requirements are fulfilled) (Joshi et al. 2017).

In Figure 23.2, S represents new possible sources of food, R represents possible foragers, and A represents the already existing sources.

The fundamental properties on which self-organization depends, in the situation of honey bees, are as follows:

1. **Positive feedback:** As the quantity of nectar in a source increases, the number of onlooker bees seeking them also increases.

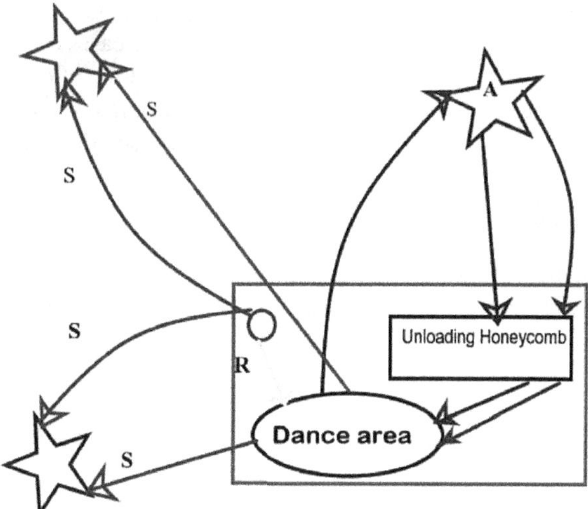

FIGURE 23.2 The behavior of the honey bee looking for nectar.
Source: Joshi et al., 2017.

2. **Negative feedback:** The food source which has been marked as abandoned is not included in the exploration process.
3. **Fluctuations:** The scouts conduct a random discovery process to locate new sources.
4. **Multiple interactions:** Bees share their data on the locations of new food sources and the amount of nectar available in their dancing area for their swarm members.

23.3.3 Algorithm

1. Generate the first population of n scout bees.
2. Evaluate the fitness of all scout bees.
3. Select m top sites for vicinity search.
4. Select a proper value for patch size (*ngh*) for all sites.
5. Employ some bees (*neh*) for some best e sites.
6. Elect the best bees from the sites as new sites.
7. Send the remainder n-m bees for random search.
8. Form the new generation of scout bees.
9. Check if termination criterion is satisfied.
   a. If no, then restart the process from step 2
   b. Else print the fittest bee as the final solution.

23.4 CUCKOO APPROACH

23.4.1 Introduction

The Cuckoo Search (CS) algorithm is also an algorithm inspired by nature, based on cuckoo birds' evolutionary approach to improve their population. CS is more efficient than other algorithms inspired by evolution, though. Differential evolution and particle swarm optimization (PSO) are simply unique cases of the CS algorithm, so it is not unexpected that they are surpassed by it. In terms of convergence speed, the CS algorithm outpaced the DE algorithm to find an optimal solution. Furthermore, the CS algorithm was stated to be more effective in computation than the PSO.

23.4.2 Cuckoo Reproductive Method

Cuckoos belong to that family of birds which practice a special breeding tactic that is more hostile as compared to other species of birds. Many varieties of cuckoo birds place their eggs in common nests; however, they might get rid of the eggs of others to increase the prospects of hatching their own eggs. Many species lay eggs in the nests which belong to other birds or hosts using the brood parasitism process. In new nests, where eggs have only just been placed, the parasitic cuckoos are fine, and their time for the breeding of eggs is very exact.

We laid only a single egg at a time in the nest of the host that will usually hatch faster than all other eggs. When this occurs, by removing those eggs out of the nest, the foreign cuckoo can expel those eggs which have not yet produced the offspring. This action is intended to reduce the possibility of the hatching of real eggs. In addition, the alien cuckoo's chicks will try acquiring more nutrition by imitating the host chicks' cry. Often the owner cuckoo will find one or more alien eggs. The cuckoo either removes the egg in that situation or absolutely abandons the nest and goes somewhere else to build a new nest.

23.4.3 The Lévy Flights Mechanism

In nature, species hunt for food in an unsystematic or quasi-random way. An animal's foraging route is essentially an arbitrary walk since both the current position/state and the likelihood of change to the next position are the basis for the next step. The probability of the chosen paths was mathematically modeled. Various studies have proved that most of the animals as well as insects' flight activity shows the characteristics typical of a phenomenon known as Lévy flights. It is a random walk which constitutes the length if each step is computed using probability distribution. The expanse from the origin of the walk tends towards a steady gap after a significant count of moves (Figure 23.3).

23.4.4 Algorithm

In the natural world, if the egg of a cuckoo is somewhat close to the egg of a host, so the egg of that cuckoo is less likely to be found, fitness can be attributed to the disparity in solutions. It is also a smart idea to do a random walk with certain random parameter values in a skewed fashion.

With this algorithm we can consider the host's eggs in the nest as the already prevalent solutions and the cuckoo egg as the new solution. There are three idealized rules that are followed in this technique:

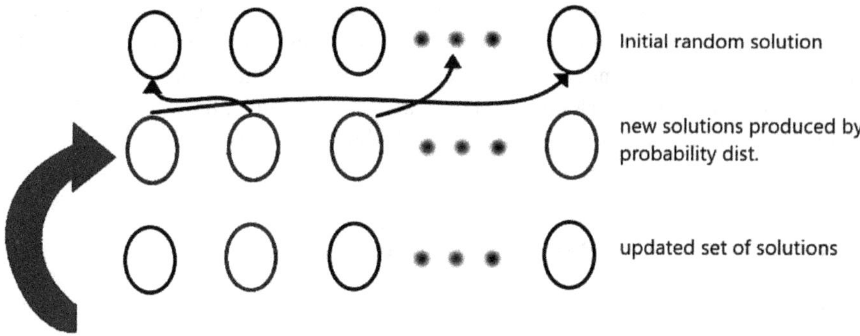

FIGURE 23.3 Schematic illustration of the CS method.

1. Each one of the cuckoo lays only one egg at one time, in a nest which is chosen at random.
2. That nest is passed on to the next generation with a higher quality.
3. The number of available host nests is fixed, and the egg laid by a cuckoo is discovered by the host bird with a probability of the cuckoo egg being discovered by the host as p_a.

The discovered eggs are considered to be part of the worst nests and are therefore dumped.

The pseudo-code with respect to CS for a global optimization (Bonyadi & Michalewicz, 2017) is as given below:

Objective function *f (a), a= (a1, a2... ad)q*

Original population production of a host nest is (*m = 1, 2...x*)

While (*q* < Generation Max) or (discontinue criteria)

Find a cuckoo egg (say l) arbitrarily;

Assess *Fi* for its quality/fitness;

Choose a nest among *k (say n*) randomly;

Evaluate its quality/fitness F_j;

If $(F_i > F_j)$

Replace *n* by the new solution;

Close

A certain segment of (P_a) of inferior nests are deserted while

fresh are constructed at different location via Lévy flights mechanism;

Store the finest solutions for next steps;

Rank all solutions and find the current best;

End while

At first sight, with regard to some large scale randomization, it seems like there are some parallels between CS and hill-climbing. In essence, however, these two algorithms are very distinct. Second, CS is a population-based algorithm identical to PSO, but it uses some form of classism and/or filtering similar to that used in the search for harmony. Second, since the step length is heavy-tailed, the randomization is more successful and every large step is feasible. Ultimately, the number of tuning parameters is fewer than in PSO, and CS will thus be much easier to conform to a broader variety of optimization issues.

23.5 PARTICLE SWARM OPTIMIZATION (PSO)

23.5.1 Introduction

In engineering issues, maximizing earnings or minimizing losses has always been a concern. The complexity of optimization problems increases for different fields of knowledge as science and technology grow. Examples of engineering problems that may involve an approach to optimization are often in the conversion and distribution of electricity, mechanical design, logistics, and nuclear reloading.

There are many methods that one could perform in order to optimize or minimize a function to find the optimum. Despite a broad variety of optimization algorithms that could be used, there is not a primary algorithm that is considered the best for any scenario. One optimization method that is suitable for a problem might not be so for another one; it depends on several features, for example, whether the function is differentiable and its concavity (convex or concave). In order to solve a problem, one must understand different optimization methods, so this person is able to select the algorithm that best fits the features' problem.

The algorithm for PSO is a stochastic swarm-based optimization strategy, given by Kennedy and Eberhart (1995). The PSO algorithm is driven by the social animal behavior of the actions of insects, herds, birds, and fish. These groups find food in a unique way, where each participant seeks food and cooperates with each other. The swarms keep changing the pattern of the hunt according to the learning interactions of its participants.

The PSO algorithm is closely connected with two academic findings: one evolutionary algorithm; a swarm mode is also used by PSO that allows it to check for a wide area at the same time in the space of the solution (functional optimization). The other is artificial stimulation of life, and analysis of artificial structures of life traits. In the study of the artificial life behavior of social animals, of how to create the artificial life systems of the swarm, Millonas gave five base principles (van den Bergh 2001):

- Proximity: The digitally made swarm must be capable of doing easy space-time computations.
- Quality: The swarm should feel the consistency of the environmental change and react.
- Diverse response: The swarm should never restrict ways of getting the sources to a small reach.
- Stability: With each environmental shift, the swarm does not change its mode of action.
- Adaptability: If this improvement is worthy, the swarm should change its mode of behavior.

The fourth and fifth principles are the reverse of each other. The key features of artificial life systems are found in these five concepts and they direct the creation of the swarm's artificial life in the digital world. In PSO, particles change and update their locations and speeds due to environmental changes, meeting proximity, and consistency requirements. Furthermore, the PSO swarm does not restrict its travel, but constantly looks for the most efficient solution out of all possible solutions present in the space. PSO particles can maintain their movement in the search room, thus respond to the environmental changes and move accordingly. So the above five criteria are fulfilled by particle swarm networks.

23.5.2 The History of Particle Swarm Optimization

Agents are autonomous entities in the natural programming model, typically with little understanding of the high goal in pursuit, but model complex structures in this real world. Multiple low-level agendas that facilitate constructive concerted action arising from this apparently not intelligent and not influential singular actors allow this to be possible.

Fletcher and Reeves (1964) argued that such a model is capable of representing, by classical surface-based depictions, the dynamics and shape of natural systems that are complicated. In the Boid model (1986), Reynolds introduced basic rules. That improved particle action autonomy and defined the basic rules that these particles can follow to trigger the evolving behavior. Reynolds proposed three distinct flocking laws that all particles can obey: separation, alignment, and cohesion. Where separation theory allows each particle to move from one another in different directions in order to prevent gathering, the alignment and cohesion rules enable spatial changes to allow movement to average heading and direction respectively, with neighboring flock members.

There is $O(n^2)$ complexity in the case where each boid knows where other boids are located, causing computational difficulty. However, Reynolds proposed a neighborhood model with the exchanging of information between boids in common vicinity, thus reducing $O(n)$ complexity and speeding the algorithm's implementation.

This algorithm was first given in 1995 by Eberhart and Kennedy as one expansion of Reynold's work. The flock or swarm converged prematurely in a collective manner by incorporating local exchanging of information by the closest matching of adjacent velocity. A spontaneous disruption or craziness in the particle velocities was then added, resulting in important variations and then ensuring life-like dynamics of the group. Both criteria were eliminated, because the flock converged equally well on attractors after their elimination also. And this model acquired an agent population that was more in line with the dynamics of a colony than a flock (Figure 23.4) (Wang et al., 2018).

23.5.3 Algorithm

A group of particles known as a swarm exists in a space of solutions, a base version of this algorithm's functions. The motions of these particles are directed by their best-known self-location in this searchable space and by the best-known position of the entire swarm. Once an enhanced place is established for each, they will then

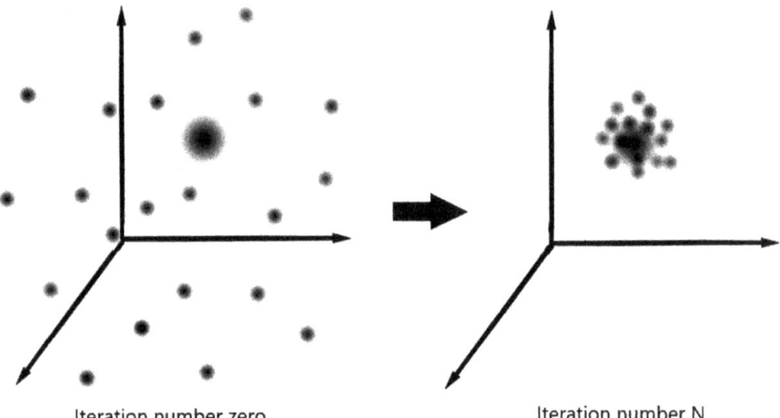

FIGURE 23.4 Pictorial representation of PSO.

direct the movements of the swarm. The procedure is replicated, and so an acceptable solution is supposed to eventually be discovered, but not assured.

Let f: Rn-R formally as the cost function that should be at minimum. A function with arguments of solution space in the vector form of real numbers, generates a real number as an output that indicates the solution by giving an objective function value. The f gradient is not established. The aim is finding a solution where $f(x) < f(y)$ representing x as a global minimum for all y in the search-space.

Let K be the number of particles in the swarm, each with the search-space location ai and the velocity Rn. And let xi be the optimal position of I particle and q be the optimal position of the group in its entirety.

Then a fundamental PSO algorithm is:

For $j = 1... K$ do (j=particles)
  Initialize the position of particle with a distributed random vector: $a1 \sim U1 (blo1, bup1)$
 Initialize best position of particle as initial: $x1 \leftarrow a1$
  If $f(x1) < f(q1)$ then
    Change best known position of swarm: $q1 \leftarrow x1$
  Set velocity of particle: $v1 \sim U1 (-|bup1\text{-}blo1|, |bup1\text{-}blo1|)$
While end condition is not true do:
  For every $j = 1... K$ (particle) do
    For $d1 = 1... l$ (dimension) do
      Select any number: $re, re \sim U(0, 1)$
      Change velocity of particle:
    $v1, d1 \leftarrow \omega\, v1, d1 + \varphi e\, re\, (p1, d1\text{-}x1, d1) + \varphi e\, re\, (gd\text{-}x1, d1)$
    Change position of particle: $a1 \leftarrow a1 + v1$
    If $f(a1) < f(x1)$ then
      Change best position of particle: $x1 \leftarrow a1$
    If $f(x1) < f(q1)$ then
      Change best position of swarm: $q1 \leftarrow x1$

23.6 COMPARATIVE STUDY OF PSO AND CS

In Sengupta et al. (2019) we see a comparison between particle search and cuckoo search algorithms by means of distance functions. These are some measurement sets which help us to compare the similarities and differences among the present optimal solutions and the remainder of the solutions. This study has focused mainly on the problem specific distance function (PSDF), as its measurements are associated directly to the problem that is to be solved by the evolutionary method. One more reason to select this method is that we have already discerned the ideal values, which are X equals 0 and Y equals 0.

Our problem equation is:

$$\text{Min} \quad F(X,Y) = X^2 + Y^2 \quad \text{Where,} -5.12 < \text{X,Y} < 5.12$$

23.6.1 Experiments

The main tool used in this experiment was MATLAB. The problem equation was used with the following values: population size is 50, gen = 300, and runs = 100. In Cuckoo Search, $p_a = 0.25$. In particle search, 0.4 is the inertia weight and 2 is the value of both confidence aspects.

Figure 23.5 shows the difference between *X* values for both CS and PSO programs for the circle function.

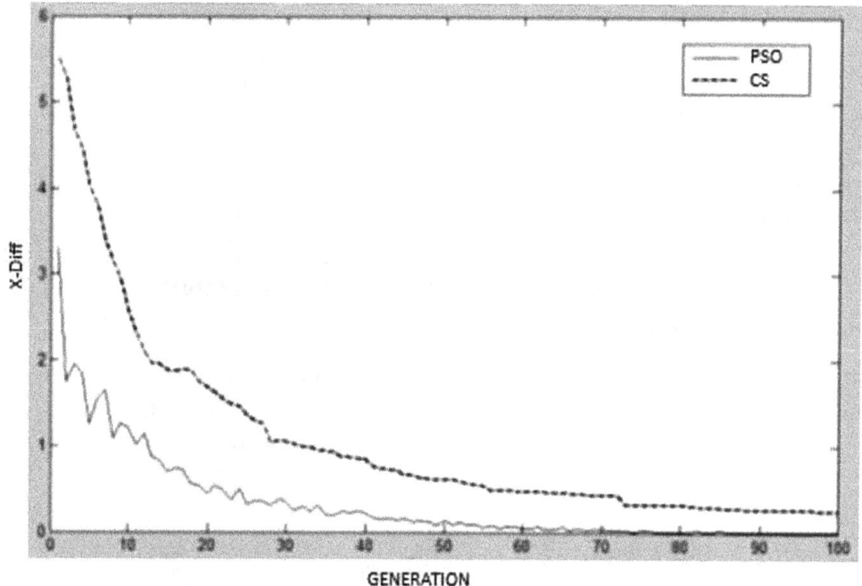

FIGURE 23.5 Difference between *X* values of CS and PSO.

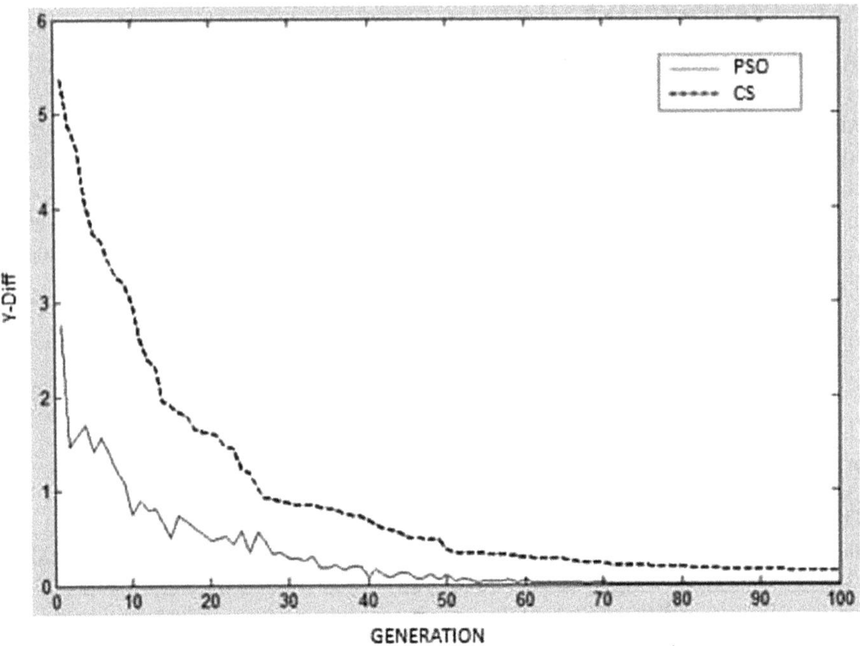

FIGURE 23.6 Difference between *Y* values of CS and PSO.

Figure 23.6 shows the difference between *Y* values for both CS and PSO programs for the circle function.

As both figures show the same nature, we can assume that the PSO population was not as diverse as the CS population.

Figure 23.7 shows the average fitness for both the algorithms.

We can see that CS started with a higher fitness value, more search area was navigated, and then both reached their best value.

Thus, we can come to the conclusion that CS has more advantage over PSO, with the main advantage being adjustment of only one parameter p_a.

23.7 COMPARATIVE STUDY BETWEEN ANT COLONY OPTIMIZATION AND CUCKOO SEARCH OPTIMIZATION

Slow convergence and an easy drop into a local optimum are two drawbacks of the ACO. Since the ACO has several constraints, it can easily result in a lack of preliminary pheromone and a high level of complexity. Furthermore, the algorithm is vulnerable to being stuck and falling into a local optimum, which is inconvenient for finding the global best solution.

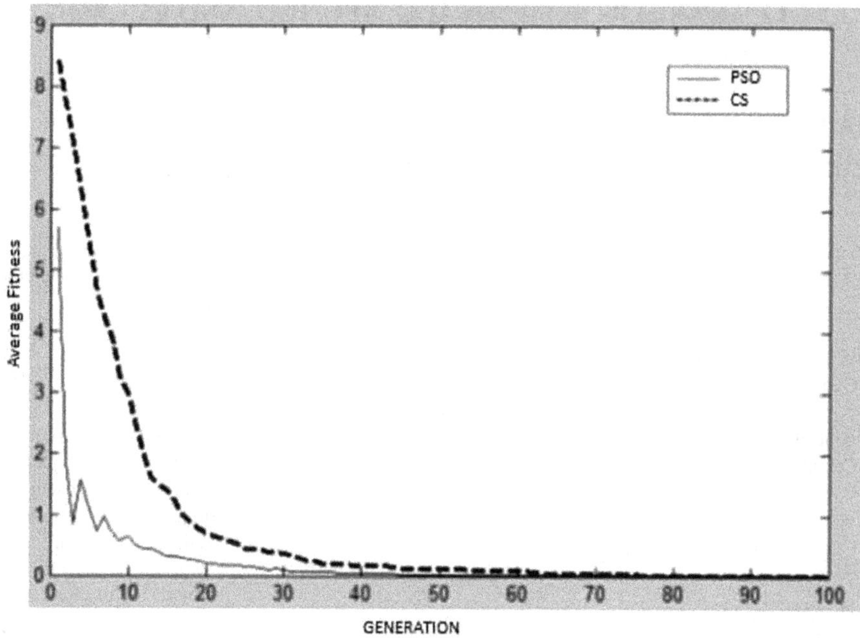

FIGURE 23.7 Average fitness for CS and PSO.

It is possible to measure it several times using random initial values, but it is time intensive. As a result, figuring out how to seek the balance point is a challenge that must be addressed.

Cuckoo search, on the other hand, is a modern meta-heuristic search algorithm with a lot of testing potential. Through comparing performance metrics such as time and space complexity, robustness, readability, and platform independence, it can be concluded that the cuckoo search has significant advantages in terms of overall robustness, readable logics, portability, and platform independence. The algorithm also has good global search capabilities, a small subset of chosen parameters, an impressive search path, and a strong ability to solve multi-objective problems (Table 23.1).

TABLE 23.1
Comparison between CS and ACO

	Cuckoo Search	Ant Colony Optimization
Advantages	More robust, few parameters, enhanced global search ability, versatile	Positive feedback mechanism, reliable
Disadvantages	Low convergence rate	Takes longer time to search, not suitable for large-scale problems
Scope of application	Continuous optimization, large-scale problems	Discrete optimization; small-scale problems

23.8 COMPARISON BETWEEN ACO, ABC, PSO, AND CUCKOO SEARCH

TABLE 23.2
Comparison between ACO, ABC, PSO, and CS

	Parameters	Working Mechanism	Features
Particle Swarm Optimization	Learning factors; no dimension and range of particles, maximum change of particle velocity	Optimal solution obtained by the movements of the particle in the problem space	Simple calculations, no overlapping and mutation calculation
Ant Colony Optimization	No. of ants, initial pheromone value, weight and evaporation rate	Ants use "pheromone deposition" concept for finding the optimal route	Used in dynamic applications
Artificial Bee Colony Optimization	Employed bees, onlooker bees and scout bees	Collected information is shared by performing a dance	Intelligent behavior of honey bees for solving optimization problem
Cuckoo Search Optimization	No. of nests, Levy flight solution	Use the new and better solutions to replace a worst solution in the nest	Simple implementation, reduced number of parameters in comparison to PSO, ACO, and ABC

23.9 IMPLEMENTATION OF CUCKOO SEARCH ALGORITHM

1. Used for secure data key encryption and efficient authentication.
2. The intrusion detection is a type of security management system for computer or networks. Now the feature selection is used in intrusion detection systems (IDSs) by using the cuckoo search algorithm.
3. Cloud computing security frameworks such as gathering information, network mapping, vulnerabilities exploration, audits, and penetration tests can be solved with the heuristic approach of a cuckoo search algorithm.

The cuckoo approach is an important algorithm that can be used for security purposes and we intend to work on this algorithm further with the main aim of securing data in wireless networks using data key encryption.

23.10 CONCLUSION

Throughout the available literature, many evolutionary computational algorithms have been explored in order to optimize classification efficiency and minimize the computational time of the intrusion detection system. We plan to use this knowledge for a more effective and productive approach in handling any attacks on data protection while avoiding them all the time. Further research is required to identify the behavior of the proposed algorithms in the field of security.

When studying the methods of optimization, we encountered only a few drawbacks of each, which may impede our future work. The low convergence rate of PSO, poor local search ability of ABC optimization, and difficult theoretical analysis of ACO is to name but a few.

Therefore, with the aid of the cuckoo search method, we plan to move forward with our research in the field of data protection in wireless networks because it is simple, flexible, robust, and easy to implement.

BIBLIOGRAPHY

Adnan, M. A., & Razzaque, M. A. (2013, March). A comparative study of particle swarm optimization and Cuckoo search techniques through problem-specific distance function. In *2013 International Conference of Information and Communication Technology (ICoICT)* (pp. 88–92). IEEE.

Akhtar, A. (2019). Evolution of ant colony optimization algorithm–a brief literature review. In *arXiv: 1908.08007.*

Bai, Q. (2010). Analysis of particle swarm optimization algorithm. *Computer and Information Science*, *3*(1), 180.

Bansal, J. C., Gopal, A., & Nagar, A. K. (2018). Stability analysis of artificial bee colony optimization algorithm. *Swarm and Evolutionary Computation*, *41*, 9–19.

Beckers, R., Deneubourg, J. L., & Goss, S. (1993). Modulation of trail laying in the ant Lasius niger (Hymenoptera: Formicidae) and its role in the collective selection of a food source. *Journal of Insect Behavior*, *6*(6), 751–759.

Bergh, F. V. D., & Engelbrecht, A. P. (2001, July). Effects of swarm size on cooperative particle swarm optimisers. In *Proceedings of the 3rd annual conference on genetic and evolutionary computation* (pp. 892–899).

Bonyadi, M.R., & Michalewicz, Z. (2017). Particle swarm optimization for single objective continuous space problems: A review *Evolutionary Computation, 25* (1): 1–54.

Chen, W. N., & Zhang, J. (2010). A novel set-based particle swarm optimization method for discrete optimization problem In *IEEE Transactions on Evolutionary Computation.*

Chen, W. N., & Zhang, J. (2012, October). A set-based discrete PSO for cloud workflow scheduling with user-defined QoS constraints. In *2012 IEEE International Conference on Systems, Man, and Cybernetics (SMC)* (pp. 773–778). IEEE.

Chen, W.N., Zhang, J., Chung, H.S., Zhong, W.L., Wu, W.G. and Shi, Y.H., 2009. A novel set-based particle swarm optimization method for discrete optimization problems. *IEEE Transactions on Evolutionary Computation*, *14*(2), pp.278–300.

Dorigo, M., Maniezzo, V., & Colorni, A. (1996). Ant system: optimization by a colony of cooperating agents. *IEEE Transactions on Systems, Man, and Cybernetics, Part B (Cybernetics)*, *26*(1), 29–41.

Eberhart, R., & Kennedy, J. (1995, November). Particle swarm optimization. In *Proceedings of the IEEE International Conference on Neural Networks* (Vol. 4, pp. 1942–1948).

Fletcher, R., & Reeves, C. M. (1964). Function minimization by conjugate gradients. *The computer journal*, *7*(2), 149–154.

Fidanova, S., Luque, G., Roeva, O., Paprzycki, M., & Gepner, P. (2018, September). Hybrid ant colony optimization algorithm for workforce planning. In *2018 Federated Conference on Computer Science and Information Systems (FedCSIS)* (pp. 233–236). IEEE.

Gou, J., Lei, Y. X., Guo, W. P., Wang, C., Cai, Y. Q., & Luo, W. (2017). A novel improved particle swarm optimization algorithm based on individual difference evolution. *Applied Soft Computing*, *57*, 468–481.

Joshi, A. S., Kulkarni, O., Kakandikar, G. M., & Nandedkar, V. M. (2017). Cuckoo search optimization-a review. *Materials Today: Proceedings*, *4*(8), 7262–7269.

Karaboga, D., & Basturk, B. (2007). A powerful and efficient algorithm for numerical function optimization: artificial bee colony (ABC) algorithm. *Journal of Global Optimization, 39*(3), 459–471.

Mareli, M., & Twala, B. (2018). An adaptive Cuckoo search algorithm for optimisation. *Applied Computing and Informatics, 14*(2), 107–115.

Nasrinpour, H. R., Bavani, A. M., & Teshnehlab, M. (2017). Grouped bees algorithm: A grouped version of the bees algorithm. *Computers, 6*(1), 5.

Pandey, V. A., Pulastiya, P., Kaur, I., & Rastogi, S. (2020). A survey on key management using particle swarm optimization in MANET. *Available at SSRN 3570308.*

Sengupta, S., Basak, S., & Peters, R. A. (2019). Particle swarm optimization: A survey of historical and recent developments with hybridization perspectives. *Machine Learning and Knowledge Extraction, 1*(1), 157–191.

Shi, Y., & Eberhart, R.C. (1998). A modified particle swarm optimizer *in Proceedings of IEEE International Conference on Evolutionary Computation.*

Skaraboga, D. (2005). *An idea based on honey bee swarm for numerical optimization* (Vol. 200, pp. 1–10). Technical report-tr06, Erciyes University, engineering faculty, computer engineering department.

Sun, L., Chen, T., & Zhang, Q. (2018). An artificial bee colony algorithm with random location updating. *Scientific Programming, 2018*, 1–4.

Wang, D., Tan, D., & Liu, L. (2018). Particle swarm optimization algorithm: An overview. *Soft Computing, 22*(2), 387–408.

Wen, X., Huang, M., & Shi, J. (2012, October). Study on resources scheduling based on ACO algorithm and PSO algorithm in cloud computing. In *2012 11th International Symposium on Distributed Computing and Applications to Business, Engineering & Science* (pp. 219–222). IEEE Computer Society.

Zhao, G. (2014). Cost-aware scheduling algorithm based on PSO in Cloud Computing environment. *International Journal of Grid and Distributed Computing, 7*(1), 33–42.

Index

A

ABC algorithm, 375–376
accelerometer, 362–363, 365
access, 88–92, 94, 97
a critical survey of autonomous vehicles, 235
action-effectiveness evaluation, 153
active cases, 325
acute respiratory distress syndrome, 331
advanced metering infrastructure, 29, 32
agricultural industry, 170
agriculture, 222–224, 232
  industry, 164
air pollution, 332
algorithms, 270–273, 275, 277–280
algorithm selection problem, 256
amalgamation, 99–101
Amazon Web Services, 45
analysis, 222–223, 226–228, 231–233
ant colony optimization, 372, 384, 386
application layer, 52
approvals, 245
ARIMA, 283, 285, 288, 290–291, 293–295
artificial intelligence, 179, 182, 192, 200, 202, 224, 270, 280, 372
artificial neural network, 187–188
ASD, 269–271, 278–280
assessment and evaluation, 122
atmospheric condition, 332
attack, 94–95, 97, 99, 298–299, 303–307, 309–310
augmented reality, 182, 189
authentication token, 119–120
authorize, 92–94, 96
autism spectrum disorder, 269, 271
automated robots, 182, 188
autonomous vehicle, 190–192, 360–361, 364, 366–367
autoregressive, 283, 285, 287, 290–291

B

back propagation (BP), 227
blockchain, 96–101
Bonnie++, 60

C

cameras for accidental situations, 241–242
capacitated vehicle routing problem, 257–258
capacity, 91, 97
carbon computing, 283, 285–287, 289–291
carbon emission, 283, 285, 287–290
carbon footprints, 283
checkpoint/restore userspace (CRIU), 61
China, 311–313
CIA triangle, 118
classification, 222, 226, 232
client, 92–94, 96
cloud auditor, 43
cloud broker, 43
cloud carrier, 43
cloud computing, 39, 57, 59–61, 68, 88–95, 99–101, 106–107, 135–136, 181, 186
  security, 92, 100
cloud computing reference architecture, 41
cloud consumers, 43
cloud data center, 59–62
cloud developer, 43
cloud provider, 42
cloud services provider, 40
CNN, 74, 76–77
CO, 332–333, 335–336
collaborative learning, 123
communication protocol, 200, 204, 208, 210–211, 218
computing technologies, 283–285, 287, 289–291
confidentiality, 92–94, 96, 99
connectivity protocol, 203, 206, 208–209, 212
container auto-scaling, 68
container-based virtualization, 58–60
container failure management, 68
containerization, 58–62
container resource allocation, 68
continuous measurement, 148
continuous-time hidden Markov model, 148, 150–153, 156, 159
controller, 361, 366
convolutional neural networks, 150, 188–190
coronavirus, 312–314, 326
correlation, 334–338
cost of vehicles, 243–245
Covid-19, 331–333, 337–338
CRISP-DM, 315–316
CSP, 88–89, 92–95
cuckoo search, 377, 383–386
cyber attack, 2, 6, 8, 14–15
cyber-physical system, 2, 117, 121, 130
cyber security, 2–4, 7–8, 14–16, 20–21, 117–118, 121–122, 124–126, 128–130, 181, 185
cyber threat, 3

D

data analysis, 222–223, 228, 233
data analysis and prediction, 312
database, 88, 94
data breaches, 44
data center, 46
data integrity, 92–94
data mining process, 332
data security, 164
datasets used for autonomous driving, 243
DDOS, 51
death cases, 312, 324
decentralization, 96–99
decision tree, 186–187, 270, 272
decreasing cases, 328
Deendayal Upadhyaya Gram Jyoti Yojana, 30
deep learning, 147, 153, 159, 179, 185–186
deep neural networks, 148
demand response program, 26–27, 32–34, 36
demand side management, 28, 32–33, 35
denial of service, 3, 17, 120
differential evolution, 298–300, 302, 308
digital watermarking, 297–300, 309
distributed denial of service, 3, 17
distributed generation, 26
distribution center, 147, 152, 159
Docker, 59–61
downtime, 66
DWT, 298–302

E

EDA, 311, 317
edge computing, 108–109
    vs fog computing, 111–112
e-governance, 40, 134–135, 143
EKF, 350, 352–353, 355, 359, 367
e-learning, 118, 123
encryption, 140, 386
energy conservation, 26, 32
energy consumption, 68
energy efficiency, 26, 30–33
energy efficiency services limited, 30
ergonomics, 148, 159–160
evolutionary algorithm, 372, 380
exploratory data analysis, 311, 317

F

FANN, 312
fast artificial neural network, 312
firewall and access control, 139
fitness, 376–379, 384
FNN, 226–228
fog computing, 110, 144
forms of camera, 240

G

GCP, 47
Google earth engine, 333, 338
GPU, 80–82
granulometry, 74–76
green computing, 283–287, 294–295
green design, 284–285
green manufacturing, 284–285

H

handedness indicator, 149, 156
hand speed indicator, 149, 156, 159
healthcare industry, 164, 171
high performance computing, 60
HIPAA, 53
history of autonomous vehicles, 236–239
hover-over effect, 322
human action recognition, 150–151, 153
human effort, 147–148
human pose estimation, 149–151
humidity, 225–233
hypervisor-based VM migration, 58

I

IAM, 50
IBM, 45
identity related service, 201
IDS/IPS, 139, 142
image segmentation, 74–76, 82–83
imperceptibility, 297–299, 309
India smart grid forum, 29–31
India smart grid task force, 29–31
Industrial Internet of Things, 10, 202
Industry 4.0, 10, 14–16, 18, 21, 163, 180–186, 188, 190–191
Industry 5.0, 163–164
inertial measurement unit, 345, 364
infected cases, 328
infected individuals, 314
information security, 2, 14, 20
infrastructure layer, 51
integrated power development scheme, 30, 32
integration, 99–100, 185–186
intelligent irrigation, 165
interference of containers, 68
internet of things, 4, 10, 200, 202, 221
intrusion detection, 166
IOT, 73–76, 79, 81–84, 144, 163–166, 171–173, 176, 221–224, 232
    architecture, 4, 12–13, 22
    device, 4–6, 8–9, 14, 17, 21, 204–206
    security, 8–9, 16, 18, 21
    service, 200–201, 218
    system, 200–202, 211–212, 216–218

ipy notebooks, 316
iron ore, 74–75, 77, 79, 82, 84
ISO/IEC 72001, 53

J

Javascript libraries, 327
Jetson, 76, 81, 84
Johns Hopkins University Github repository, 316

K

kernel-compile, 63–64
k-nearest neighbor, 187–188, 269–270, 272, 275, 278–279
KVM, 59, 62

L

labor productivity metrics, 154
learning management systems, 117, 121
Levy flight, 378–379, 386
limitations and challenges to implementations, 243–247
litigation, liability, and general opinion, 246
localization, 342, 344–345, 349–350, 352–353, 359, 364, 366–367
lockdown measures, 332, 334–337
lockdowns, 312
logistic regression, 269–270, 272–274, 278–279
logistics, 147
LXC, 59–62, 64
Lxdbr0 bridge, 64
LXD containers, 61–62, 64

M

machine learning, 185–186, 270–271, 280, 288–289, 294–295, 311–312
macro-action recognition, 147, 151, 153, 156–157
macro actions, 148–155, 157, 159
magnetometer, 362–363
manual-labor performance, 147–150, 159
manufacturing, 147–148
MAPE, 318–319
MATLAB, 174
medical thing, 202–203, 206
meta-features, 260
meta-label, 260
meta-learning, 259
  techniques, 263
MFA, 50
micro-action recognition, 147, 149–150, 153–154
micro actions, 148–155, 157, 159
Microsoft Azure, 47
migration times, 66
ML, 270–273, 275–280
moisture, 222, 224, 226
monitoring system, 204–205, 215, 219–220, 222–232
multilayer perceptron (MLP), 227
multi-tenancy, 46

N

naïve Bayes, 269–271, 276, 278–279
naive bayesian, 187–188
National Smart Grid Mission, 29–30, 32, 35–36
Nbench, 60
NC, 303–305, 308–309
Netperf, 60
network connectivity protocol, 208
networking layer, 51
neural network, 226–227, 232
NIST, 40, 135
NO_2, 332–334, 336–337
no free lunch theorems, 256
novel framework, 311, 313, 315, 317, 319, 321, 323, 325, 327

O

O_3, 332–335
odometry, 343, 345, 348, 354–355, 364
oil and gas industry, 172–173
oil industry, 164
omega, 333, 336
omnidirectional cameras, 240–241
optimization, 372, 374–376, 379–381, 384, 386
overall labor effectiveness, 149, 155, 158

P

pandemic, 312–313, 329
particle filter, 350, 352–355, 359, 367
path planning, 342–343, 356, 359
perception, 342, 355, 367
pH, 222, 224–228, 231–233
phasor measurement unit, 28, 32
platform, 87, 90, 93, 100
PM10, 332, 337
pollutant, 332, 334, 336–338
population, 372, 375–377, 379–380, 383–384
precision agriculture, 164
predetermined time system, 147
predict, 312–313, 329
principles of the SIR model, 311, 314
privacy, 248
privacy and security, 45
productivity, 148–149, 156
prototyping and outlining AI, 191–193
PSNR, 299, 303–305, 308–309

PSO, 378, 380–384, 386
public health, 331–332

R

raspberry pie, 165
real time, 202, 206, 214–215, 217, 222, 226
recovered individuals, 314
recurrent neural network, 149–151
reinforcement learning, 186–187
renewable energy sources, 26–27, 31–33
RFID, 221
RKT, 59–61
robust, 297–299, 303, 305, 309–310
rule-based learners, 187–188

S

SARS, 312–313
SARS-Cov-2, 312–313
saving algorithm, 262
SBC, 359, 361–362, 366
scaling factor, 298–299, 302, 308–309
scipy's integrate, 320
screening, 75
secure, 93, 96, 100
security, 90–100, 247
    framework, 40
    of e-governance applications, 135
    risk, 9, 16
    threats, 137
security model for e-governance, 138
SEIR, 312
sensor, 221–226, 232
Sentinel-5 precursor, 333
server, 88, 90, 94–97
service, 88–96, 99–100
session hacking, 120
Shakti Sustainable Energy Foundation, 31
shared responsibility model, 40
simulation, 344–345, 348, 354, 359, 367
SIR model, 311–314, 319–321, 328–329
SLA, 53
smart city, 204
smart devices, 112–113
smart environment, 200, 204–205, 217–220
smart grid, 26–31, 35–36
smart healthcare, 165, 206–207, 210, 217, 219–220
smart home, 200–201, 210, 218
smart irrigation, 164
smart meter, 214–215
smart sensor, 202, 204–207, 214
smelts, 312
SO_2, 332–333, 335
social welfare maximization model, 26, 28, 32–34, 36
soil, 222–228, 231–233
spatio-temporal, 222, 227–228, 232–233
SPO2, 313
SQL injection, 118–120, 124
SSL/TSL, 51
standard times, 147, 149, 154–155, 158
state estimation, 342–343, 352, 359, 364, 366
supervised learning, 186–187
supply-chain, 14–15
support vector machine, 187–189, 269–270, 273–274, 278–279
susceptible-exposed-infected-recovered, 313
susceptible individuals, 314, 321
susceptible-infectious-removed, 312
sustainability, 92, 94–95
SVD, 298–299, 301–302, 309–310
SVG, 325, 327
sweep algorithm, 263

T

technologies used in autonomous vehicles, 239–242
technology acceptance model, 124, 132
temperature, 222, 225–233
threats to the security of e-governance services, 137
time and motion study (TMS), 147–149, 159
trends of cloud, edge and fog computing, 113–114

U

Ubuntu 16.04 servers, 64
Ujwal Discom Assurance Yojana, 30
U-Net, 74–75, 78, 80, 82–83
unsupervised learning, 186–187

V

vehicle model, 350, 364
vehicle routing problem, 255
vehicular ad hoc networks (VANET) in autonomous vehicles, 242–243
virtual instance, 59
virtualization, 57–62, 64, 68, 92–95
    and simulation, 181, 183
    layer, 51
virtual machine, 50, 58
virus, 312–313
vulnerabilities, 118, 120, 122, 124, 130–132

W

web animation, 327
web application, 322
wireless sensor network, 200, 219, 233

worker availability, 149, 155, 158
work measurement system, 147

X

XEN, 59
XSS attacks, 118

Y

Y-cruncher, 60

Z

Z file system, 60

For Product Safety Concerns and Information please contact our EU representative GPSR@taylorandfrancis.com
Taylor & Francis Verlag GmbH, Kaufingerstraße 24, 80331 München, Germany

www.ingramcontent.com/pod-product-compliance
Lightning Source LLC
Chambersburg PA
CBHW060749220326
41598CB00022B/2374

* 9 7 8 0 3 6 7 7 0 5 1 6 9 *